SPECTRAL METHODS FOR PARTIAL DIFFERENTIAL EQUATIONS

Proceedings of a Symposium
August 16–18, 1982 • NASA Langley Research Center

The symposium was sponsored by the Institute for Computer Applications in Science and Engineering (ICASE)

Spectral Methods for Partial Differential Equations

edited by
Robert G. Voigt, David Gottlieb, and M. Yousuff Hussaini

siam Philadelphia/1984

Library of Congress Catalog Card Number: 84-50634
ISBN: 0-89871-195-9

CONTENTS

PREFACE

On August 16–18, 1982, the Institute for Computer Applications in Science and Engineering (ICASE) sponsored a Symposium on Spectral Methods for Partial Differential Equations at the NASA Langley Research Center. Approximately five years had passed since the 1977 SIAM monograph by David Gottlieb and Steven Orszag had reviewed spectral methods, and it was thought that sufficient developments had taken place to merit a symposium on the subject. The intent of the organizing committee, consisting of David Gottlieb, M. Yousuff Hussaini, Milton E. Rose and Robert G. Voigt, Chairman, was to bring researchers who were advancing the theory of spectral methods together with scientists who were attempting to use spectral methods for applications. Over one hundred scientists representing universities, private industry and government research laboratories attended the three-day meeting.

The program included eleven invited talks with an additional keynote address on fluid dynamical applications of spectral methods given by Steven A. Orszag. Four of the talks dealt primarily with advances in the theory of spectral methods while the remaining seven concentrated on applications from a variety of fields related to fluid dynamics. All eleven talks are represented by papers in this volume.

In addition, an unusual feature is the inclusion of a lengthy introductory article by David Gottlieb, M. Yousuff Hussaini and Steven A. Orszag. This article primarily surveys developments not covered in the earlier Gottlieb-Orszag monograph. In particular, the pseudospectral method is presented in some detail. The article also has an extensive discussion of developments in the application of spectral methods, with almost all of the work in compressible fluid dynamics having taken place since the earlier monograph.

—ROBERT G. VOIGT

INTRODUCTION
Theory and Applications of Spectral Methods

DAVID GOTTLIEB,* M. YOUSUFF HUSSAINI,† AND STEVEN A. ORSZAG‡

The 1977 monograph by Gottlieb & Orszag reviewed the state of the art, at that time, of spectral methods. The theory presented in that work centered on the Galerkin and tau methods and the fluid dynamical applications were confined to incompressible flows. In the past several years, there has been extensive activity in both the theory and application of spectral methods. This activity has been mainly concentrated in the area of pseudospectral methods. For the theory, functional analysis has proved to be a powerful tool for obtaining useful error estimates. For spectral techniques, improved iteration methods have been developed that make possible large-scale calculations of complicated physical phenomena. For applications, first results for compressible flow problems have been obtained for rather complicated flow fields including shock waves, and significant progress has been made on transition and turbulence in incompressible flows.

The aim of this article is to review some of the major developments that have been made since 1977. We concentrate on pseudospectral techniques, as the analysis of Galerkin and tau methods is adequately covered in our 1977 monograph. The review is divided into two main parts. First, we give a summary of results that explain the nature of pseudospectral methods, the correct way to implement them and the error behavior that should be expected to be produced by them. It should be noted that some very important results in the theory of spectral methods are omitted. Our guideline has been to quote those results that are used in the applications discussed in the second part.

The part on applications contains two major subsections, namely compressible and incompressible flows. Our review is by no means exhaustive; some major applications are not discussed at all. For example, there is a large and rapidly growing literature on the applications of spectral methods in meteorology (see, e.g., Haltiner & Williams [1980]). Our goal here is to introduce the potential user to the range of applications of spectral methods and to the problems encountered in implementing them. Spectral methods provide a very sophisticated tool for scientific computing; as in the case of other sophisticated techniques, they can be dangerous if misused. A properly constructed spectral method can be used to obtain solutions where other numerical techniques fail. However, a poorly designed spectral method may perform much worse than simpler finite difference or finite element techniques. We hope that this review will serve as a useful guide to the proper implementation of spectral methods.

1. Pseudospectral Fourier method. In order to set the framework for this review, let us begin by explaining how to construct pseudospectral approximations. There are

---

*ICASE, NASA Langley Research Center, Hampton, Virginia 23665 and Tel-Aviv University, Tel Aviv, Israel. The work of this author was supported in part by NASA contracts NAS1-15810 and NAS1-17130 while he was in residence at ICASE, and in part by Air Force Office of Scientific Research grant 83-0089.

†ICASE, NASA Langley Research Center, Hampton, Virginia 23665. The work of this author was supported by NASA contracts NAS1-15810, NAS1-16394 and NAS1-17130 while he was in residence at ICASE.

‡Massachusetts Institute of Technology, Cambridge, Massachusetts 02139. The work of this author was supported in part by Office of Naval Research contract N00014-82-C-0451, in part by National Science Foundation grant ATM-8310210, and in part by Air Force Office of Scientific Research grant 83-0089.

basically two steps to obtaining a numerical approximation $u_N(x)$ to a solution $u(x)$ of a differential equation. First, an appropriate finite or discrete representation of the solution must be chosen. This may take the form of an interpolating function between the values $u(x_j)$ at some suitable points x_j, or series coefficients in the finite representation

$$u_N(x) = \sum_{k=0}^{N} a_k \phi_k(x)$$

with given expansion functions $\phi_k(x)$. The second step is to obtain equations for the discrete values $u_N(x_j)$ or the coefficients a_k from the original equation. In the case of a differential equation, this second step involves finding an approximation for the differential operator in terms of the grid point values of u_N or, equivalently, the expansion coefficients.

In pseudospectral Fourier approximation, the problem involves periodic boundary conditions. Thus, all functions appearing in the problem are periodic. Let $f(x)$ be a smoothly differentiable function with period 2π. First, we approximate $f(x)$ by the trigonometric polynomial $P_N f(x)$ that interpolates it at one of the following sets of points:

$$\text{(a)} \qquad x_j = \frac{\pi j}{N} \qquad (j = 0, \cdots, 2N-1),$$

$$(1.1) \quad \text{(b)} \qquad y_j = \frac{2\pi j}{2N+1} \qquad (j = 0, \cdots, 2N),$$

$$\text{(c)} \qquad z_j = \frac{(2j+1)\pi}{2N} \qquad (j = 0, \cdots, 2N-1).$$

As will be explained below, fast Fourier transforms (FFT) play a key role in many applications of spectral methods. Many FFT programs are most efficient when 2^P points are transformed, which is not compatible with (1.1b). In practice, the points (1.1a) are most frequently used. This case is discussed first.

When the collocation points x_j given by (1.1a) are used, the approximation $P_N f(x)$ has the form

$$P_N f(x) = \sum_{j=0}^{2N-1} f(x_j) g_j(x) \tag{1.2}$$

where $g_j(x_k) = \delta_{jk}$ and $g_j(x)$ is a polynomial in the functions sin x, cos x of degree at most N. The polynomials $g_j(x)$ are given explicitly by

$$g_j(x) = \frac{1}{2N} \sin\,[N(x - x_j)] \cot \frac{x - x_j}{2}. \tag{1.3}$$

The fact that $g_j(x)$ is a trigonometric polynomial of degree N follows from the equivalent representation

$$g_j(x) = \frac{1}{2N} \sum_{l=-N}^{N} \frac{1}{c_l} e^{il(x-x_j)} \tag{1.4}$$

where $c_l = 1$ ($|l| \neq N$), $c_N = c_{-N} = 2$. Thus we can represent $P_N f(x)$ either as

$$P_N f(x) = \frac{1}{2N} \sum_{j=0}^{2N-1} f(x_j) \sin\,[N(x-x_j)] \cot \frac{x-x_j}{2} \tag{1.5}$$

using (1.3), or as

$$P_N f(x) = \sum_{j=0}^{2N-1} f(x_j) \frac{1}{2N} \sum_{l=-N}^{N} \frac{1}{c_l} e^{il(x-x_j)} \tag{1.6}$$
$$= \sum_{l=-N}^{N} \frac{1}{c_l} e^{ilx} \frac{1}{2N} \sum_{j=0}^{2N-1} f(x_j) e^{-ilx_j}$$

using (1.4). Defining

$$a_l = \frac{1}{2Nc_l} \sum_{j=0}^{2N-1} f(x_j) e^{-ilx_j}, \tag{1.7}$$

(1.6) becomes

$$P_N f = \sum_{l=-N}^{N} a_l e^{ilx}. \tag{1.8}$$

When applying the pseudospectral Fourier method, either the explicit interpolatory formula (1.5) or the complex Fourier representation (1.7)–(1.8) may be used.

As outlined above, the next step in the pseudospectral method is to seek equations for an approximate solution u_N to a differential equation whose exact solution is u. The crucial step here is to obtain values for the derivatives $d^k u_N(x)/dx^k$ at the collocation points x_j in terms of the values $u_N(x_j)$. This is done pseudospectrally in two formally equivalent, but computationally distinct, ways.

One way is simply to differentiate (1.5) and to evaluate the resulting expression at the points x_j:

$$\frac{d^k u_N(x_j)}{dx^k} = \sum_{n=0}^{2N-1} u_N(x_n) \frac{d^k g_n(x_j)}{dx^k} = (D_k \vec{u})_j \tag{1.9}$$

where D_k is an $2N \times 2N$ matrix with elements

$$(D_k)_{j,n} = \frac{d^k g_n(x_j)}{dx^k}$$

and $\vec{u}$ is the column vector

$$\vec{u} = \begin{pmatrix} u_N(x_0) \\ \vdots \\ u_N(x_{2N-1}) \end{pmatrix}$$

Explicitly,

$$(D_1)_{jn} = \begin{cases} \frac{1}{2}(-1)^{j+n} \cot \frac{x_j - x_n}{2}, & j \neq n, \\ 0, \quad j = n; \end{cases} \tag{1.10}$$

$$(D_2)_{jn} = \begin{cases} \frac{1}{2}(-1)^{j+n+1} \frac{1}{\sin^2((x_j - x_n)/2)}, & j \neq n, \\ -\frac{2N^2+1}{6}, \quad j = n. \end{cases} \tag{1.11}$$

More generally,

$$D_k = (D_1)^k, \tag{1.12}$$

which easily follows from (1.3) and the properties of $g_j(x)$. Also, since D_1 is a real, antisymmetric matrix we note that D_{2k} is a real, symmetric matrix while D_{2k+1} is a real, antisymmetric matrix.

Computationally, the evaluation of derivatives using (1.9)–(1.12) involves the multiplication of a $2N$-component vector $\vec{u}$ by a $2N \times 2N$ matrix, D_k, which typically requires $O(N^2)$ arithmetic operations. However, since the matrix product is actually a convolutional sum, it is possible to use the FFT to evaluate (1.9)–(1.12) in only order $N \log N$ operations if N is a highly composite integer (like 2^p or 3^q). Nevertheless direct matrix multiplication can be quite efficient if N is not too large or a highly parallel computer is used.

On the other hand, it is also possible to evaluate derivatives using (1.7)–(1.8). Indeed, (1.8) gives

$$\frac{d^k u_N}{dx^k}(x_j) = \sum_{|n| \le N} (in)^k a_n e^{inx_j} \tag{1.13}$$

where a_n is given by (1.7). In this approach, a_n is first evaluated by (1.7) and then derivatives at x_j are evaluated by (1.13). If N is a highly composite integer, the two discrete Fourier transforms (1.7) and (1.13) can be efficiently evaluated by the FFT algorithm in $O(N \log N)$ operations. This, evaluation of derivatives requires just two FFTs together with the complex muliplication by $(in)^k$ in (1.13).

The above discussion concentrates on the use of the points x_j given by (1.1a). Now we summarize the results for the sets of points y_j and z_j given by (1.1b) and (1.1c), respectively. The trigonometric polynomial $Q_N f(x)$ that interpolates the periodic function $f(x)$ at the collocation points $x = y_j$ $(j = 0, \cdots, 2N)$ is given by

$$Q_N f(x) = \sum_{j=0}^{2N} f(y_j) h_j(x)$$

where

$$h_j(x) = \frac{1}{2N+1} \sin\left[\left(N + \frac{1}{2}\right)\left(x - y_j\right)\right] \frac{1}{\sin((x - y_j)/2)}. \tag{1.14}$$

The kth derivative of $Q_N f$ at the point y_j is

$$\frac{d^k Q_N f}{dx^k}(y_j) = \sum_{n=0}^{2N} f(y_n) \frac{d^k}{dx^k} h_n(y_j) = \sum_{n=0}^{2N} (\bar{D}_k)_{j,n} f(y_n)$$

where

$$(\bar{D}_k)_{j,n} = \frac{d^k h_n(y_j)}{dx^k}.$$

In particular

$$(\bar{D}_1)_{j,n} = \frac{1}{2} \frac{(-1)^{j+n}}{\sin((y_j - y_n)/2)}, \quad j \neq n, \qquad (\bar{D}_1)_{j,j} = 0. \tag{1.15}$$

Moreover, since $h_j(x)$ can be represented as

$$h_j(x) = \frac{2}{2N+1} \sum_{j=-N}^{N} e^{il(x-y_j)},$$

we obtain

$$Q_N f(x) = \sum_{l=-N}^{N} e^{ilx} \frac{2}{2N+1} \sum_{j=0}^{2N} f(y_j) e^{-ily_j}, \tag{1.16}$$

so

$$Q_N f(x) = \sum_{l=-N}^{N} \bar{a}_l e^{ilx}, \qquad \bar{a}_l = \frac{2}{2N+1} \sum_{j=0}^{2N} f(y_j) e^{-ily_j}. \tag{1.17}$$

With the points z_j defined in (1.1c), the trigonometric interpolant is given by (1.5) with z_j replacing x_j. Equivalently, (1.7)–(1.8) holds with this modification.

Some examples to illustrate the application of the pseudospectral Fourier method follow:

Example 1. Consider the linear equation

$$Lu \equiv u_{xx} + f(x)u_x + g(x)\,u = h(x) \tag{1.18}$$

with 2π-periodic boundary conditions. The pseudospectral Fourier equations for the approximation u_N based on the points x_j in (1.1a) are

$$L_N u_N(x_j) \equiv D_2 u_N(x_j) + f(x_j) D_1 u_N(x_j) + g(x_j) u_N(x_j) = h(x_j) \qquad (j = 0, \cdots, 2N-1). \tag{1.19}$$

The algebraic system (1.19) may be solved either directly by inverting the $2N \times 2N$ matrix L_N, or indirectly by the spectral iteration method explained in §8.3.

Example 2. Consider the nonlinear hyperbolic equation

$$u_t = \tfrac{1}{2}(u^2)_x \tag{1.20}$$

with 2π-periodic boundary conditions. The pseudospectral approximation using the points x_j is given explicitly by

$$\frac{\partial u_n}{\partial t}(x_j) = \frac{1}{4} \sum_{\substack{n=0 \\ n \neq j}}^{2N} (-1)^{j+n} \cot \frac{x_j - x_n}{2} u_N^2(x_n). \tag{1.21}$$

A suitable time marching technique should be used to solve this system of $2N$ ordinary differential equations in t (see §8.3).

2. Approximation theory for periodic functions. In this section we discuss various results about the quality of the Fourier approximations described in §1. Since we are interested in the application of the pseudospectral method to differential equations, it is of interest to determine how good the approximation is, not only for the solution, but also for its derivatives. The approximation results for the pseudospectral Fourier method turn out to be very similar to Galerkin approximations based on Fourier series (often called spectral approximations), so we review both kinds of results and point out the relation between them here.

If u is a periodic function with p continuous periodic derivatives, then u can be expanded in a Fourier series of the form

$$u(x) = \sum_{k=-\infty}^{\infty} \hat{a}_k e^{ikx}. \tag{2.1}$$

The spectral (Galerkin) approximation to u is given by

$$\hat{P}_N u = \sum_{k=-N}^{N} \hat{a}_k e^{ikx}, \tag{2.2}$$

which should be contrasted to the pseudospectral (collocation) approximation (1.7)–(1.8).

For each q, the following quantities are defined:

$$|u|_q^2 = \sum_{k=-\infty}^{\infty} k^{2q} |a_k|^2, \tag{2.3}$$

$$\|u\|_q^2 = \sum_{j=0}^{q} |u|_j^2. \tag{2.4}$$

The quantity defined in (2.3) is the norm of the qth derivative of u, whereas the norm defined in (2.4) measures the magnitude of u and its first q derivatives.

In Theorem 2.1 we quote the principal approximation theory result for the spectral (Galerkin) Fourier method (2.2) (see Canuto & Quarteroni [1982b] and references cited therein):

THEOREM 2.1. *For any* $0 \leqq q \leqq p$, *there exists a constant* C *independent of* N *and* u *such that*

$$\|u - \hat{P}_N u\|_q \leqq C N^{q-p} |u|_p. \tag{2.5}$$

Theorem 2.1 implies that if u has p derivatives, derivatives of $\hat{P}_N u$ (up to qth order) approximate those of u.

At this stage it is interesting to compare the spectral Fourier approximation $\hat{P}_N u$ defined by (2.2) to the pseudospectral Fourier approximation $P_N u$ given by (1.7)–(1.8). From (1.7),

$$a_l = \frac{1}{2Nc_l} \sum_{j=0}^{2N-1} u(x_j)\, e^{ilx_j}. \tag{2.6}$$

Substituting for $u(x_j)$ from (2.1) into (2.6) gives

$$a_l = \frac{1}{2Nc_l} \sum_{j=0}^{2N-1} e^{-ilx_j} \sum_{k=-\infty}^{\infty} \hat{a}_k e^{ikx_j} = \frac{1}{2Nc_l} \sum_{k=-\infty}^{\infty} \hat{a}_k \sum_{j=0}^{2N-1} e^{i(k-l)x_j}.$$

Now using the identity

$$\sum_{j=0}^{2N-1} e^{i(k-l)x_j} = \begin{cases} 0, & k - l \neq 2nN, \quad n \text{ integer}, \\ 2N, & k - l = 2nN, \end{cases} \tag{2.7}$$

we obtain

$$a_l = \frac{1}{c_l} \sum_{n=-\infty}^{\infty} \hat{a}_{l+2nN} = \frac{\hat{a}_l}{c_l} + \frac{1}{c_l} \sum_{\substack{n=-\infty \\ n \neq 0}}^{\infty} \hat{a}_{l+2nN}. \tag{2.8}$$

(2.8) establishes the relation between $P_N u$ and $\hat{P}_N u$. If u is smooth then $a_k \to 0$ rapidly as $|k| \to \infty$, so the second term on the right side of (2.8) is small.

The main approximation theory result for $P_N u(x)$ is very similar to that for $\hat{P}_N u$, namely:

THEOREM 2.2. *Let $p > \frac{1}{2}$. Then for $0 \leqq q \leqq p$, there exists a constant C independent of N and u such that*

$$\|u - P_N u\|_q \leqq C N^{q-p} |u|_p \tag{2.9}$$

(*see Canuto & Quarteroni* [1982b]).

Unfortunately, the norm defined in (2.4) is not the most natural one for the pseudospectral method. It is more convenient to measure the approximation errors in u and its derivatives at the grid points x_j given by (1.1a). Equivalently, we would like to get a bound on the discrete norm

$$\|\!|u - P_N u\|\!|_q^2 = \sum_{l=0}^{q} \frac{1}{2N} \sum_{j=0}^{2N-1} \left[\left(\frac{\partial^l u}{\partial x^l} - \frac{\partial^l P_N u}{\partial x^l} \right)(x_j) \right]^2. \tag{2.10}$$

In order to get an estimate for (2.10), we need the following lemma (Davis & Rabinowitz [1975]):

LEMMA 2.3. *Let $f(x)$ be of the form*

$$f(x) = \sum_{k=-2N+1}^{2N-1} \hat{f}_k e^{ikx}.$$

Then

$$\int_0^{2\pi} f(x)\, dx = \frac{\pi}{2N} \sum_{j=0}^{2N-1} f(x_j).$$

The proof involves a simple application of (2.7). Lemma 2.3 provides the necessary connection between (2.10) and (2.3) to establish the following theorem.

THEOREM 2.4. *Let $p > \frac{1}{2}$. Then for $0 \leqq q \leqq p$, there exists a constant C independent of N and u such that*

$$\|\!|u - P_N u\|\!|_q \leqq C N^{q-p} |u|_p. \tag{2.11}$$

The error estimates given in (2.5) and (2.11) show the power of the spectral method. The rate of decrease of the error with increasing N only depends on the smoothness of the function u being approximated. However, it is misleading to conclude from (2.5) and (2.11) that if u is not smooth then the approximation $P_N u(x)$ to $u(x)$ is poor. In many applications, one encounters functions u which are piecewise smooth. To illustrate what happens for such u, let us consider a function with a single discontinuity in each period, namely

$$u(x) = x \qquad (|x| \leqq \pi),$$

extended periodically in x. The pseudospectral approximation $P_N u$ to u is given by

$$P_N u = \sum_{k=-N}^{N} a_k e^{ikx}$$

where a_k is given by (1.7). In general $\partial P_N u/\partial x$ is a poor approximation to $\partial u/\partial x$, even away from the discontinuity at $x = \pm\pi$. However, if the expansion coefficients a_k are damped as in

$$b_k = a_k \exp\left[-\rho \left(\frac{|k| - k_0}{N - k_0} \right)^4 \right] \qquad (|k| \geqq k_0), \tag{2.12}$$

then, for suitable choice of ρ and k_0,

$$v(x) = \sum_{k=-N}^{N} b_k e^{ikx}$$

has the property that $\partial v/\partial x$ is a good approximation to $\partial u/\partial x$ away from the points of discontinuities (see Fig. 1).

3. Stability and convergence of pseudospectral Fourier methods. Consider the solution of the time-dependent initial value problem

$$u_t = Lu \tag{3.1}$$

with initial condition $u(t = 0) = u_0$ using the pseudospectral Fourier method. Here L is a linear differential operator with periodic coefficients. The pseudospectral approximate u_N

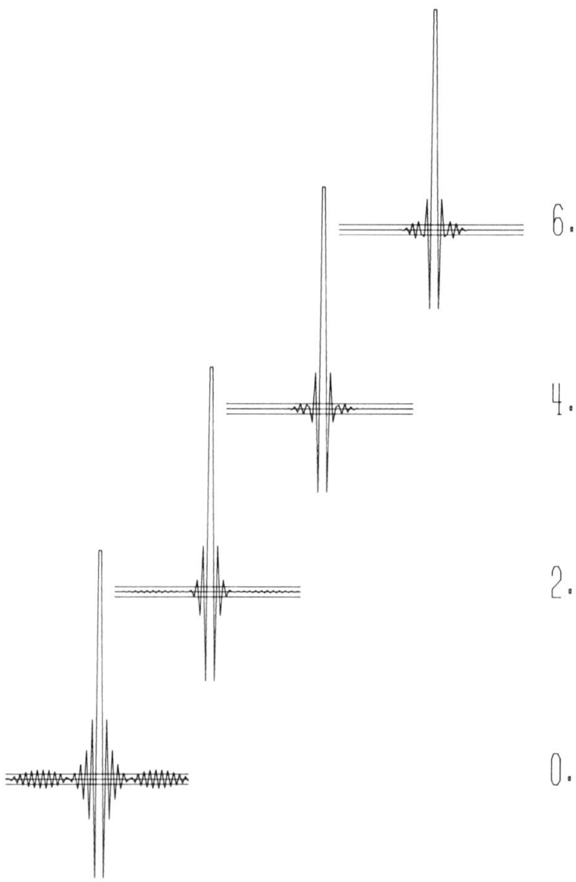

FIG. 1. *Filtered Fourier derivative of the function*

$$u(x) = \begin{cases} x, & 0 \leq x \leq \pi, \\ x - 2\pi, & \pi < x \leq 2\pi, \end{cases}$$

with $k_0 = 45$, $N = 64$, $\rho = 0, 2, 4,$ *and* 6 (*formula* (2.12)).

to u satisfies $\partial u_N/\partial t = L_N u_N$, where u_N $(t = 0)$ is the trigonometric interpolant of u_0 and L_N is a finite-dimensional approximation to L as explained in §1.

The analysis of approximation errors given in §2 yields estimates for the truncation error

$$(L - L_N)u \tag{3.2}$$

for differential operators L. It was shown that $(L - L_N)u$ decreases rapidly with N at a rate that depends only on the smoothness of u and the coefficients of the various derivatives appearing in L.

In addition to the consistency properties of a numerical method, the method must be stable to be useful. A scheme is said to be stable if

$$\|u_N(t)\| \leqq C\,\|u_N(0)\| \qquad (0 \leqq t \leqq T) \tag{3.3}$$

for some appropriate norm. Here C may depend on T, but not on N. The importance of stability derives from the Lax equivalence theorem, which implies that if (3.3) is satisfied and $(L - L_N)u \rightarrow 0$ as $N \rightarrow \infty$ in the same norm as (3.3), then u_N converges to u:

$$\|u - u_N\| \rightarrow 0 \qquad (0 \leqq t \leqq T) \quad \text{as } N \rightarrow \infty. \tag{3.4}$$

In other words, stability (3.3) and consistency $((L - L_N)u \rightarrow 0)$ imply convergence.

In this section, we review some results about the stability and convergence of the pseudospectral Fourier method for hyperbolic and parabolic problems.

Constant coefficient wave equation. Consider the problem

$$\begin{aligned} &u_t = u_x \qquad (0 \leqq x \leqq 2\pi), \\ &u(0, t) = u(2\pi, t), \qquad u(x, 0) = u_0(x). \end{aligned} \tag{3.5}$$

The pseudospectral Fourier approximation can be written

$$\frac{\partial u_N}{\partial t}(x_k) = \sum_{j=0}^{2N-1} (D_1)_{k,j}\, u_N(x_j) \tag{3.6}$$

where the antisymmetric matrix D_1 is given by (1.10).

The stability of (3.6) can be proven by the energy method. There are two ways of deriving the desired energy estimate. One method is algebraic in nature and uses the structure of the matrix D_1, while the other makes use of the fact that the differential equation is satisfied exactly at the collocation points x_j. These two techniques are extensively used in the theory of spectral methods, so we demonstrate them in detail as applied to (3.5)–(3.6). In the first case, we multiply (3.6) by $u_N(x_k)$ and sum on k $(0 \leqq k \leqq 2N - 1)$ to get

$$\sum_{k=0}^{2N-1} u_N(x_k) \frac{\partial u_N}{\partial t}(x_k) = \sum_{k=0}^{2N-1} \sum_{j=0}^{2N-1} u_N(x_k)\,(D_1)_{k,j}\, u_N(x_j),$$

so

$$\frac{1}{2}\frac{d}{dt} \sum_{k=0}^{2N-1} u_N^2(x_k) = \frac{1}{2} \sum_{k=0}^{2N-1} \sum_{j=0}^{2N-1} [(D_1)_{k,j} + (D_1)_{j,k}] u_N(x_k) u_N(x_j) = 0 \tag{3.7}$$

since D_1 is an antisymmetric matrix. (3.7) implies energy conservation:

$$\sum_{k=0}^{2N-1} u_N^2(x_k, t) = \sum_{k=0}^{2N-1} u_N^2(x_k, 0). \tag{3.8}$$

Therefore the scheme is stable in the sense of (3.3) in the usual L_2 norm.

The second method to establish stability is based on the fact that (3.5) is satisfied at the points x_j:

$$\frac{\partial u_N}{\partial t} = \frac{\partial u_N}{\partial x} \quad \text{at } x = x_j = \frac{\pi j}{N} \quad (j = 0, \cdots, 2N-1). \tag{3.9}$$

Therefore

$$\frac{1}{2N}\sum_{j=0}^{2N-1} u_N(x_j)\frac{\partial u_N}{\partial t}(x_j) = \frac{1}{2N}\sum_{j=0}^{2N-1} u_N(x_j)\frac{\partial u_N}{\partial x}(x_j). \tag{3.10}$$

The right side of (3.10) is a trigonometric polynomial of degree $2N - 1$, so Lemma 2.3 gives

$$\frac{1}{2N}\sum_{j=0}^{2N-1} u_N(x_j)\frac{\partial u_N}{\partial x}(x_j) = \frac{1}{2\pi}\int_0^{2\pi} u_N(x)\frac{\partial u_N}{\partial x}\,dx = 0,$$

and (3.8) holds.

Similar arguments demonstrate stability for the problem

$$\vec{u}_t = A\,\vec{u}_x \tag{3.11}$$

where A is a diagonalizable $m \times m$ matrix and $\vec{u}(x, t)$ is a vector with m components. Using the stability result (3.8), it is straightforward to establish an error estimate of the form

$$\|\vec{u}(t) - \vec{u}_N(t)\|_0 \leqq C(1 + N^2)^{(1-q)/2}\|\vec{u}_0\|_q \tag{3.12}$$

where the norms are as defined in §2.

Variable coefficient wave equation. Consider the problem

$$u_t = a(x)\,u_x \quad (0 \leqq x \leqq 2\pi), \quad u(x, 0) = u_0(x), \quad u(0, t) = u(2\pi, t) \tag{3.13}$$

with $a(x)$ 2π-periodic. If $a(x) > 0$ for $0 \leqq x \leqq 2\pi$, then the argument given above shows that

$$\frac{d}{dt}\sum_{j=0}^{2N-1}\frac{u_N^2(x_j)}{a(x_j)} = 0,$$

so the same stability and convergence results hold.

However, if $a(x)$ changes sign in $(0, 2\pi)$, $u_N(x, t)$ may grow without bound as $t \rightarrow \infty$. The following example (Gottlieb, Orszag & Turkel [1981]) illustrates the nature of the instability that may occur. Consider (3.13) with

$$a(x) = \alpha \sin x + \beta \cos x + \gamma. \tag{3.14}$$

For this problem one may derive two different energy estimates. The first one is

$$\|u_N(t)\| \leqq e^{(|\alpha| + |\beta|)t/2}\|u_N(0)\| + 2\sigma_N(e^{(|\alpha| + |\beta|)t/2} - 1) \tag{3.15}$$

where

$$\|u_N(t)\|^2 = \frac{1}{2N}\sum_{j=0}^{2N-1} u_N^2(x_j, t) \quad \text{and} \quad \sigma_N = \sum_{j=0}^{2N-1} (-1)^j u(x_j, 0).$$

The second estimate is

$$\|u_N(t)\|_1 \leqq e^{(|\alpha|+|\beta|)t/2}\|u_N(0)\|_1. \tag{3.16}$$

Relations (3.15) and (3.16) prove stability; however, the constant C in (3.3) involves $\exp[(|\alpha| + |\beta|)T/2]$, which grows in time.

The trouble here is the behavior of the solution itself. If $\alpha = 1$, $\beta = \gamma = 0$ in (3.14), then

$$u(x, t) = u_0\left[2\tan^{-1}\left(e^t \tan\frac{x}{2}\right)\right].$$

For large t, the exact solution changes rapidly near $x = 0$ so large values of N are required to resolve this steep gradient. In the literature on numerical methods, this kind of behavior is often referred to as "nonlinear instability," as it often leads to spectacular growth of errors in a nonlinear problem. The problem here can be quite serious in practice, but it is due not to lack of stability in the formal sense, but rather to the lack of spatial resolution. It can be treated either by refining the mesh as the solution develops smaller scales, or by treating the region of rapid change as a discontinuity (see below).

Wave equation with discontinuity. If the initial condition $u_0(x)$ has a finite number of discontinuities in $[0, 2\pi]$ but is otherwise smooth, the pseudospectral method is still applicable if some smoothing procedure is used (see Majda, McDonough & Osher [1978] and Mercier [1981]). Typically, one has to modify (1.8) by defining

$$\tilde{u}_N = \sum_{k=-N}^{N} a_k \rho_k e^{ikx} \tag{3.17}$$

where

$$\rho_k = \begin{cases} 1 & (0 \leqq |k| \leqq k_0), \\ \exp\left[-\rho\left(\dfrac{|k| - k_0}{N - k_0}\right)^4\right], & |k| > k_0. \end{cases}$$

It may be shown that, for any q, there is a constant $C_{s,q}$ depending only on $u(t)$ $(0 \leqq t \leqq T)$ such that

$$\|u(t) - u_N(t)\|_s \leqq C_{s,q} N^{-q} \qquad (0 \leqq t \leqq T). \tag{3.18}$$

Moreover, if the modification (3.17) is applied to the spectral representation of $u_0(x)$,

$$u_0(x) = \sum_{k=-N}^{N} \hat{a}_k \rho_k e^{ikx},$$

then the rate of convervence of u_N to u is arbitrarily fast in any compact region in which $u(x, t)$ is smooth. This shows that the high-order accuracy of the pseudospectral method holds even for nonsmooth problems (Osher, this volume).

Nonlinear equations. It is possible to find energy estimates for pseudospectral approximations to nonlinear equations. For example, consider the inviscid Burgers equation (1.20) with initial condition $u(x, 0) = u_0(x)$. The pseudospectral approximation (1.21) does not conserve energy, but this property can be recovered by rewriting (1.20) as

$$u_t = \tfrac{1}{3}(u^2)_x + \tfrac{1}{3}uu_x. \tag{3.19}$$

The pseudospectral approximation to (3.19) is given by

$$\frac{\partial u_N}{\partial t}(x_k) = \frac{1}{3}\sum_{j=0}^{2N-1}(D_1)_{k,j}u_N^2(x_j) + \frac{1}{3}\sum_{j=0}^{2N-1}u_N(x_k)(D_1)_{k,j}u_N(x_j). \tag{3.20}$$

Multiplying by $u_N(x_k)$ and summing on k gives

$$\frac{1}{2}\frac{d}{dt}\sum_{k=0}^{2N-1}u_N^2(x_k) = 0$$

since D_1 is antisymmetric. Thus, stability is ensured for this pseudospectral scheme.

It is interesting to note that the procedure described above to prove stability may be applied to the Euler equations of gas dynamics. It is known (see Harten [1983a] and Tadmor [1983]) that the Euler equations can be expressed in the form

$$H(\vec{v})\vec{v}_t = A(\vec{v})\vec{v}_x \tag{3.21}$$

where $\vec{v}$ is a three-component vector, and $H(\vec{v})$ and $A(\vec{v})$ are 3×3 symmetric matrices. Also, $H(\vec{v})$ is positive definite and H and A satisfy

$$H(\vec{v})\vec{v}_t = [H(\vec{v})\vec{v}]_t, \qquad A(\vec{v})\vec{v}_x = [A(\vec{v})\vec{v}]_x. \tag{3.22}$$

(See Harten [1983a] for the explicit formulas for H and A.) Using (3.22) one can rewrite (3.21) as

$$H(\vec{v})\vec{v}_t + [H(\vec{v})\vec{v}]_t = A(\vec{v})\vec{v}_x + [A(\vec{v})\vec{v}]_x. \tag{3.23}$$

For this form of the Euler equations, it follows easily that

$$\frac{d}{dt}\sum_{k=0}^{2N-1}\vec{v}^*(x_k)H(\vec{v}(x_k))\vec{v}(x_k) = 0, \tag{3.24}$$

which implies stability in the norm induced by H.

Heat equation. Consider the problem

$$u_t = u_{xx} \quad (0 \leqq x \leqq 2\pi), \quad u(0, t) = u(2\pi, t), \quad u(x, 0) = u_0(x). \tag{3.25}$$

The pseudospectral Fourier approximation is

$$\frac{\partial u_N}{\partial t}(x_k) = \sum_{j=0}^{2N-1}(D_2)_{k,j}u_N(x_j).$$

Then

$$\frac{\partial u_N}{\partial t} = \frac{\partial^2 u_N}{\partial x^2} \quad \text{at } x = x_k = \frac{\pi k}{N} \quad (k = 0, \cdots, 2N-1) \tag{3.26}$$

so that

$$\frac{\pi}{N}\sum_{k=0}^{2N-1}u_N(x_k)\frac{\partial u_N}{\partial t}(x_k) = \frac{\pi}{N}\sum_{k=0}^{2N-1}u_N(x_k)\frac{\partial^2 u_N(x_k)}{\partial x^2}$$

$$= \int_0^{2\pi}u_N\frac{\partial^2 u_N}{\partial x^2}\,dx = -\int_0^{2\pi}\left(\frac{\partial u_N}{\partial x}\right)^2 dx \leqq 0,$$

proving the stability of this pseudospectral approximation.

4. Pseudospectral Chebyshev method. When a function $f(x)$ is not periodic, a trigonometric interpolation polynomial does not provide a good enough approximation to yield accurate approximations to the derivatives of $f(x)$. It is better to approximate $f(x)$ by polynomials in x. However, it is well known that the Lagrange interpolation polynomial based on equally spaced points does not give a satisfactory approximation to general smooth f. In fact, as the number of collocation points increases, interpolant polynomials typically diverge. This poor behavior of polynomial interpolation can be avoided for smoothly differentiable functions by removing the restriction to equally spaced collocation points. Good results are obtained by relating the collocation points to the structure of classical orthogonal polynomials, like Chebyshev or Legendre polynomials.

In the most common pseudospectral Chebyshev method, the interpolation points in the interval $(-1, 1)$ are chosen to be the extrema

$$x_j = \cos \frac{\pi j}{N} \qquad (j = 0, \cdots, N) \tag{4.1}$$

of the Nth order Chebyshev polynomials $T_N(x)$. Here the Chebyshev polynomial of degree n is defined by

$$T_n(x) = \cos (n \cos^{-1} x). \tag{4.2}$$

It follows that

$$T_n(x_j) = \cos \frac{\pi j n}{N}, \tag{4.3}$$

which indicates a close relation between the pseudospectral Chebyshev and the pseudospectral Fourier method. In order to construct the interpolant of $f(x)$ at the point x we define the polynomials

$$g_j(x) = \frac{(1 - x^2)T'_N(x)(-1)^{j+1}}{\bar{c}_j N^2(x - x_j)} \qquad (j = 0, \cdots, N) \tag{4.4}$$

with $\bar{c}_0 = \bar{c}_N = 2, \bar{c}_j = 1$ $(1 \leqq j \leqq N - 1)$. It is readily verified that

$$g_j(x_k) = \delta_{jk}.$$

The Nth degree interpolation polynomial, $P_N f(x)$, to $f(x)$ is given by

$$P_N f(x) = \sum_{j=0}^{N} f(x_j) g_j(x). \tag{4.5}$$

A different way of representing $P_N f(x)$ is to use the identity

$$\sum_{n=0}^{N} \frac{T_n(x_j)T_n(x)}{\bar{c}_n} = \frac{(1 - x^2)T'_N(x)}{2N(x - x_j)}(-1)^{j+1},$$

giving

$$\begin{aligned}\sum_{j=0}^{N} f(x_j) g_j(x) &= \frac{2}{N} \sum_{j=0}^{N} \frac{f(x_j)}{\bar{c}_j} \sum_{n=0}^{N} \frac{T_n(x_j)T_n(x)}{\bar{c}_n} \\ &= \frac{2}{N} \sum_{n=0}^{N} T_n(x) \frac{1}{\bar{c}_n} \sum_{j=0}^{N} \frac{f(x_j)T_n(x_j)}{\bar{c}_j}.\end{aligned}$$

Thus

$$P_N f(x) = \sum_{n=0}^{N} a_n T_n(x) \tag{4.6}$$

where

$$a_n = \frac{2}{N}\frac{1}{\bar{c}_n}\sum_{j=0}^{N}\frac{f(x_j)T_n(x_j)}{\bar{c}_j}. \tag{4.7}$$

It should be noted that the coefficients a_n in (4.7) can be evaluated using the FFT. In fact, using (4.3) in (4.7) gives

$$a_n = \frac{2}{N}\frac{1}{\bar{c}_n}\sum_{j=0}^{N}\frac{f(x_j)}{\bar{c}_j}\cos\frac{\pi jn}{N}. \tag{4.8}$$

The second step in getting a pseudospectral approximation is to express the derivatives of $P_N f$ in terms of $f(x)$ at the collocation points x_j. This can be done by differentiating either (4.5) or (4.6). With (4.5) we obtain

$$\frac{d^p P_N f(x)}{dx^p} = \sum_{j=0}^{N} f(x_j)\frac{d^p}{dx^p}g_j(x), \tag{4.9}$$

so that

$$\frac{d^p P_N f(x_k)}{dx^p} = \sum_{j=0}^{N} f(x_j)(D_p)_{k,j} \tag{4.10}$$

where

$$(D_p)_{k,j} = \frac{d^p}{dx^p}g_j(x)\bigg|_{x=x_k}. \tag{4.11}$$

For example

$$\begin{aligned} (D_1)_{kj} &= \frac{\bar{c}_k}{\bar{c}_j}\frac{(-1)^{j+k}}{x_k - x_j} \qquad (k \neq j), \\ (D_1)_{jj} &= -\frac{x_j}{2(1-x_j^2)}, \qquad (D_1)_{00} = \frac{2N^2+1}{6} = -(D_1)_{NN} \end{aligned} \tag{4.12}$$

and

$$D_p = (D_1)^p. \tag{4.13}$$

It should be noted from the explicit formula (4.12) that the matrix D_1 is not antisymmetric; also D_2 is not symmetric. These facts introduce both theoretical and practical difficulties in the pseudospectral Chebyshev method.

A different way to obtain an expression for the derivative of $P_N f$ is to differentiate (4.6) to get

$$\frac{d^p}{dx^p}P_N f = \sum_{n=0}^{N} a_n T_n^{(p)}(x) \tag{4.14}$$

where the coefficients a_n are given by (4.7). For example,

$$\frac{dP_N f}{dx} = \sum_{n=0}^{N} a_n T_n'(x) = \sum_{n=0}^{N} b_n T_n(x) \tag{4.15}$$

where

$$b_N = 0, \qquad b_{N-1} = 2Na_N$$

and

$$\bar{c}_n b_n = b_{n+2} + 2(n+1)a_{n+1} \qquad (0 \leq n \leq N-2). \tag{4.16}$$

In evaluating the first derivative at the collocation points x_j, the FFT is used to evaluate a_n by (4.8) and then

$$\frac{d}{dx} P_N f(x_j) = \sum_{n=0}^{N} b_n \cos \frac{\pi n_j}{N}.$$

More generally,

$$\frac{d^p(P_N f)}{dx^p} = \sum_{n=0}^{N} b_n^{(p)} T_n(x)$$

where $b_b^{(0)} = a_n$ $(0 \leq n \leq N)$, $b_n^{(1)} \equiv b_n$, and

$$\bar{c}_n b_n^{(p)} = b_{n+2}^{(p)} + 2(n+1)b_{n+1}^{(p-1)}, \tag{4.17}$$

with

$$b_{N-1}^{(p)} = 0, \qquad b_{N-1}^{(p)} = 2Nb_N^{(p-1)}.$$

The set of points x_j defined in (4.1) is not the only set used with pseudospectral Chebyshev approximations. For hyperbolic problems, a convenient alternative set of collocation points is

$$y_j = \cos \frac{\pi j}{N+1} \qquad (j = 0, \cdots, N). \tag{4.18}$$

The interpolation polynomial at the points y_j is of degree N and is given by

$$Q_N f = \sum_{j=0}^{N} f(y_j) h_j(x)$$

where

$$h_j(x) = \frac{(-1)^{j+1}}{\bar{c}_j(N+1)^2} \frac{(1-x)(1+y_j)T'_{N+1}(x)}{(x-y_j)}. \tag{4.19}$$

Upon differentiation, we obtain

$$\frac{d(Q_N f)}{dx}(y_k) = \sum_{j=0}^{N} (\hat{D}_1)_{kj} f(y_j) \qquad (k = 0, \cdots, N) \tag{4.20}$$

where

$$(\hat{D}_1)_{k,j} = \frac{1+y_j}{1+y_k} (D_1)_{k,j}$$

and D_1 is given by (4.12). It is clear from (4.4) and (4.19) that, for every function $f(x)$ $(-1 \leq x \leq 1)$,

$$(1+x)\, Q_N f(x) = P_{N+1}[(1+x) f(x)].$$

Moreover, if $Q_N f(x)$ is expanded in terms of Chebyshev polynomials,

$$Q_N f = \sum_{n=0}^{N} \bar{a}_n T_n(x), \tag{4.21}$$

there is a simple relation between $\{\bar{a}_n\}$ and $\{a_n\}$ defined in (4.7) with $N \leftarrow N + 1$. In fact,

$$Q_N f = Q_N(P_{N+1} f) = P_{N+1} f + \frac{a_{N+1}}{N+1}(1 - x)T'_{N+1}(x), \tag{4.22}$$

so, recalling the derivations of (4.6)–(4.7),

$$\bar{a}_n = a_n + 2(-1)^{n+N+1} a_{N+1} \qquad (1 \leqq n \leqq N), \tag{4.23}$$

which shows that FFT techniques can also be used with the set of points y_j.

The other two sets of points that are sometimes used are

$$z_j^{(1)} = \cos \frac{2\pi j}{2N+1} \qquad (j = 0, \cdots, N) \tag{4.24}$$

and

$$z_j^{(2)} = \cos \frac{\pi(2j+1)}{2N+1} \qquad (j = 0, \cdots, N). \tag{4.25}$$

5. Approximation theory of the pseudospectral Chebyshev method. In order to measure the approximation error between f and $P_N f$ defined in §4, we introduce the norms

$$\|f\|^2 = \int_{-1}^{1} \frac{f^2(x)}{\sqrt{1-x^2}}\, dx, \tag{5.1}$$

$$\|f\|_q^2 = \left[\|f\|^2 + \left\|\frac{\partial f}{\partial x}\right\|^2 + \cdots + \left\|\frac{\partial^q f}{\partial x^q}\right\|^2\right]. \tag{5.2}$$

These norms are useful since we are interested in applying the pseudospectral method to the solution of differential equations, and hence we need to measure the error between the derivatives of the solution and those of the approximation.

The main approximation result is summarized in the following theorem.

THEOREM 5.1. *Let $f(x)$ be a function with s continuous derivatives and let $P_N f$ be defined as in* §4. *Then there is a constant C independent of $f(x)$ and N such that*

$$\|f - P_N f\|_q \leqq C N^{2q-s} \|f\|_s \qquad (0 \leqq q \leqq \tfrac{1}{2} s). \tag{5.3}$$

The difference between (5.3) and (2.8) should be noted. Each derivative on the left side of (5.3) introduces a new factor of N^2 in the error estimate instead of the factor N appearing in (2.8).

As before, the norms defined in (5.1) and (5.2) should be modified in order to reflect the errors at the collocation points. To do this, we use the following two lemmas:

LEMMA 5.2 (see Rivlin [1974]). *Let $g(x)$ be a polynomial of degree $2N - 1$ and let $x_j = \cos \pi j/N$ $(j = 0, \cdots, N)$. Then*

$$\frac{\pi}{N} \sum_{j=0}^{N} \frac{g(x_j)}{\bar{c}_j} = \int_{-1}^{1} \frac{g(x)}{\sqrt{1-x^2}}\, dx \tag{5.4}$$

where $\bar{c}_0 = \bar{c}_N = 2$, $\bar{c}_j = 1$ $(1 \leqq j \leqq N-1)$.

LEMMA 5.3. *Let $f(x)$ be a polynomial of degree N. Then*

$$\int_{-1}^{1} \frac{f^2(x)}{\sqrt{1-x^2}}\,dx \leqq \frac{\pi}{N}\sum_{j=0}^{N} \frac{f^2(x_j)}{\bar{c}_j} \leqq 2\int_{-1}^{1} \frac{f^2(x)}{\sqrt{1-x^2}}\,dx. \tag{5.5}$$

This lemma is proven by using Lemma 5.2 after representing $f^2(x)$ as a mixture of $T_{2N}(x)$ plus a polynomial of degree $2N-1$. Using Lemma 5.3 we get the following error estimate:

THEOREM 5.4. *Let*

$$h(x) = f(x) - P_N f(x).$$

Then

$$\frac{\pi}{N}\sum_{l=0}^{q}\sum_{j=0}^{N}\left(\frac{\partial^l h(x_j)}{\partial x^l}\right)^2 \leq CN^{2q-s}\|f\|_s \tag{5.6}$$

where C is independent of f and N.

When $f(x)$ is discontinuous at one point (or a finite number of points), we can still

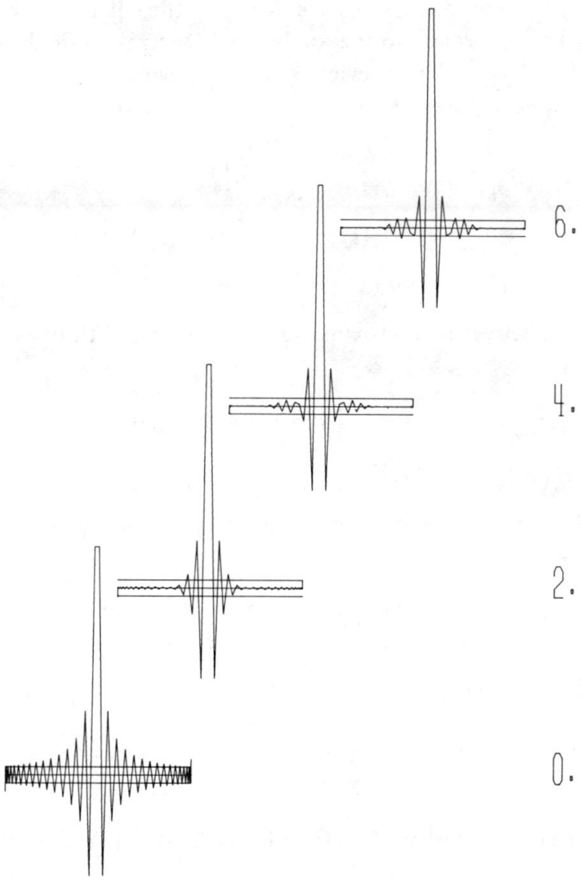

FIG. 2. *Filtered Chebyshev derivative of a step function with $n_0 = 45$, $N = 64$, $\rho = 0, 2, 4,$ and* 6 (*formula* (5.8)).

obtain a good approximation for the derivative far away from the discontinuity. In order to do so, we modify the expression for $P_N f$ as in

$$\tilde{P}_N f = \sum_{n=0}^{N} a_n \rho_n T_n(x) \tag{5.7}$$

where

$$\rho_n = \begin{cases} 1 & (n \leqq n_0), \\ \exp\left[-\rho\left(\dfrac{n-n_0}{N-n_0}\right)^4\right] & (n_0 < n \leqq N). \end{cases} \tag{5.8}$$

In Fig. 2 we plot the derivative of a step function with and without the smoothing described in (5.7)–(5.8).

6. Stability and convergence of pseudospectral Chebyshev methods. In this section we review some results concerning the stability and convergence of the pseudospectral Chebyshev method. The most common way to prove stability is to use energy estimates similar to those used to demonstrate the well-posedness of the differential equation. This approach is not systematic; one has to guess a suitable norm in which the energy is bounded. An alternative approach is to solve explicitly the equation for the error and then to verify stability. This approach to the analysis of spectral methods was introduced by Dubiner in unpublished work that is reviewed in the Appendix.

Results for parabolic equations. Consider the equation

$$\text{(6.1)}\quad \begin{aligned} &\text{(a)} && \frac{\partial u}{\partial t} = a(x)\frac{\partial^2 u}{\partial x^2} && (|x| \leqq 1), \\ &\text{(b)} && u(-1,t) = g(t), && u(1,t) = h(t), \\ &\text{(c)} && u(x,0) = u_0(x), && a(x) \geqq a_0 > 0. \end{aligned}$$

The pseudospectral Chebyshev approximation $u(x, t)$ is an Nth degree polynomial in x that satisfies

$$\begin{aligned} &\frac{\partial u_N}{\partial x} = a(x)\frac{\partial^2 u_N}{\partial x^2} \qquad \left(x = x_j \equiv \cos\frac{\pi j}{N}, 0 < j < N\right), \\ &u_N(-1,t) = g(t), \qquad u_N(1,t) = h(t). \end{aligned} \tag{6.2}$$

The stability of this method is easily demonstrated by an energy argument. In fact, (6.2) gives (with $g \equiv h \equiv O$)

$$\frac{\pi}{N}\sum_{j=0}^{N} \frac{u_N(x_j,t)}{a(x_j)}\frac{\partial u_N}{\partial t}(x_j,t) = \frac{\pi}{N}\sum_{j=0}^{N} u_N(x_j)\frac{\partial^2 u_N}{\partial x^2}(x_j). \tag{6.3}$$

Then Lemma 5.2 implies that

$$\frac{\pi}{2N}\frac{d}{dt}\sum_{j=0}^{N}\frac{u_N^2(x_j,t)}{a(x_j)} = \int_{-1}^{1}\frac{\partial^2 u_N}{\partial x^2}\frac{dx}{\sqrt{1-x^2}}. \tag{6.4}$$

The right side of (6.4) is nonpositive (see Gottlieb & Orszag [1977, p. 82]) so that

$$\sum_{j=0}^{N}\frac{u_N^2(x_j,t)}{a(x_j)} \leqq \sum_{j=0}^{N}\frac{u_N^2(x_j,0)}{a(x_j)}. \tag{6.5}$$

The stability of the scheme (6.2) gives the following error estimate:

THEOREM 6.1 (Canuto & Quarteroni [1981]). *Let* $\sigma > \frac{1}{2}, s > \sigma + 2, 0 < t < T$, *and suppose that* $u(x, t)$ *has s spatial derivatives. Then there is a constant C independent of u and N such that*

$$\frac{\pi}{N}\sum_{j=0}^{N}\frac{1}{\bar{c}_j}|u_N(x_j, t) - u(x_j, t)|^2 \leq CN^{2\sigma+4-s}.$$

A similar proof may be applied to (6.1) when the boundary conditions are of Neumann type, i.e. $u_x(\pm 1, t)$ are given. For the Neumann problem we get the following stability estimate.

THEOREM 6.2. *Let* $u_N(x, t)$ *be the pseudospectral Chebyshev approximation to the solution u of* (6.1) (*with* $a(x) = 1$) *and boundary conditions* $u_x(\pm 1, t) = 0$. *Suppose that* u_N *has a Chebyshev expansion of the form*

$$u_N(x, t) = \sum_{n=0}^{N-1} w_k T_k(x) + w_N(t) T_N(x).$$

Then

$$\frac{d}{dt}\left[\int_{-1}^{1}\sqrt{1-x^2}\left(\frac{\partial u_N}{\partial x}(x, t)\right)^2 dx + N^2 w_N^2(t)\right] \leq 0. \tag{6.6}$$

Note that the norm used in (6.6) differs from the one used in (6.5).

If $a(x) \equiv 1$ (and $g(t) = h(t) = 0$), then it is possible to gain more information about the pseudospectral Chebyshev approximation to (6.1) by explicitly studying the eigenvectors and eigenvalues of the finite-dimensional pseudospectral approximation. Since $u_N(x, t)$ is a polynomial of degree N in x, so is $E_N(x, t)$, defined by

$$E_N(x, t) = \left(\frac{\partial}{\partial T} - \frac{\partial^2}{\partial x^2}\right) u_N.$$

Moreover, since $E_N(x, t)$ vanishes at the points x_j (which are the roots of T'_N), it follows that

$$E_N(x, t) = (\hat{A}(t) + \hat{B}(t)x)\, T'_N(x)$$

for appropriate functions $\hat{A}(t), \hat{B}(t)$. Assume that

$$u_N(x, t) = e^{\lambda t} v_N(x). \tag{6.7}$$

Then $\hat{A}(t) = Ae^{\lambda t}, \hat{B}(t) = Be^{\lambda t}$ so $v_N(x)$ satisfies

$$\lambda v_N(x_j) - \frac{\partial^2 v_N(x_j)}{\partial x^2} = (A + Bx)\, T'_N. \tag{6.8}$$

Therefore

$$v_N(x) = A\,\phi_N(x, \lambda) + B\chi_N(x, \lambda) \tag{6.9}$$

where

$$\phi_N(x, \lambda) = \sum_{k=0}^{\infty} \lambda^{-k-1} \frac{\partial^{2k+1}}{\partial x^{2k+1}} T_N(x) \tag{6.10a}$$

and

$$\chi_N(x, \lambda) = \sum_{k=0}^{\infty} \lambda^{-k-1} \frac{\partial^{2k}}{\partial x^{2k}} \left(x \frac{\partial T_N}{\partial x} \right). \tag{6.10b}$$

We assume for simplicity that N is even, and therefore $\phi_N(x, \lambda)$ is an odd function of x and $\chi_N(x, \lambda)$ is an even function of x. For the Dirichlet problem with homogeneous boundary conditions, the eigenvalues λ are obtained by solving

$$\phi_N(1, \lambda)\chi_N(1, \lambda) = 0. \tag{6.11}$$

Upon setting $\mu = 1/\lambda$, it follows that ϕ_N is a polynomial of degree $N/2 - 1$ in μ while χ_N is a polynomial of degree $N/2$ in μ. It has been shown (Gottlieb & Lustman [1982]) that the roots of $\phi_N(1, 1/\mu)$ are real, negative and distinct, as are those of $\chi_N(1, 1/\mu)$. Therefore, (6.11) yields $N - 1$ real, negative and distinct eigenvalues, which is the required number of roots since there are $N - 1$ interior points in the scheme.

In conclusion we have now explicitly $N - 1$ eigenfunctions

$$\phi_N(x, \lambda_1), \cdots, \phi_N(x, \lambda_{N/2-1}), \chi_N(x, \nu_1), \cdots, \chi_N(x, \nu_{N/2}). \tag{6.12}$$

where $\lambda_1, \cdots, \lambda_{N/2-1}$ are the roots of $\phi_N(1, \lambda)$ and $\nu_1, \cdots, \nu_{N/2}$ are those of $\chi_N(1, \lambda)$.

Consider now the more general case of boundary conditions

$$\alpha U(1) + \beta u_x(1) = 0, \qquad \gamma u(-1) + \delta u_x(-1) = 0. \tag{6.13}$$

There are as yet no stability proofs for the pseudospectral Chebyshev method for (6.1) with (6.13). The difficulty in finding an energy estimate is that the norm used for Dirichlet boundary conditions is different from that used for Neumann boundary conditions. However, it is still possible to find the eigenvalues and eigenvectors for this problem. As a matter of fact, the eigenvalues are given by the roots η_i of the polynomial equation

$$\begin{aligned} &\left[[\alpha\phi_N(1, \eta) + \beta \frac{\partial \phi_N}{\partial x}(1, \eta)\right]\left[\gamma\chi_N(1, \eta) - \delta \frac{\partial \chi_N}{\partial x}(1, \eta)\right] \\ &\quad + \left[\alpha\chi_N(1, \eta) + \beta \frac{\partial \chi_N}{\partial x}(1, \eta)\right]\left[\gamma\phi_N(1, \eta) - \delta \frac{\partial \phi_N}{\partial x}(1, \eta)\right] = 0. \end{aligned} \tag{6.14}$$

It may be shown that the roots η_i $(i = 1, \cdots, N - 1)$ of (6.15) are real, negative and distinct if α and β are of the same sign and γ and δ are of the opposite sign.

The results for the eigenvalue analysis indicate that, for fixed number of mesh points N, the pseudospectral approximation converges to a steady state as $t \to +\infty$, as does the analytic solution to (6.1). The analysis does not prove convergence of u_N to u as N increases. However, it is important to know the structure of the eigenvalues and eigenvectors to fully understand the behavior of the approximation.

Results for scalar hyperbolic problems. Let us start by considering the scalar equation

$$\begin{aligned} &u_t = u_x \qquad (|x| < 1), \\ &u(x, 0) = f(x), \qquad u(1, t) = g(t). \end{aligned} \tag{6.15}$$

For this simple equation all the methods described in §4 are applicable. We shall review here the theory for the pseudospectral approximation based on the points y_j defined in (4.18). The use of other points will be discussed later.

Let u_N be the pseudospectral Chebyshev approximation to u. Then u_N satisfies

$$\frac{\partial u_N}{\partial t} = \frac{\partial u_N}{\partial x} \qquad (x = y_j, j = 1, \cdots, N), \tag{6.16}$$
$$u_N(1,t) = u_N(y_0, t) = g(t), \qquad u_N(x,0) = P_N f(x).$$

Then the following inequality holds (Gottlieb [1981]):

$$\frac{\pi}{N}\frac{d}{dt}\sum_{j=1}^{N}(1+y_j)u_N^2(y_j,t) + \frac{\pi}{8N}u_N^2(-1,t) \leqq \frac{\pi N}{6}g^2(t), \tag{6.17}$$

which gives the following error estimate:

$$\frac{\pi}{N}\sum_{j=0}^{N}(1+y_j)[u(y_j,t) - u_N(y_j,t)]^2 \leqq CN^{2(1+\sigma)-s} \tag{6.18}$$

where $\sigma > 1/2$ and $s > 2(1+\sigma)$, s is the number of derivatives possessed by the initial conditions and the constant C is independent of u and N. Similar results hold for the points $z_j^{(1)}$, $z_j^{(2)}$ defined in (4.24). Unfortunately, there is still no estimate of the form (6.17) and (6.18) for the set of points x_j defined in (4.1), which is the set most commonly used in applications.

If the differential equation is of the slightly more general form

$$u_t = c(x)\, u_x \qquad (|x| < 1)$$

where $c(x)$ does not change sign in the domain, the above results still hold. The case in which $c(x)$ changes sign introduces some difficulty. The following examples illustrate the kinds of problems that may arise. Consider the equations

$$u_t = xu_x \qquad (|x| < 1), \tag{6.19}$$
$$u(x,0) = f(x), \quad u(1,t) = f(e^t), \quad u(-1,t) = f(-e^t)$$

whose analytic solution is $u(x,t) = f(xe^t)$, and

$$u_t = -xu_x \quad (|x| < 1), \qquad u(x,0) = f(x), \tag{6.20}$$

with solution $u = f(xe^{-t})$. The pseudospectral Chebyshev approximation to both problems has been shown to be stable (Gottlieb [1981]). However, there is a noticeable difference in the behavior of the error between (6.19) and (6.20). For (6.19), there is a significant loss of accuracy as t increases. This is due to the fact that the gradient of the solution grows with time as e^t and, therefore, with a fixed number of grid points, the numerical approximation cannot resolve the steep gradients. This phenomenon does not occur when (6.20) is solved.

When the differential equation being solved is nonlinear, behavior similar to that encountered for (6.19) may occur in the numerical approximation, except now the solution may diverge explosively in a finite time rather than having just gradual loss of accuracy in time. Smoothing, done by cutting high modes, should be applied in these cases to avoid such "nonlinear instabilities."

Results for systems of hyperbolic equations. Consider the symmetric 2×2 system

$$\frac{\partial}{\partial t}\begin{pmatrix} u \\ v \end{pmatrix} = \begin{pmatrix} \frac{1}{2} & 1 \\ 1 & \frac{1}{2} \end{pmatrix} \frac{\partial}{\partial x}\begin{pmatrix} u \\ v \end{pmatrix},$$
$$u(-1) = f(t), \qquad u(1) = g(t), \tag{6.21}$$
$$\begin{pmatrix} u \\ v \end{pmatrix}_{x=0} = \begin{pmatrix} \psi_1(x) \\ \psi_2(x) \end{pmatrix}.$$

For such a nondiagonal system, the points $z_j^{(1)}$, $z_j^{(2)}$ are not easily applicable. Therefore, we shall discuss the pseudospectral method for the set of collocation points $x_j = \cos \pi j/N$ $(j = 0, \cdots, N)$. (Similar results hold for the points y_j.) Assume that u and v are known at some time t_1 and that we would like to advance to the time level t_2. First, we get the values of $u_x(x, t_1)$ and $v_x(x, t_1)$ by the pseudospectral procedure described in §4. Then we multiply by the matrix

$$\begin{pmatrix} 1 & \frac{1}{2} \\ \frac{1}{2} & 1 \end{pmatrix}$$

and advance in time to t_2. This algorithm gives values for u and v at the time t_2 at each of the collocation points x_j, which we shall denote by $u^c(x_j, t_2)$ and $v^c(x_j, t_2)$. Then we set

$$u(x_j, t_2) = u^c(x_j, t_2) \qquad (j = 1, \cdots, N-1),$$
$$v(x_j, t_2) = v^c(x_j, t_2) \qquad (j = 0, \cdots, N), \tag{6.22}$$
$$u(-1, t_2) = f(t_2), \qquad u(1, t_2) = g(t_2).$$

It turns out that the pseudospectral method just described is unstable, although the differential equation is well posed. Instead of (6.22), improved boundary conditions may be shown to lead to a stable scheme. The form of these conditions is suggested by using characteristic variables which are, in this example, $u + v$ on the characteristic $dx/dt = -\frac{3}{2}$ and $u - v$ on the characteristic $dx/dt = \frac{1}{2}$. Ideally, the quantity $u - v$ on the incoming characteristic should be specified at $x = -1$, whereas $u + v$ at $x = -1$ is carried on an outgoing characteristic and should be determined by the time marching scheme. Similarly, $u + v$ should be given at $x = 1$, while $u - v$ at $x = 1$ should be given by the time marching scheme. The procedure described by (6.22) is unstable because it uses the numerical scheme (at $x = -1$) to find v rather than $u + v$. The scheme is stabilized by requiring that $u + v$ at $x = -1$ and $u - v$ at $x = 1$ have the values computed by the time marching scheme from the previous time level, even after imposition of the boundary conditions on $u(\pm 1, t_2)$. Thus (6.22) is replaced by

$$u(x_j, t_2) = u^c(x_j, t_2) \qquad (j = 1, \cdots, N-1),$$
$$v(x_j, t_2) = v^c(x_j, t_2) \qquad (j = 1, \cdots, N-1),$$
$$u(x_0, t_2) = g(t_2), \qquad u(x_N, t_2) = f(t_2), \tag{6.23}$$
$$u(x_0, t_2) - v(x_0, t_2) = u^c(x_0, t_2) - v^c(x_0, t_2),$$
$$u(x_N, t_2) + v(x_N, t_2) = u^c(x_N, t_2) + v^c(x_N, t_2).$$

In general, when a hyperbolic system of equations is solved one should use the time marching scheme to determine the boundary values of the outgoing characteristic variables. This procedure yields stable results. Applications to multidimensional problems are given by Gottlieb, Lustman & Streett, this volume.

Eigenvalues and eigenvectors for hyperbolic equations. As in the case of parabolic equations, it is possible to find explicitly the eigenvectors and the characteristic equations for the Chebyshev approximations to (6.15). Let u_N denote the approximation to the solution of (6.15) based on the points x_j, and let v_N denote the approximations based on the points y_j. Then u_N and v_N satisfy exactly

$$\frac{\partial u_N}{\partial t} = \frac{\partial u_N}{\partial x} + \tau \frac{(1+x)T'_N}{2N^2}, \qquad \frac{\partial v_N}{\partial t} = \frac{\partial v_N}{\partial x} + \bar{\tau} \frac{T'_{N+1}}{N^2} \tag{6.24}$$

for suitable $\tau, \bar{\tau}$. Assuming that $u_N = e^{\lambda t}\hat{u}_N$, $v_N = e^{\mu t}\hat{v}_N$ and substituting into (6.24), one obtains

$$\frac{\partial \hat{u}_N}{\partial x} = \lambda u_N + \alpha \frac{(1+x)T'_N}{2N^2}, \qquad \frac{\partial \hat{v}_N}{\partial x} = \mu \hat{v}_N + \beta \frac{T'_{N+1}}{N^2} \tag{6.25}$$

for suitable α, β. Therefore

$$\hat{u}_N(x) = \sum_{k=0}^{\infty} \lambda^{-k-1} \frac{d^k}{dx^k} \frac{(1+x)T'_N}{2N^2}, \qquad \hat{v}_N(x) = \sum_{k=0} \mu^{-k-1} \frac{d^{k+1}}{dx^{k+1}} \frac{T_{N+1}}{N^2} \tag{6.26}$$

where λ and μ are determined by the eigenvalue conditions

$$\hat{u}_N(1) = 0, \qquad \hat{v}_N(1) = 0.$$

All the eigenvalues satisfy $\operatorname{Re} \lambda < 0$, $\operatorname{Re} \mu < 0$.

7. Other polynomial expansions. Although most of the actual computations with spectral methods involve expansion with trigonometric functions and Chebyshev polynomials, there are other basis functions that are currently being used. Here we briefly review some of those methods.

Legendre polynomials. An attractive alternative to Chebyshev polynomial expansions is Legendre polynomial expansions. It suffices to explain how to construct a pseudospectral Legendre polynomial approximation to a derivative.

Let $x_0 = -1$, $x_N = 1$, and let x_i $(i = 1, \cdots, N-1)$ be the roots of $q'_N(x)$, where $q_N(x)$ is the Legendre polynomial of degree N. Given the values of any function $f(x)$ at the points x_j $(j = 0, \cdots, N)$, we construct the interpolating polynomial

$$P_N f = \sum_{j=0}^{N} f(x_j) g_j(x) \tag{7.1}$$

where

$$g_j(x) = -\frac{1}{\alpha_N q_N(x_j)} \frac{(1-x^2)q'_N(x)}{x - x_j}$$

with $\alpha_N = N(N+1)$. Therefore,

$$\left. \frac{d^l}{dx^l} P_N f \right|_{x=x_k} = \sum_{j=0}^{N} f(x_j) \frac{\partial^l g_j(x_k)}{\partial x^l} = \sum_{j=0}^{N} (D_l)_{kj} f(x_j). \tag{7.2}$$

For example,

$$(7.3)\qquad \begin{aligned} &(D_1)_{kj} = \frac{q_N(x_k)}{q_N(x_j)}\frac{1}{x_k - x_j} \qquad (k \neq j),\\ &(D_1)_{00} = \tfrac{1}{4}\alpha_N = -(D_1)_{NN},\\ &(D_1)_{kk} = 0 \qquad (k \neq 0, k \neq N) \end{aligned}$$

The difference between the Chebyshev and Legendre methods is evident here. The matrix D_1 for Legendre polynomials is nearly antisymmetric, in contrast to the Chebyshev matrix given in (4.12).

By the same method, we obtain

$$(7.4)\qquad \begin{aligned} &(D_2)_{kj} = -2\frac{q_N(x_k)}{q_N(x_j)}\frac{1}{(x_k - x_j)^2} \qquad (1 \leqq k,j \leqq N-1, k \neq j),\\ &(D_2)_{jj} = -\frac{1}{3}\frac{N}{1 - x_j^2} \qquad (1 \leqq j \leqq N-1). \end{aligned}$$

This shows that $D_2 = \Lambda S \Lambda^{-1}$, where Λ is a diagonal matrix and S is symmetric.

The following result indicates that the accuracy of Legendre approximations is comparable with the accuracy of Chebyshev approximations:

THEOREM 7.1. *Let $f(x)$ have σ smooth derivatives for $|x| \leqq 1$, and define*

$$\|f\|^2 = \sum_{l=0}^{\sigma} \int_{-1}^{1} \left(\frac{d^l f}{dx^l}\right)^2 dx.$$

Then there is a constant C independent of $f(x)$ and N such that

$$(7.5)\qquad \sum_{l=0}^{\mu}\left[\sum_{j=0}^{N}\left(\frac{d^l f}{dx^l}(x_j) - \frac{d^l P_N f(x_j)}{dx^l}\right)^2\right]^{1/2} \leqq C N^{2\mu-\sigma}\|f\|_\sigma$$

for $\mu < \tfrac{1}{2}\sigma$ and where $\|f\|_\sigma$ is defined in (5.2).

Stability results are much easier to obtain here than with Chebyshev approximations. For example, consider the pseudospectral Legendre method for the approximation of

$$(7.6)\qquad u_t = u_x \quad (|x| < 1), \qquad u(1, t) = 0.$$

From (7.3) it follows that

$$(7.7)\qquad \begin{aligned} \frac{1}{2}\frac{d}{dt}\sum_{k=0}^{N}\frac{1}{[q_N(x_k)]^2}u_N^2(x_k) &= \sum_{k=0}^{N}\sum_{j=0}^{N}\frac{1}{[q_N(x_k)]^2}(D_1)_{kj}u_N(x_j)u_N(x_k)\\ &= \frac{1}{2}\sum_{k=0}^{N}\sum_{j=0}^{N}\left[\frac{(D_1)_{kj}}{[q_N(x_k)]^2} + \frac{(D_1)_{jk}}{[q_N(x_j)]^2}\right]u_N(x_j)u_N(x_k)\\ &= -\frac{1}{2}\alpha_N\frac{u^2(x_N)}{q_N^2(x_N)} < 0, \end{aligned}$$

showing stability.

A similar proof holds for parabolic equations.

Jacobi polynomials. An interesting use of a different set of expansion functions was suggested by Wray & Leonard [1982] for the case of three-dimensional flow in a pipe.

The velocity vector $\vec{u}$ is expanded

$$\vec{u}(r, \theta, x, t) = \sum_{n,k,l} a_{m,k,l}(t)\vec{\chi}_{m,k,l}(r) \exp(ikx + il\theta). \tag{7.8}$$

The vector $\vec{\chi}(r)$ is expressed in terms of q_n^l where

$$q_n^l = r^{|l|}(1 - r^2)^2 g_n^{|l|}(r^2),$$

where $g_n^l(y)$ is the shifted Jacobi polynomial

$$g_n^l(y) - P_n^{(0,l)}(2y - 1).$$

With this choice of basis function, one gets the correct behavior of $\vec{u}$ as $r \to 0$. Moreover, since $q_n^{(l)}$ satisfy the orthogonality condition

$$\int_0^1 y q_m^{(l)}(y) q_n^{(l)}(y)\, dy = c_m^l \delta_{m,n},$$

one gets simple expressions for the derivatives. For example, one can show that

$$\nabla^2[q_n^l(r) \exp(il\theta)] = r^l[b_n^l q_{n-1}^{(l)} + c_n^l q_n^{(l)} + d_n^l q_{n+1}^{(l)}] \exp(il\theta)$$

for certain constants b_n^l, c_n^l, d_n^l.

8. Time discretization and iterative methods.

8.1. Time discretization. One of the major difficulties in the application of spectral methods to flow problems is time marching methods. We may divide flow problems into three categories. First, we have problems for which we are interested only in steady state solutions. These problems require a time marching technique which may be inaccurate, but should converge fast to the steady state solution. Discussion of these methods will be given in §8.3. The second class of problems includes those in which the temporal behavior of the solution occurs on a much slower scale than the spatial one. Here it is reasonable to use low order accurate finite difference schemes in order to advance the solution in time. The problem that one faces here is that spectral methods require severe time step limitations when an explicit time marching technique is applied. In addressing this problem, Gottlieb & Turkel [1980] suggested a second-order-in-time, unconditionally stable scheme. This scheme seems to perform satisfactorily for Fourier methods, but their success in the case of Chebyshev methods is still questionable. For parabolic problems the modification of the Du-Fort–Frankel scheme was suggested by Gottlieb & Gustafsson [1976] and Gottlieb & Lustman [1981], and has been investigated by Funaro [1983a].

The third class of flow problems involves phenomena in which the temporal and spatial evolution occur on the same scale. It is still possible to use finite differencing techniques for time marching, but the time step must be small. Another alternative is to use spectral discretization in time; however, this approach usually leads to a very large and complicated set of equations. Morchoisne [1979] suggested simplifying the set of equations thus obtained by using the idea of approximate factorization. Several other methods are discussed by Deville et al. [1981].

8.2. Time stepping methods for incompressible flow problems. Here, we survey time stepping methods and boundary condition procedures for incompressible Navier–Stokes equations which are incompletely parabolic and time singular. The equations are

$$\frac{\partial \vec{v}}{\partial t} + \vec{\nabla} p = -\vec{v} \cdot \vec{\nabla} \vec{v} + \nu \nabla^2 \vec{v} + \vec{f} \qquad (\vec{x} \in D), \tag{8.1}$$

$$\vec{\nabla} \cdot \vec{v} = 0 \qquad (\vec{x} \in D), \tag{8.2}$$

where $\vec{v}(\vec{x}, t)$ is the velocity field at $\vec{x}, t$, $p(\vec{x}, t)$ is the pressure, ν is the kinematic viscosity and $\vec{f}(\vec{x}, t)$ is an external force. The (constant) density is assumed to be 1. If the boundaries of the region D in which the flow satisfying (8.1)–(8.2) occurs are stationary, then for $\nu > 0$, the appropriate boundary conditions are the no-slip conditions

$$\vec{v}(\vec{x}, t) = 0 \qquad (\vec{x} \in \partial D). \tag{8.3}$$

The pressure in (8.1) may be considered a Lagrange multiplier that ensures satisfaction of the (kinematical) incompressibility constraint (8.2) everywhere in D. The most obvious way to obtain an equation for p is to take the divergence of (8.1) and apply (8.2), which gives the Poisson equation

$$\nabla^2 p = -\vec{\nabla} \cdot (\vec{v} \cdot \vec{\nabla})\vec{v} + \vec{\nabla} \cdot \vec{f} \qquad (\vec{x} \in D). \tag{8.4}$$

A warning of possible trouble in the numerical solution is given when boundary conditions for (8.4) are sought (Orszag & Israeli [1974]). Applying (8.3) to (8.1) gives

$$\vec{\nabla} p = \nu \nabla^2 \vec{v} + \vec{f} \qquad (\vec{x} \in \partial D), \tag{8.5}$$

so both Dirichlet and Neumann conditions for p are available if $\vec{v}$ is known. In numerical integrations of (8.1)–(8.3), it is not obvious which of these boundary conditions to use.

The ambiguity regarding boundary conditions on p can be avoided if the pressure is eliminated from (8.1)–(8.3). Applying the operator $\vec{\nabla} \times (\vec{\nabla} \times \cdots)$ to (8.1) gives, using (8.2),

$$\frac{\partial}{\partial t} \nabla^2 \vec{v} = \vec{\nabla} \times (\vec{\nabla} \times (\vec{v} \cdot \vec{\nabla})\vec{v}) + \nu \nabla^4 \vec{v} - \vec{\nabla} \times (\vec{\nabla} \times \vec{f}) \qquad (\vec{x} \in D). \tag{8.6}$$

For flow in a plane channel, implicit time stepping of the linear terms in (8.6) gives a fourth-order equation for the component v_n of $\vec{v}$ in the direction $\vec{n}$ of the normal to the channel walls. The boundary conditions on v_n are $v_n = \partial v_n / \partial n = 0$. Once v_n is known, the other components of $\vec{v}$ are obtained using incompressibility and the equation for the component of vorticity in the direction $\vec{n}$. The numerical implementation of this scheme is discussed in Orszag & Patera [1983, App. II].

A one-dimensional linear model that embodies the essential features of the incompressibility and viscous terms of the Navier–Stokes equations is obtained by considering a solution to the two-dimensional Stokes equations of the form

$$\vec{v} = (u(y, t)e^{ikx}, v(y, t)e^{ikx}), \qquad p = p(y, t)e^{ikx}$$

for some real wavenumber k. Further details of the following analyses are given by Orszag, Israeli & Deville [1983]. The equations satisfied by (u, v, p) are

$$\frac{\partial u}{\partial t} = -ikp + \nu(u_{yy} - k^2 u), \quad \frac{\partial v}{\partial t} = -\frac{\partial p}{\partial y} + \nu(v_{yy} - k^2 v), \quad iku + \frac{\partial v}{\partial y} = 0 \tag{8.7}$$

for $-1 \leq y \leq 1$. The boundary conditions are

$$u(\pm 1, t) = v(\pm 1, t) = 0, \tag{8.8}$$

which simulate rigid no-slip boundary conditions. In general three-dimensional geometries, these equations are coupled and are not easily solved.

The solution to an initial value problem for (8.7)–(8.8) can be expressed in terms of normal modes, for which

$$(u, v, p)(y, t) = e^{\bar{\sigma} t}(\hat{u}, \hat{v}, \hat{p})(y) \tag{8.9}$$

with symmetric modes

$$v(y) = \cos \bar{\mu} \cosh ky - \cosh k \cos \bar{\mu} y \tag{8.10}$$

or antisymmetric modes

$$v(y) = \sin \bar{\mu} \sinh ky - \sinh k \sin \bar{\mu} y. \tag{8.11}$$

Here

$$\bar{\mu} = \left(-\frac{\bar{\sigma}}{\nu} - k^2\right)^{1/2} \tag{8.12}$$

satisfies the eigenvalue relations

$$k \tanh k = -\bar{\mu} \tan \bar{\mu} \tag{8.13}$$

for (8.10) and

$$k \coth k = \bar{\mu} \cot \bar{\mu} \tag{8.14}$$

for (8.11). These eigenmodes are complete on the interval $|y| \leqq 1$.

Here we outline several schemes for the solution of (8.7)–(8.8) in which time t is discretized by differences, but space x is discretized spectrally.

Scheme 1. *Full implicit time differencing.* The system (8.7)–(8.8) is approximated by backwards Euler time differencing:

$$\begin{aligned} \frac{u^{n+1} - u^n}{\Delta t} &= -ikp^{n+1} + \nu(D^2 - k^2)u^{n+1}, \\ \frac{v^{n+1} - v^n}{\Delta t} &= -Dp^{n+1} + \nu(D^2 - k^2)v^{n+1}, \qquad iku^{n+1} + Dv^{n+1} = 0 \end{aligned} \tag{8.15}$$

where the superscript index n indicates time level $n\Delta t$ and D indicates a spectral-y derivative. It follows that

$$\frac{1}{\Delta t}(D^2 - k^2)u^{n+1} - \nu(D^2 - k^2)^2 u^{n+1} = -\frac{ik}{\Delta t} Dv^n - \frac{k^2}{\Delta t} u^n \tag{8.16}$$

with $u^{n+1} = Du^{n+1} = 0$ at $y = \pm 1$. It may be shown that this scheme is stable and spectrally accurate in y and first order in Δt. Methods to solve (8.16) for the spectral representation of u^{n+1} are given by Orszag, Israeli & Deville [1983]. Higher order results in Δt are easily obtained by using higher order time differencing methods.

Scheme 2. *Implicit time splitting.* Here the time differencing scheme involves two split (or fractional) time steps (Orszag & Kells [1980]). The first step involves solution of the inviscid equations

$$\frac{u^* - u^n}{\Delta t} = -ikp^*, \quad \frac{v^* - v^n}{\Delta t} = -Dp^*, \quad iku^* + Dv^* = 0 \tag{8.17}$$

for $|y| \leq 1$ with boundary condition

$$u^* = 0 \qquad (y = \pm 1).$$

The second step involves the solution of the viscous equations

$$\text{(8.18)} \qquad \frac{u^{n+1} - u^*}{\Delta t} = \nu(D^2 - k^2)u^{n+1}, \qquad \frac{v^{n+1} - v^*}{\Delta t} = \nu(D^2 - k^2)v^{n+1}$$

for $|y| \leq 1$ with boundary conditions

$$u^{n+1} = v^{n+1} = 0 \qquad (y = \pm 1).$$

In this scheme, u^n and v^n do not satisfy the incompressibility constraint (8.2), although the intermediate variables u^* and v^* do. In comparison with Scheme 1, Scheme 2 has the advantage of only involving the solution of second-order Helmholtz (or Poisson) equations. In general geometries, it offers significant simplifications compared to unsplit schemes. However, Scheme 2 suffers from large time stepping errors if $\nu\Delta t$ is large; these can only be removed at considerable computational expense by extrapolation (see Orszag, Israeli & Deville [1983]).

It is by no means obvious that Scheme 2 has solutions consistent (and convergent) to the solutions of the Stokes equations. For example, it follows from (8.17) and the boundary conditions that

$$Dp^* = 0 \qquad (y = \pm 1),$$

but (8.10) gives

$$Dp = \nu D^2 u \qquad (y = \pm 1),$$

which is, in general, not vanishing. It seems that there is an $O(1)$ error in Dp as $\Delta t \to 0$ at $y = \pm 1$. Nevertheless, Orszag, Israeli & Deville [1983] show that, away from the boundaries $y = \pm 1$, the global error estimates

$$\text{(8.19)} \qquad u^n(y) - u(y, n\Delta t) \sim u_1(y)\Delta t + O(\Delta t^{3/2}),$$

$$\text{(8.20)} \qquad v^n(y) - v(y, n\Delta t) \sim v_1(y)\Delta t + O(\Delta t^{3/2}),$$

$$\text{(8.21)} \qquad p^n(y) - p(y, n\Delta t) \sim p_1(y)\Delta t + O(\Delta t^{3/2})$$

hold for fixed y with $|y| < 1$ as $\Delta t \to 0$, where u_1, v_1, p_1 are finite functions of y. Near the boundaries $y = \pm 1$, the error estimates (8.19), (8.20) hold together with

$$Du^n(y) - Du(y, n\Delta t) = O(\Delta t^{1/2}),$$

$$D^2u^n(y) - D^2u(y, n\Delta t) = O(1),$$

$$D^2v^n - D^2v(y, n\Delta t) = O(\Delta t^{1/2}),$$

$$p^n(y) - p(y, n\Delta t) = O(\Delta t^{1/2})$$

for $|y| - 1 = O((\nu\Delta t)^{1/2})$, while the error in Dp is, as noted above, $O(1)$ in this region.

Scheme 3. *Boundary-divergence-free implicit time splitting.* Marcus, Orszag & Patera [1982] and Marcus [1983] use a modification of Scheme 2 that removes many of the boundary errors analyzed above. The idea is that splitting error induces large boundary errors, because a large divergence of (u, v) develops near the boundaries $|y| = 1$ on the viscous step. The modified method is based on the observation that the normal flow boundary condition $v = 0$ is applied twice each time step in Scheme 2, so that by relaxing these conditions on v it may be possible to reduce the error in the boundary divergence. The modified scheme is given by dropping $v^* = 0$ at $y = \pm 1$ in favor of the condition

$$iku^{n+1} + Dv^{n+1} = 0 \qquad (y = \pm 1) \tag{8.22}$$

while applying (8.17) and (8.18). Thus, normal flow is allowed at the boundary during the inviscid pressure step in order to ensure that (8.22) holds. Analysis of this modified scheme shows that it is uniformly first order in Δt as a function of y, while higher-order time differencing methods (e.g. Crank–Nicolson differencing) leads to correspondingly higher-order results. However, in more than two space dimensions, inversions of large matrices are required by this method.

8.3. Iterative methods. The system of algebraic equations which results from a pseudospectral approximation to elliptic problems or evolution problems (with implicit time discretization) may be written as

$$L(u) = f \tag{8.23}$$

where L is an operator, possibly nonlinear, f is a known vector and u is the approximate solution. As shown in §§2 and 4, even in the linear case the matrices which represent the operator are full and are difficult to invert directly. Various basic iterative schemes which are available to solve (8.23) can be discussed within the framework of the defect correction procedure. Suppose there is a sparse, efficiently invertible matrix H which approximates the Jacobian J_L of L. Let u^n be a guess for a solution u of (8.23). Taylor expansion about u^n gives

$$0 = L(u) - f = L(u^n) - f + J_L(u - u^n) + O([u - u^n]^2) \cong L(u^n) - f + H(u - u^n).$$

Thus, the simplest defect correction iterative scheme is

$$H(u^{n+1}) = H(u^n) - [L(u^n) - f],$$

or

$$u^{n+1} = u^n - H^{-1}[L(u^n) - f]. \tag{8.24}$$

The various iterative methods differ in the choice of the preconditioning operator H. For instance, in the case of Jacobi's method, H is the diagonal of J_L, and for the Gauss–Seidel method, H is the lower triangular part of J_L (Hageman & Young [1981]).

Morchoisne [1979] has applied the pseudospectral method in space and time to the unsteady two-dimensional Navier–Stokes equations—in the form of a fourth-order equation in the stream function. As a preconditioning operator, he used an approximate factorization of the Navier–Stokes operator. More specifically, the operators L and H were defined as in the usual notation with $\Delta \equiv \nabla^2$:

$$L(\psi^n) = -\frac{\partial}{\partial t}\Delta\psi^n - \frac{\partial\psi^n}{\partial y}\frac{\partial}{\partial x}\Delta\psi^n + \frac{\partial\psi^n}{\partial x}\frac{\partial}{\partial y}\Delta\psi^n + \nu\Delta\psi^n,$$

with x, y, t $[-1, 1] \times [-1, 1] \times [0, T]$, and $H \equiv \mathcal{L}_y\mathcal{L}_x\mathcal{L}_t$ where

$$\mathcal{L}_t = \frac{1}{\varepsilon\eta^\sigma}\left(1 + \varepsilon\frac{\partial}{\partial t}\right), \quad \mathcal{L}_x = \left(1 + \varepsilon u^n\frac{\partial}{\partial x} - \varepsilon\nu\frac{\partial^2}{\partial x^2}\right)\left(1 - \eta\frac{\partial^2}{\partial x^2}\right),$$

$$\mathcal{L}_y = \left(1 + \varepsilon v^n\frac{\partial}{\partial y} - \varepsilon\nu\frac{\partial^2}{\partial y^2}\right)\left(1 - \eta\frac{\partial^2}{\partial y^2}\right)$$

and $u^n = \partial\psi^n/\partial y$, $v^n = -\partial\psi^n/\partial x$. This scheme awaits more general testing.

The nonstationary analogue of the basic iterative scheme (8.24) is the preconditioned Richardson's method

$$u^{n+1} = u^n - \alpha_n H^{-1} [L(u^n) - f], \tag{8.25}$$

which has faster convergence. The convergence rate can further be accelerated by applying polynomial acceleration (Hageman & Young [1981]) to (8.24):

$$u^{n+1} = \omega_n u^n - \alpha_n \omega_n H^{-1}(L(u^n) - f) + (1 - \omega_n) u^{n-1}.$$

This is usually called a nonstationary second degree iterative scheme, and it includes Chebyshev and conjugate gradient methods (Hageman & Young [1981]).

Orszag [1980] discusses the Richardson, Chebyshev and conjugate gradient iterative schemes in the linear case, and suggests that the preconditioning operator H should be a suitably chosen low-order finite difference approximation H_{FD} to L. For a differential equation dominated by second derivative terms, Orszag [1980] shows that

$$m \leqq \|H_{FD}^{-1} L\| \leqq M$$

with $m = 1$ and $M \sim 2.5$. Unfortunately, $\|H_{FD}^{-1} L\|$ is unbounded if L is a first derivative operator and H_{FD} is a central difference approximation. In order to keep $\|H_{FD}^{-1} L\|$ bounded in this case, it is necessary to have more degrees of freedom (grid points) for H_{FD} than for L, and then to modify suitably higher frequencies in the spectral approximation. An alternate remedy is discussed in Hussaini & Zang (this volume) where a semiimplicit Fourier–Chebyshev pseudospectral method of solution (involving a finite difference preconditioning) is described for solving the unsteady two-dimensional incompressible Navier–Stokes equations.

In multidimensional problems it may prove expensive to evaluate the inverse of H_{FD}. Efficiently invertible approximate versions of H_{FD} such as approximate factorizations and incomplete LU decomposition of H_{FD} are the obvious choices. Zang et al. [1982] and Wong et al. [1983] discuss several types of LU decompositions. Application of these preconditioning operators is typically far less expensive than the spectral evaluation of the residue. Thus, there is a large payoff for improving the effectiveness of the preconditioning and in using iterative schemes with better convergence rates.

The minimum residual (MR) method (Wong et al. [1983]) is a robust parameter-free scheme. The only requirement for convergence is that all the eigenvalues of $H^{-1}L$ be strictly in the right half of the complex plane. It can be put in the form of the classical Richardson iterative scheme (8.3) where

$$\alpha_n = \frac{(r^n, Lv^n)}{(Lv^n, Lv^n)}, \qquad r^n = Lu^n - f, \quad v^n = H^{-1} r^n.$$

Figure 3 shows the convergence history of MR solutions to this Chebyshev pseudospectral approximation to the two-dimensional Poisson's equation with Dirichlet boundary conditions (Wong et al. [1983]). The labels A and B refer to different LU decompositions; the label I indicates no preconditioning.

Spectral multigrid methods are very much in the initial stages of development. The work of Zang et al. [1982] and Hussaini & Zang (this volume) suggests that multigrid procedures, which have significantly improved the rates of convergence for finite difference and finite element methods, may do equally well in the case of spectral methods. A glimpse of this promise is seen in Table 1, which compares the convergence

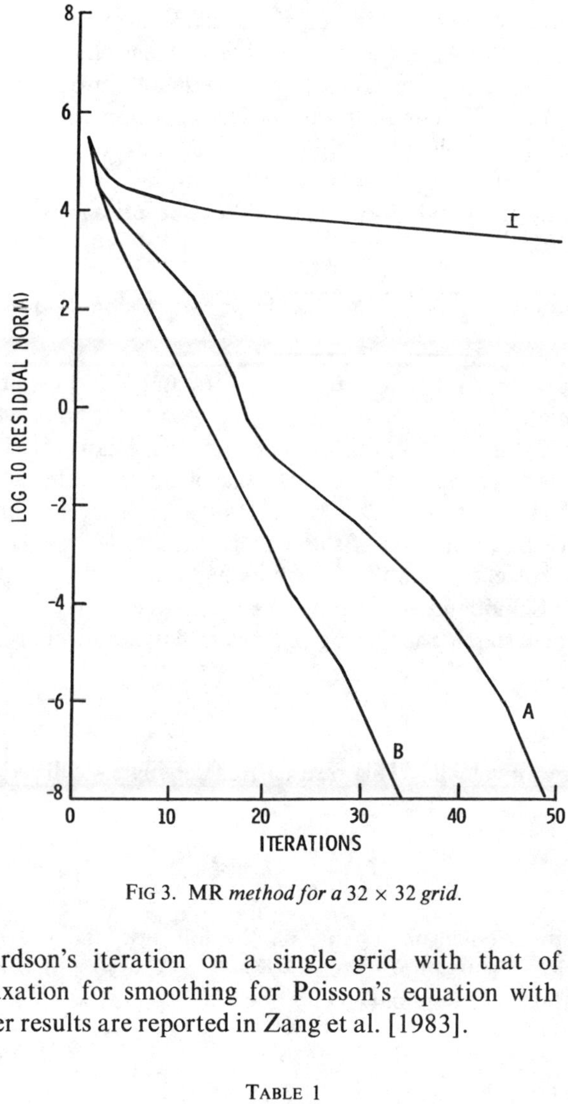

FIG 3. MR *method for a* 32 × 32 *grid.*

history of Richardson's iteration on a single grid with that of a multigrid with Richardson's relaxation for smoothing for Poisson's equation with periodic boundary conditions. Further results are reported in Zang et al. [1983].

TABLE 1
RMS *residual*

Relaxation	Single grid	Multigrid
3	2.73 (1)	2.82 (7)
6	2.08 (1)	2.42 (−1)
9	1.63 (1)	6.35 (−3)
12	1.31 (1)	4.56 (−4)
15	1.07 (1)	8.30 (−5)

9. Incompressible flows.

9.1. Three-dimensional numerical algorithms. The spectral simulation of viscous incompressible (external and internal) flows up to 1974 were reviewed by Orszag & Israeli [1974].

We present here a detailed description of a spectral algorithm for the solution of the Navier–Stokes equations (8.4) in the case of three-dimensional plane shear flows. The physical domain for these flows extends to infinity in the horizontal directions and is assumed periodic in those directions; in the vertical direction, the flow is either confined between two parallel rigid walls $|y| \leq 1$ (plane channel flow) or bounded by a wall on the one side and extends to infinity in the other direction (boundary layer flow). In this case, it is natural to expand the flow field $\vec{v} = (u, v, w)$ in Fourier series in x and z, and Chebyshev polynomials in y:

$$\vec{v}(\vec{x}, t) = \sum_{|k_x|<N_1} \sum_{|k_z|<N_2} \sum_{k=0}^{N_3} \tilde{\vec{v}}(k_x k_z, k, t) e^{i(k_x\alpha x + k_z\beta z)} T_k(y),$$

where $\alpha = 2\pi/L_1$, $\beta = 2\pi/L_2$, $0 < x < L_1$, $0 < z < L_2$ and $T_k(y)$ denotes the kth degree Chebyshev polynomial. The solution techniques of Orszag & Kells [1980], Moin & Kim [1980] and Kleiser & Schumann (this volume) use the Adams–Bashforth method for time discretization and are essentially similar except for the different treatment of the incompressibility condition. Orszag & Kells employ a three-level fractional time step in which the divergence-free condition of the flow field is satisfied at the second fractional step. Moin & Kim solve the continuity equation directly along with the momentum equations. Kleiser & Schumann use an influence matrix technique.

The first fractional step in the Orszag & Kells [1980] algorithm is

$$\frac{\hat{\vec{v}}^{n+1} - \tilde{\vec{v}}^{n}}{\Delta t} = \frac{3}{2} F^n - \frac{1}{2} F^{n-1} - \frac{i}{2} k_x \alpha U(y) \left[\hat{\vec{v}}^{n+1} - 2\hat{\vec{v}}^{n} + \hat{\vec{v}}^{n-1}\right],$$

where $U(y)$ is the mean velocity. This results from solving the advection part, rewritten as

$$\frac{\partial \vec{v}}{\partial t} + U(y) \frac{\partial \vec{v}}{\partial x} = \vec{F} + U(y) \frac{\partial \vec{v}}{\partial x},$$

by applying the Crank–Nicolson scheme on the left side and the Adams–Bashforth scheme on the right. The nonlinear terms $\vec{F} = (\vec{v} \times \nabla \times \vec{v})$ are computed using the pseudospectral method. No boundary conditions are applied at this stage. The second fractional step consists of

$$\hat{\hat{\vec{v}}}^{n+1} - \hat{\vec{v}}^{n} = -\Delta t\, \nabla \hat{\pi}^{n+1}, \qquad \nabla \cdot \hat{\hat{\vec{v}}}^{n+1} = 0$$

where $\hat{\pi}$ is the total pressure. The impermeability conditions

$$v = 0 \quad \text{at } y = \pm 1$$

are imposed at this stage. The third fractional step includes the viscous effects and the no-slip boundary conditions

$$\vec{v}^{n+1} - \hat{\hat{\vec{v}}} = \nu \Delta t \cdot \nabla^2 \vec{v}^{n+1}, \qquad \vec{v}^{n+1} = 0 \quad \text{at } y \pm 1.$$

In summary, this algorithm consists of a spectral tau method with the nonlinear terms computed by the pseudospectral method. It has a global error of order $O(\Delta t^2 + \nu \Delta t)$ in time discretization. The time steps are formally restricted only by the convective stability limit of the perturbed velocity field.

Moin & Kim [1980] use the Crank–Nicolson method on ∇P and $\nabla^2 \vec{v}$, and the

Adams–Bashforth on the remaining terms:

$$\nabla^2 \vec{v}^{n+1} - \beta \vec{v}^{n+1} - \tfrac{1}{2}\nabla P^{n+1} = -\beta \vec{v}^{n} - \tfrac{1}{2}[3\vec{F}^{n} - \vec{F}^{n-1}] + \tfrac{1}{2}\nabla P^{n} + \nabla^2 \vec{v}^{n} \equiv \vec{R}^{n},$$

$$\nabla \cdot \vec{v}^{n+1} = 0,$$

where $\beta = 2/\nu\Delta t$ and $\vec{F} = \vec{v} \cdot (\nabla \cdot \vec{v})$. Fourier transforming the above equations in x and z gives

$$\begin{aligned} &\frac{\partial^2 \tilde{\vec{v}}}{\partial y^2} - [\lambda + (k_x^2 + k_y^2)]\,\tilde{\vec{v}}^{n+1} - \frac{1}{2}\nabla \tilde{P}^{n+1} = \tilde{\vec{R}}^{n}, \\ &ik_x \tilde{u}^{n+1} + \frac{\partial \tilde{v}^{n+1}}{\partial y} + ik_z \tilde{w}^{n+1} = 0 \end{aligned} \tag{9.1}$$

where

$$\nabla \tilde{P}^{n+1} = \begin{bmatrix} ik_x \tilde{P}^{n+1} \\ \dfrac{\partial \tilde{P}^{n+1}}{\partial y} \\ ik_z \tilde{P}^{n+1} \end{bmatrix}.$$

For every pair of wavenumbers k_x and k_z, there is a linear system of four ordinary differential equations for u^{n+1}, v^{n+1}, w^{n+1} and P^{n+1}. The y-derivatives in the above system are approximated by either central differences or Chebyshev spectral approximations. Both cases yield block tridiagonal matrix operators which are inverted by conventional methods.

Although Moin & Kim [1980] discussed an algorithm with Chebyshev polynomials in the vertical direction, all of their published results thus far have used an alternative algorithm which employs finite differences in the vertical direction, but retains the Fourier spectral representation in the horizontal direction.

The algorithm of Kleiser & Schumann [1980] differs from that of Moin & Kim [1980] in the solution of (9.1). Instead of using the block-tridiagonal matrix inversion method, Kleiser & Schumann (this volume) obtain the solution of this system solving by a sequence of one-dimensional scalar Helmholtz equations of the type

$$\frac{\partial^2 \tilde{\vec{v}}}{\partial y^2} - (\lambda + k_x^2 + k_z^2)\,\tilde{\vec{v}} = \vec{b}, \qquad \tilde{\vec{v}}(\pm 1) = \tilde{\vec{v}}_1 .$$

Use of the influence matrix technique (Buzbee et al. [1971]) yields pressure boundary conditions which ensure a divergence-free solution within round-off errors.

Like Moin & Kim, Wray & Hussaini [1980] also employ central-difference discretization in the vertical direction. However, they address boundary layer rather than channel flow. The boundary layer is assumed to be "parallel," and a key part of their approach is to treat the mean flow as the solution of the diffusion equation

$$\frac{\partial U}{\partial t} = \nu \frac{\partial^2 U}{\partial y^2}$$

with the Blasius profile as the initial condition. The numerical method is only utilized to calculate the perturbations from this profile. A second major difference is the use of fourth-order Runge–Kutta time discretization. They also satisfy the divergence-free condition within round-off errors.

We warn the reader that the formulation of the spectral tau method to the solution of (9.1) requires some care to avoid instabilities (see Gottlieb & Orszag [1977, §13]).

9.2 Applications to hydrodynamic stability theory. The eigenvalue problems of hydrodynamic stability theory are often very difficult, because they involve the solution of nearly singular boundary value problems. For example, the solutions to the Orr–Sommerfeld equation that describes the stability of plane parallel shear flows typically have boundary layers of thickness $(\alpha R)^{-1/2}$ and an internal layer structure of thickness $(\alpha R)^{-1/3}$, where R is the Reynolds number and α is the streamwise disturbance wavenumber. In this case, spectral methods based on polynomial expansions offer the attractive advantage of high boundary resolution. As discussed by Orszag and Israeli [1974] and Gottlieb & Orszag [1977], only order $1/\sqrt{\delta}$ polynomials are required to resolve a solution with a boundary layer of thickness δ. This property allows the accurate solution of hydrodynamic stability problems with a minimum of spatial degrees of freedom. The general stability analysis code SALLY (Srokowski & Orszag [1977]) is a flexible, accurate code for the analysis of the stability of the boundary layers on aircraft.

The key ideas involved in the generation of modern spectral codes for stability analysis are the use of collocation (pseudospectral) methods, the use of matrix algebra for the setup of the matrix equations to be solved, and the use of the spectral iteration method for the efficient solution of these large matrix equations. The use of matrix algebra is particularly convenient in encountering a new problem for the first time. Here, as we described in §1, the spectral derivative operators are represented by matrices D, nonconstant coefficient terms are represented by diagonal matrices, and boundary conditions are imposed by replacing suitable rows of the matrix. The resulting scheme for generation of the spectral differential operator is both easy and accurate; it permits the efficient solution of a wide variety of boundary and eigenvalue problems, especially of the kind encountered in hydrodynamic stability theory.

Thus, the pseudospectral matrix representation of the Orr–Sommerfeld equation is simply

$$P\{(D^2 - \alpha^2 I)^2 - iR[(\alpha\bar{U} - \omega)(D^2 - \alpha^2 I) - \alpha\bar{U}''I]\}\phi = 0, \qquad (I - P)B\phi = 0,$$

where ϕ is the eigenvector, P is the projection operator onto the interior collocation points (so $I - P$ projects onto the boundary), and B is the operator representing the boundary conditions. The setup of the matrix operator is easily done by algebraic operations applied to the spectral derivative matrix D.

9.3 Applications to transition and turbulence. Spectral methods have proven to be the key element in obtaining basic new physical insights in the mechanisms of transition to turbulence. Transition is the process in which a flow first becomes chaotic and random. Because spectral methods are not subject to phase errors, it is possible to use them to follow in detail the nonlinear interaction waves that would be hopelessly diffused by more conventional finite-difference methods. Several studies of transition phenomena are noteworthy here.

Orszag & Kells [1980] studied the transition to turbulence in plane channel flows and found that one could achieve good agreement with gross experimental features of the flows. In particular, they found that transition typically occurs in plane Poiseuille and plane Couette flows at Reynolds numbers of the order of 1000, in agreement with available experimental observations. Their key conclusion, which confirms the experimental observations of Kovasznay et al. [1962] and Klebanoff et al. [1962] which were stressed previously by Stuart [1965] and Tani [1969], is the essentially three-dimensional character of transition. In two dimensions, no evidence of real fluid chaos was found.

Orszag [1976] and Wray & Hussaini [1980] studied the transition to turbulence in boundary layers. The latter authors made comparisons between their numerical flow profiles and available experimental data; the basic flow was taken as parallel, assuming that the spatial instability of the laboratory boundary layer corresponds to the temporal growth of the numerical boundary layer with periodic boundary conditions. The plots in Fig. 4 (from Wray & Hussaini) show that the events which constitute the incipient stages of transition (up to the so-called "two-spike" stage) are remarkably similar to the corresponding events observed in the laboratory. A similar set of comparisons between laboratory and numerical experiments has been made for plane Poiseuille flow by Kleiser & Schumann (this volume).

More recently, Orszag & Patera [1980], [1981a], [1983], Pierrehumbert and Widnall [1982], Herbert [1983] and Brachet & Orszag [1983] have used spectral methods to isolate a three-dimensional instability that seems to be a prototype of that responsible for the onset of chaos in shear flows. They have found that the two-dimensional (or axisymmetric) flow-states that result from primary, classical, shear layer instability saturate into ordered finite-amplitude flow-states and that, while these finite-amplitude states are stable to two-dimensional disturbances, they are strongly unstable to three-dimensional disturbances. It seems that the three-dimensional flows do not saturate into ordered states and instead lead directly to chaos.

Marcus et al. [1982] and Marcus [1983] have made an extensive study of the transition process in circular Couette flow. Marcus [1983] finds that the onset of wavy Taylor vortex states can be understood in terms of a vortex pairing process, not unlike that encountered in free shear layers and jets. In this work, the flow between rotating cylinders was studied by a high-resolution spectral code using Chebyshev polynomial series in r and Fourier series in θ and z.

There has been a large amount of research on convecting flows based on the application of spectral methods. The extensive classification of the instabilities of Benard layers by Busse and his collaborators has in effect been based on spectral techniques. Busse's pioneering work has clarified both the finite amplitude stability and roles, and their three-dimensional instabilities. Siggia & Zippelius [1981] have used spectral methods to study the dynamics of defects in Benard convection. McLaughlin and Orszag [1982] used spectral methods to study the onset of chaos in a convecting layer subject to rigid boundary conditions in the vertical direction and periodic boundary conditions in the horizontal. They found substantial agreement with the Ruelle–Takens picture of turbulence in which chaos ensues after three incommensurate frequencies become excited. Curry et al. [1983] made a similar study of the onset of chaos in a convecting layer subject to free-slip boundary conditions on all walls. One of their principal conclusions is that, while two-dimensional convection may be periodic in time, there is as yet no evidence for chaotic motions of the full fluid equations in two dimensions, in contrast to the chaotic motions exhibited by low-order Galerkin (spectral) approximations to the dynamic equations. Goldhirsch & Orszag [1983] have studied the onset theory of chaos in a convecting layer bounded by the three-dimensional rigid walls. In the latter study, three-dimensional Chebyshev expansions are used to resolve the fluid flow.

There are several aspects of these studies of transition that are interesting from the

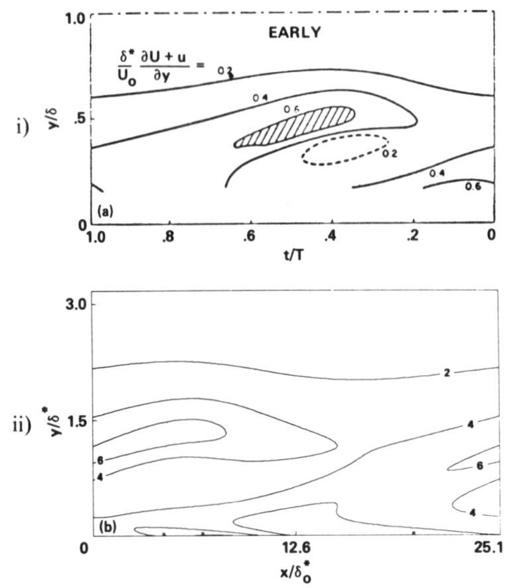

a) *Contours of $\partial u/\partial y$ at peak position; early development.* i) *Kovasznay et al.* [73, Fig. 10a]. ii) *Numerical simulation.*

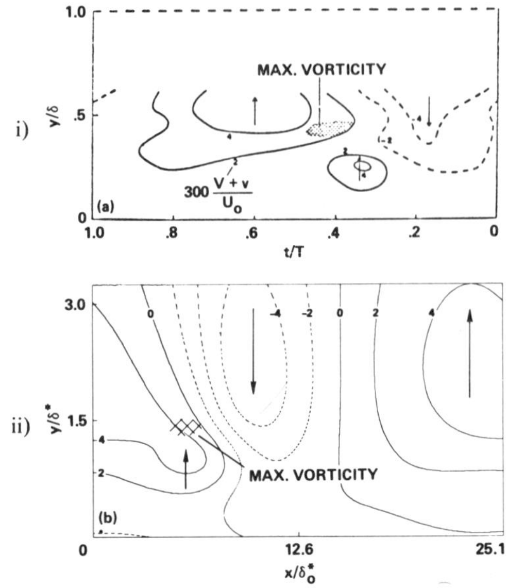

b) *Velocity normal to plate at peak position; early development.* i) *Kovasznay et al.* [73, Fig. 13a]. ii) *Numerical simulation.*

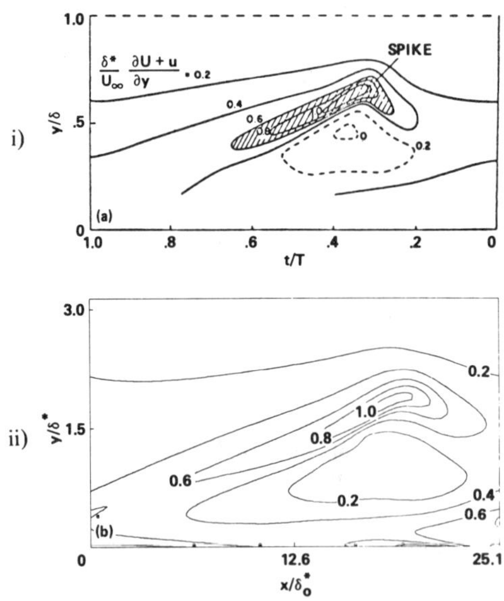

c) *Contours of $\partial u/\partial y$ at peak position; one-spike stage.* i) *Kovasznay et al.* [73, Fig. 10b]. ii) *Numerical simulation.*

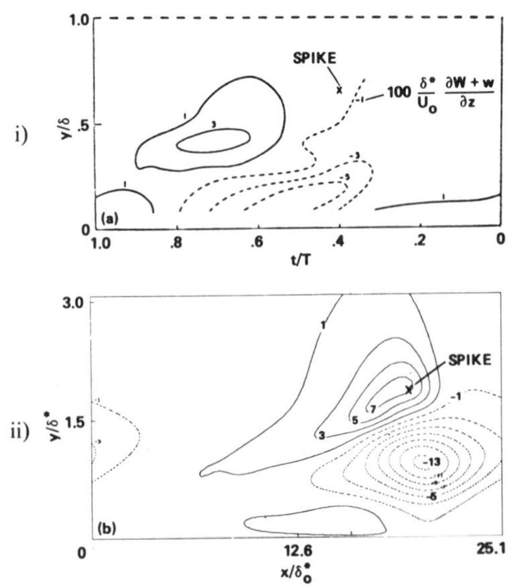

d) *Lateral vortex stretching rate $\partial w/\partial z$ at peak position; one-spike stage.* i) *Kovasznay et al.* [73, Fig. 12b]. ii) *Numerical simulation.*

FIG. 4.

numerical point of view. As mentioned above, Curry et al. [1983] and Orszag & Kells [1980] find that inadequately resolved two-dimensional flows exhibit time dependence, while adequately resolved two-dimensional flows exhibit much less time dependence than real three-dimensional flows. Toomre, Gough & Spiegel [1977] found that, provided an adequate amount of vertical resolution was used, spectral models of convection in which a single horizontal mode was retained exhibited no time dependence. Marcus [1981] has found a similar result in a study of convection in spherical geometry; he also found that the threshold for time dependence increases with increasing horizontal resolution.

Another important point concerning numerical simulations of transition is that, in contrast to the turbulence simulations discussed below, transition simulations typically require only modest spatial resolution but extremely high time resolution. It is not uncommon for a transition calculation to require 100,000 or more time steps, while typical turbulence calculations require at most 1,000–10,000 time steps, even though the latter involve higher-resolution spectral codes. The reason is that transition studies often involve relatively weak interactions between modes that require long-time integrations to resolve.

Spectral methods have been used to simulate a variety of turbulent flows. Perhaps the most distinguishing characteristic of high Reynolds number turbulent flows is their large range of excited space and time scales. In homogeneous turbulence, dissipation-scale eddies are of order R times smaller than energy-containing eddies, where R is the Reynolds number. In order to solve the Navier–Stokes equations for such a turbulent flow, it would be necessary to retain order $R^{3/4}$ spatial degrees of freedom and to perform order $R^{3/4}$ time steps to calculate for a significant evolution time of the flow. Even if only $O(1)$ arithmetic operations were required per degree of freedom per time step, the total computational work would scale as R^3, while the computer storage requirement would be order $R^{9/4}$. In this case, mere doubling of R requires an order of magnitude improvement in computer capability. In this environment, it may in fact be surprising that significant computations can be performed for turbulent flows; nevertheless, it has been possible to use spectral methods to calculate some key features.

Homogeneous turbulence is simulated by fluid motions within a box with periodic boundary conditions applied. The periodic boundary conditions allow Fourier series representation of the field variables. Starting with the work of Orszag & Patterson [1972], it has been possible to upgrade the spatial resolution of the flows being simulated to the point where it is now possible to simulate flows at Reynolds numbers comparable to those achievable in low-turbulence laboratory experiments. Herring et al. [1974] and Orszag [1976] report simulation of two-dimensional inertial range dynamics, while Brachet et al. [1983] report the first, albeit crude, calculation of a three-dimensional inertial range spectrum. The work of Brachet et al. [1983] is based on a numerical simulation of the Taylor–Green vortex, which is a special, highly symmetrical three-dimensional flow that is a prototype of motions that produce vorticity. With this flow, they were able to achieve an effective spatial resolution of $256 \times 256 \times 256$ Fourier modes for each of the velocity components using the Cray-1A computer. This allowed enough resolution to obtain a spectrum close to $k^{-5/3}$, where k is the magnitude of the wave-number vector, and to obtain the dissipation-fluctuation corrections to this spectrum.

For inhomogeneous turbulent shear flows, Orszag & Patera [1981a], [1981b] report the first direct numerical simulation of a turbulent channel flow at $R = 5000$ by using a spectral simulation with $64 \times 64 \times 65$ spatial resolution. This calculation was performed by imposing laminar, finite-amplitude two- and three-dimensional Orr–Sommerfeld eigenmodes on an initially laminar velocity profile. They found that, in time evolution, the flow became turbulent and achieved a state consistent with the structure of wall turbulence. In particular, they found that the von Karman "law of the wall" velocity profile was achieved, with a von Karman constant $\kappa = 0.46 \pm 0.05$ (which is good agreement with experiment).

If the resolution of the direct simulation of turbulence is inadequate for the Reynolds numbers that must be simulated, it is possible to use a sub-grid-scale turbulence closure, called a large-eddy simulation, following the pioneering finite-difference work of Deardorff [1970], [1971], [1972] and Schumann [1975]. Here, excitations on scales smaller than those resolvable on the numerical grid are modelled, usually by an eddy viscosity coefficient. Such a sub-grid-scale (SGS) eddy coefficient represents the dissipative effect of motions on scales smaller than the effective grid on the large eddies (defined as those motions adequately represented on the numerical grid). The most common form for this SGS eddy viscosity coefficient is due to Smagorinsky [1963]:

$$\nu_{\text{eddy}} = (c\Delta)^2 \left| \left(\frac{\partial v_i}{\partial x_j} + \frac{\partial v_j}{\partial x_i} \right)^2 \right|^{1/2} \tag{9.2}$$

where Δ is the grid scale and $\vec{v}$ is the large-eddy velocity. In spectral simulations, it is possible to use either (9.2) or some more exotic wave-space filter to remove SGS components. In the absence of walls, either method seems equally satisfactory. For wall-bounded shear flows, Deardorff calculates only up to the edge of the buffer layer between the viscous sublayer and the logarithmic region of the velocity profile. A boundary condition is imposed at this point based on the von Karman theory of the wall layer in which the turbulent fluctuating stress is assumed known. Molecular viscosity plays no role in Deardorff's calculations, which are performed at least formally, at infinite R. Clearly such a simulation does not give a faithful representation of the wall region and accompanying bursts.

If one is interested in the physics of wall turbulence, neglect of the wall region is unjustified. More recent work by Moin & Kim [1982] integrates up to the rigid wall, the increased sophistication giving a so-called transport eddy simulation. However, the method (as currently implemented) has, in effect, destroyed some of the original incentive for the use of large-eddy simulation. This point is easily demonstrated. As presently practiced, transport-eddy sub-grid-scale simulations use uniform horizontal resolution, independent of distance from the wall (Moin & Kim [1982]). If such a simulation is to capture scales down to those of the bursts, the degrees of freedom required scale as R_*^3, where R_* is the wall Reynolds number derived from friction and velocity. This estimate is based on the fact that streak structure scales with the inner variables (Kline et al. [1967]), and that streamwise and spanwise correlations scale with outer variables, i.e. channel width (Comte-Bellot [1965]). Sub-grid-scale modelling does allow somewhat higher Reynolds numbers than those that can be achieved without modelling (Moin & Kim [1982] achieve $R_* = 640$ by transport-eddy modelling while the direct numerical simulations of Orszag & Patera [1981a] are restricted to $R_* = 200$), but bursts at mean Reynolds numbers of 20,000–100,000 (in plane Poiseuille flow) are beyond any sub-grid-scale model as now implemented. In other words, for a sub-grid-scale large-eddy simulation to do significantly better than a direct simulation, the dependence of required degrees of freedom must scale less rapidly than R_*^3 as R_* becomes large. At present, no such method exists.

10. Compressible flows. In recent years, spectral methods have begun to be applied to the solution of both steady and time-dependent compressible flow problems. However, in contrast to the sophisticated applications of spectral methods to incompressible flow problems, the applications to compressible problems are, to date, less developed.

There are two unique features of compressible flow problems that have required additional theoretical and practical developments in the spectral technique. At first glance, the most difficult problem is that inviscid compressible flows typically develop discontinuous solutions. It is by no means clear that these shocked solutions are amenable to numerical solution by techniques of high formal accuracy like spectral methods. On the one hand, it has been proven (see §3) that, for linear problems, high accuracy can be maintained within spectral methods far away from a discontinuity; on the other hand, it may be thought that for nonlinear problems the overall accuracy in the

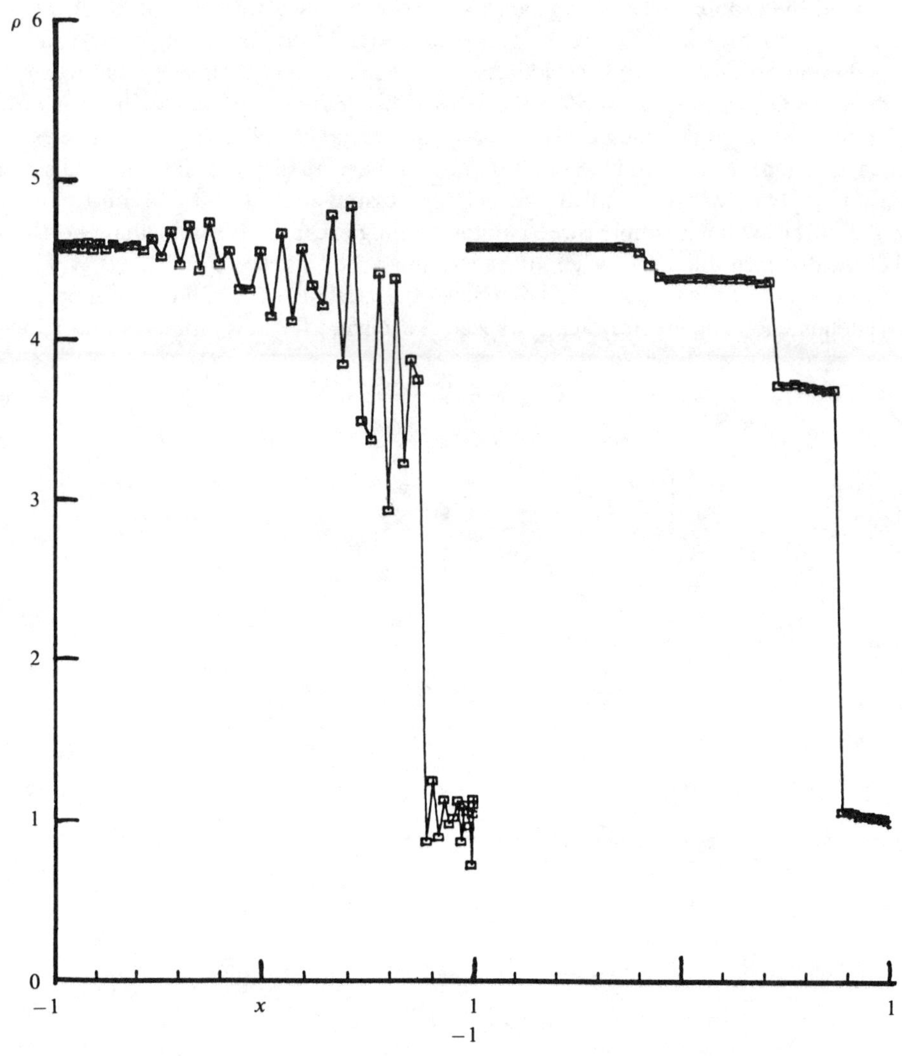

FIG. 5. *Spectral capturing of shock coalescence without and with post-processing.*

presence of discontinuities is limited to first order. However, Lax [1978] has argued that more information about the solution is contained within high resolution schemes, which may be usefully extracted in obtaining high resolution results.

The other major problem with compressible flows is associated with the boundary conditions that must be applied. For inviscid problems that are hyperbolic in nature, the analytic solution is driven by the data entering at inflow boundaries, and the numerical technique should minimize the influence of conditions imposed on outflow boundaries. Because of the high accuracy of spectral methods and their global nature, correct implementation of boundary conditions is crucial to obtaining stable and accurate results.

10.1. One-dimensional and quasi one-dimensional flows. Gottlieb, Lustman & Orszag [1981] considered the solution of one-dimensional, inviscid compressible flows with shocks by spectral methods. In particular they discussed methods capable of capturing a shock wave over one grid interval. In addition to the conventional shock tube problem, they treated the problem of a strong shock wave overtaking a weak shock, resulting in a single shock and a rarefaction wave. To demonstrate resolution of this shock and the contact discontinuity over a single grid interval, the density plot after shock coalescence is reproduced in Fig. 5. Figure 5a shows the density distribution after the application of the stabilizing low-pass filter which consists of an exponential cut-off of high frequencies (see Gottlieb, Lustman & Orszag [1981] for details). Figure 5b displays the density distribution after applying what is called the post-processing filter (Gottlieb, Lustman and Orszag [1981]) which is simply three-point averaging except in the immediate neighborhood of the discontinuity itself, where the averaging is one-sided.

These simple filters appear to be sufficient to stabilize the solution, suppress the Gibbs phenomenon and possibly achieve spectral accuracy for flow problems consisting of

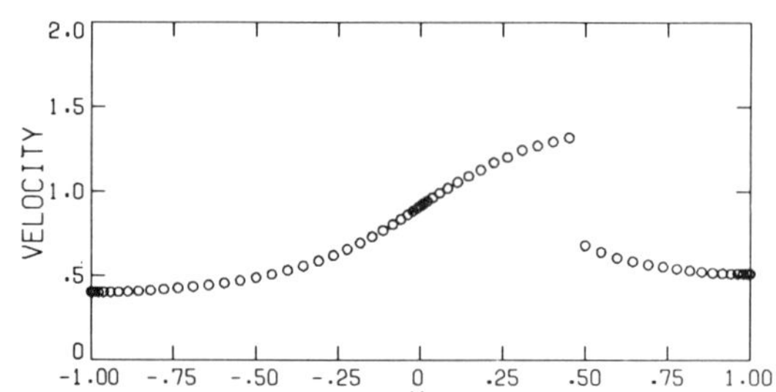

a) *Quasi one-dimensional nozzle flow.*

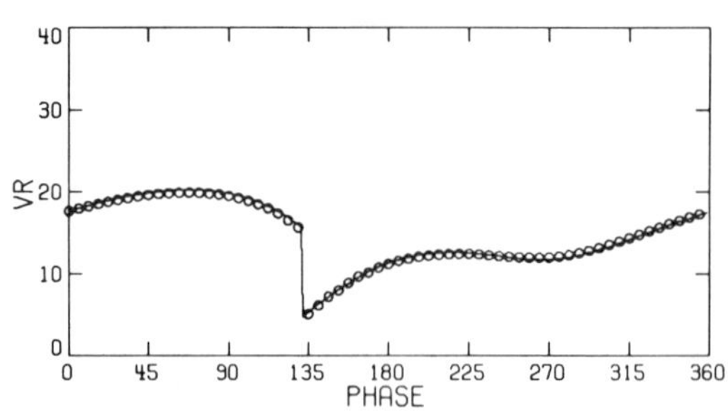

b) *Astrophysical problem.*

FIG. 6.

piecewise linear profiles. It remains to be seen if spectral accuracy can be maintained for more structured flows. It should also be noted that in this method, although shock waves are not fitted as discontinuities, the post-processing filters employed do require knowledge of the shock locations, which can be obtained in either physical or spectral space. Even assuming knowledge of the shock locations, there is a serious need for development of spectral filtering techniques to remove global oscillations without loss of spectral accuracy.

Zang & Hussaini [1980] present pseudospectral solutions to more complicated flow problems. Their results show that a highly structured flow field is well represented, along with the sharp front of the shock; see Fig. 6. However, the rate of convergence to the inviscid solution is only finite. Other spectral work on one-dimensional shock problems has been done by Taylor et al. [1981] and by Cornille [1982].

10.2. Two-dimensional flows.

10.2.1. Euler equations. Gottlieb, Lustman & Streett (this volume) study simple variants of the problem of a regular reflection of an oblique shock wave from a solid surface by solving the Euler equations in conservation form. The pseudospectral technique employed is a straightforward extension of the method developed for the one-dimensional case (Gottlieb, Lustman & Orszag [1981]). The smoothing and filtering methods which work well in the one-dimensional shock tube problem yield similar results in the two-dimensional shock inflection problem if the flow involved is piecewise uniform.

A careful treatment of the boundary conditions is extremely important, as the spectral method is not as forgiving as the finite difference method in this aspect. Imposition of the boundary conditions in the characteristic form is found to work satisfactorily, as predicted by the linear theory of §6. Figure 7 shows the results of

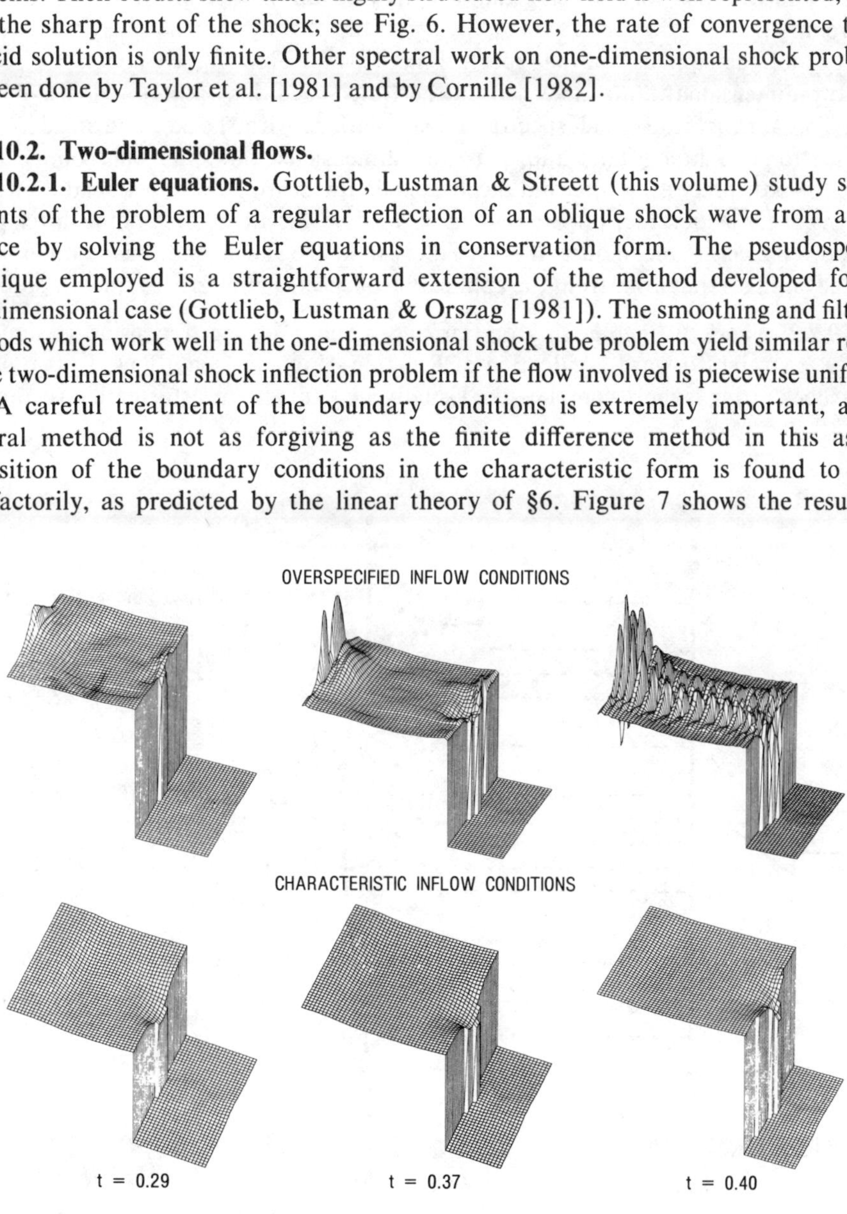

FIG. 7. *Pseudospectral pressure for Mach* 1.3 *Karman vortex sheet.*

calculations described in Salas et al. [1982] for the interaction of a Mach 1.3 shock wave with an idealized Karman vortex sheet, and it illustrates the sensitivity of spectral methods to boundary treatment. The only difference between the two calculations is the treatment of the left subsonic inflow boundary. For the calculations shown in the top row, all the variables were prescribed. This overspecification of boundary conditions leads to the eventual contamination of the entire solution. In the second calculation, only three of the four variables were specified at the left boundary. The remaining variable, the pressure, was computed from the interior solution along a linearized characteristic. The details of the method for imposing the boundary conditions are given by Gottlieb et al. (this volume).

It should be emphasized that spectral methods for even slightly complex discontinuous two-dimensional flows governed by the Euler conditions require further development. To date the best pseudospectral Euler solutions with shock capturing have been obtained for one-dimensional and quasi one-dimensional flows; as was pointed out in §10.1, the post-processing filter did involve explicitly locating the shock, and even so the accuracy of the structured, smooth regions of the flow was affected. Pending development of improved filtering techniques, it is expedient to fit the shock, and thus to maintain the usual advantages that arise from the application of spectral methods to smooth problems.

10.2.2. Shock fitting. Fitting or tracking a shock in one dimension is simple. It involves a straightforward application of Rankine–Hugoniot conditions. In two dimensions, shock fitting is more complicated. Usually a coordinate transformation is employed

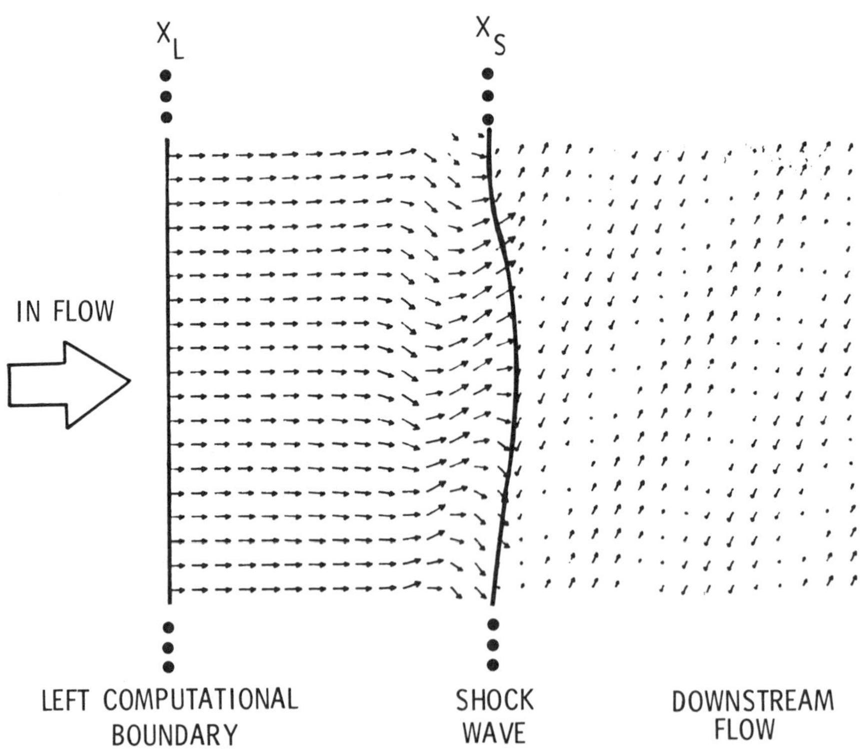

FIG. 8. *Physical plane.*

so that the shock wave becomes a coordinate boundary. To be specific, let $x_L \leqq x \leqq x_s$, $-\infty \leqq y \leqq \infty$ (see Fig. 8) be the region of computation bounded on the right by the shock wave

$$x = x_s(y, t).$$

Define the coordinate transformation

$$X = \frac{x - x_L}{x_s(y, t) - x_L}, \qquad Y = y, \quad T = t,$$

where (x,y) is the Cartesian coordinate system in the physical space, and t, the physical time. With the flow field in front of the shock known, the flow field immediately behind the shock can be evaluated via the Rankine–Hugoniot conditions if the shock velocity is known.

Let the shock speed be defined by

$$\dot{x}_s = \frac{\partial x_s}{\partial T},$$

and the unit normal $\hat{N}$ to the shock by

$$\hat{N} = \alpha \hat{x} + \beta \hat{y}$$

where $\hat{x}$ and $\hat{y}$ are unit vectors in the x- and y-direction respectively. Then the shock velocity is

$$\vec{\dot{x}} = \dot{x}_s \hat{x},$$

and its component in the direction $\hat{N}$ is

$$w_s = \vec{\dot{x}}_s \cdot \hat{N}.$$

If the velocity field in front of the shock is

$$\vec{V} = u\hat{x} + v\hat{y},$$

the velocity component relative to the shock in the direction of the normal $\hat{N}$ is

$$\tilde{U}_{\text{rel}} = \vec{V} \cdot \hat{N} - w_s = u\alpha + v\beta - w_s,$$

and the Mach number relative to the shock is

$$M_{\text{rel}} = \frac{\tilde{U}_{\text{rel}}}{a}$$

where a is the sound speed in front of the shock. With the relative Mach number known, the flow variables behind the shock can be evaluated using the Rankine–Hugoniot relations. Then the compatibility relation for the characteristic reaching the shock wave in the x, T-plane from the high pressure side is obtained by combining the equation of continuity and the x-momentum equation. If the time derivatives appearing in this expression are replaced by those obtained by differentiating the Rankine–Hugoniot relations with respect to T, the equation for shock acceleration $\partial \vec{\dot{x}}_s / \partial T$ is found. The required shock speed is obtained by integrating this equation.

While solving the equations of motion by explicit time discretization, the shock velocity and position are computed at every step. It should be noted that the shock boundary in the computational domain is a supersonic inflow boundary, and the Rankine–Hugoniot relations provide the boundary conditions for all of the variables once the unknown shock velocity is computed. The shock fitting strategy described here has been used with the pseudospectral method to study the classical blunt body problem (Hussaini & Zang (this volume)), and the more complex problem of a shock wave interacting with turbulence (Zang, Kopriva & Hussaini [1983]).

10.2.3. Potential equations. It is well known that, even with weak shocks, a good approximation to transonic flows is obtained by using the compressible potential equation. The most distinguishing, and also the most difficult, feature of transonic flows is their mixed subsonic-supersonic character, which was treated computationally by Murman & Cole [1971], who introduced type-dependent differencing schemes for steady plane transonic flow.

The extension of type-dependent schemes to employ pseudospectral methods was applied to transonic potential flow past a parabolic arc by in Gottlieb, Lustman & Streett (this volume). Streett [1983] has obtained pseudospectral solutions for arbitrary lifting airfoils. The airfoil is mapped onto a cylinder using conformal mapping. The full potential equations in the computational plane are solved by using the cylindrical polar coordinate system with Fourier representation in the azimuthal direction and Chebyshev polynomial representation in the radial direction. Shocks are captured by using the second order artificial density method, as in the finite difference codes. For supercritical cases, an iterative scheme combining AF1 (Douglas–Gunn) and AF2 (see Holst [1979]) is found to be fairly efficient. For subcritical flows, four decimal place accuracy in surface pressure and lift is obtained with a grid of 12×40 points, and a typical calculation requires about 45 seconds on CDC CYBER 175. The well-known finite difference multigrid code FLO-36 (Jameson [1979]) requires about 40 seconds on the same computer on a 32×192 grid for the same accuracy. The supercritical cases require more resolution, and the iterative scheme then suffers from slow convergence.

The multigrid procedure given in Streett et al. [1983] is found to accelerate substantially the convergence rate in both the subcritical and supercritical cases. For example, converged solutions for subcritical flow conditions are obtained in less than 15 seconds on the CYBER 175, using a grid of 16×32 points. Even more dramatic improvements are obtained in the case of supercritical flow (Fig. 9), where the machine time is reduced by a factor between 20 and 50 over a single grid iteration scheme.

In summary, the iterative schemes developed for finite difference discretization can be applied equally well to spectral discretizations. Since the residual calculation is an order of magnitude more expensive than the approximate factorization or approximate inverse calculation, reducing the number of iterations required for a given convergence criterion is the major requirement for a good spectral method.

The artificial density technique using finite differences yields satisfactory results. However, from the standpoint of accuracy, this technique for capturing shocks requires improvement. The advantage of a reduced number of points for a given accuracy is offset by the greater cost per point of the spectral method, to the extent that the total cost of a spectral solution may be of the same order as that for the finite difference solution. Improvements in iterative methods, like the multigrid method, promise to give spectral methods that are more efficient, in both storage and work, than the best available finite difference methods.

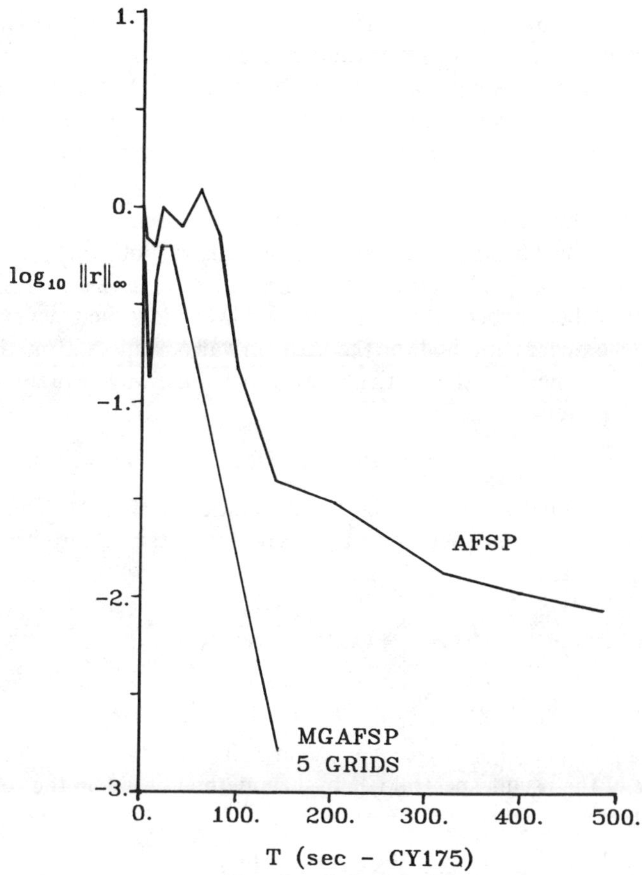

FIG. 9. *Residual versus machine time.*

10.3. Three-dimensional flows. Lambiotte et al. [1982] have developed a three-level time-split spectral/finite difference method for the numerical solution of the three-dimensional compressible Navier–Stokes equations. The first fractional step includes the effect of the advection terms, using a variable-step second-order Adams–Bashforth method. The pseudospectral method is used for calculating x- and y-derivatives. The second fractional step is an implicit pressure correction to avoid numerical instabilities due to sound waves. In the third fractional step, viscous corrections are made to the inviscid solution obtained in the previous two steps. Use of the pseudospectral technique for the evaluation of the compressible viscous terms involves a large amount of computation. Since viscosity is small at large Reynolds numbers, it is reasonable to limit the computational work by discretizing the viscous terms using central differences. The truncation error is on the order of $\nu(\Delta t + \Delta x^2 + \Delta y^2 + \Delta z^2)$ where ν is the maximum kinematic viscosity. To avoid severe time step restrictions resulting from small z in the stretched z-mesh, the z-derivatives are treated implicitly. This algorithm is being used to study compressible shear flows at high Reynolds numbers, such as the incipient stages of transition to turbulence, and receptivity of laminar boundary layers to external disturbances.

Feiereisen, Reynolds & Ferziger [1981] have carried out a numerical simulation of low Reynolds number homogeneous turbulent shear flow using the three-dimensional compressible Navier–Stokes equations. Following Rogallo [1977], they apply a coordinate transformation to the Navier–Stokes equations which permits imposition of periodic boundary conditions. Thus they were able to use a pseudospectral Fourier method in space.

Appendix A. The error bounds presented in §§3 and 5 for the pseudospectral methods were all derived using energy estimates which are global in space. Consequently, special features of the local behavior of the error may be lost in this analysis.

In an unpublished paper, Dubiner [1977] gave a detailed asymptotic analysis, independent of the energy method, of the error in various spectral methods. His novel methods are reviewed here for the particular example of the wave equation.

Consider the system

$$u_t = u_x \quad (|x| < 1), \quad u(x, 0) = f(x), \quad u(1, t) = q(t). \tag{A1}$$

Let u_N be the polynomial approximation to the solution u of (A1). This approximation may be obtained either by the pseudospectral, Galerkin or tau-Chebyshev methods. In all the above cases, u_N satisfies

$$\begin{gathered} \frac{\partial u_N}{\partial t}(x, t) = \frac{\partial u_N}{\partial x}(x, t) + \tau_N(t) q_N(x), \\ u_N(x, 0) = P_N f = f_N(x), \qquad u_N(1, t) = g(t) \end{gathered} \tag{A2}$$

where $q_N(x)$ is a polynomial of degree N in x and τ_N is a function of t.

In the case of the pseudospectral Chebyshev method based on the collocation points x_j given in (4.1),

$$q_N(x) = \frac{(1 + x)T'_N}{2N^2}. \tag{A3}$$

For the pseudospectral method based on the points y_j given in (4.18),

$$q_N(x) = \frac{T'_{N+1}}{N^2}, \tag{A4}$$

while for the Galerkin method

$$q_N(x) = \sum_{k=1}^{N} T_k(x) + 1 \tag{A5}$$

and for the tau method

$$q_N(x) = T_N(x) \tag{A6}$$

(Gottlieb & Orszag [1977]). In all cases $\tau_N(t)$ is determined by the condition that $u_N(x, t)$ be a polynomial in x of degree at most N. The quality of the approximation depends on the behavior of the unknown function $\tau_N(t)$. The basic idea of Dubiner's method is to find an expression for $\tau_N(t)$ in terms of $q_N(x)$, $P_N f(x)$ and $g(t)$ using only the fact that u_N is a polynomial in x.

To do this we Lapace transform (A2) with respect to time to get

$$s\hat{u}_N(x, s) - f_N(x) = \frac{\partial \hat{u}_N(x, s)}{\partial x} + \hat{\tau}_N(s) q_N(x), \qquad \hat{u}_N(1, s) = \hat{g}(s) \tag{A7}$$

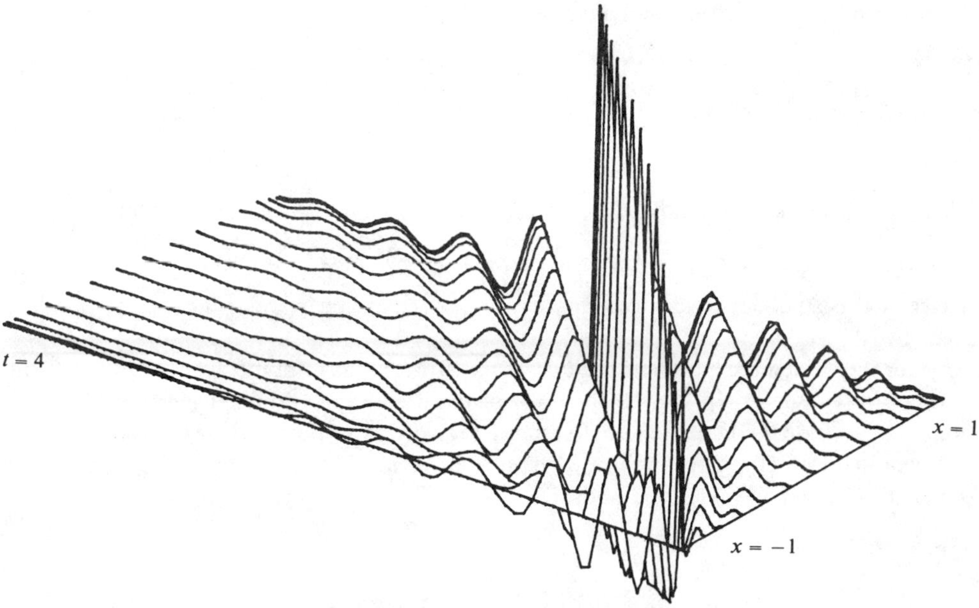

FIG. 10. *Errors in Chebyshev collocation approximations to* $u_t + u_x = 0$.

where $u_N(x, s)$ is the Laplace transform of $u_N(x, t)$. (A7) is an ordinary differential equation with solution

(A8) $$\hat{u}_N(x, s) = e^{s(x-1)}\left\{\hat{g}(s) - \hat{\tau}(s)\int_1^x e^{-s(\xi-1)}q_N(\xi)\,d\xi - \int_1^x e^{-s(\xi-1)}f_N(\xi)\,d\xi\right\}.$$

Observe now that the left side of (A8) is a polynomial in x, whereas the first term on the right side is growing exponentially with increasing x. Therefore as $x \to \infty$ the term in the brackets appearing in the right side of (A8) must vanish, giving

(A9) $$\hat{\tau}_N(s)\int_1^\infty e^{-s(\xi-1)}\,q_N(\xi)\,d\xi = -\int_1^\infty e^{-s(\xi-1)}f_N(\xi)\,d\xi + \hat{g}(s).$$

(A9) yields the desired expression for the function $\hat{\tau}_N(s)$ that governs the behavior of the error in the spectral approximation to the wave equation.

Dubiner proceeds by considering a large class of polynomials $q_N(x)$ rather than the particular forms (A3)–(A6). This class is characterized as follows:

Let

$$Z(x) = x[1 + \sqrt{1 - x^{-2}}]$$

and let $D(Z)$ be an analytic function (except for algebraic singularities at $x = \pm 1$) that grows less than exponentially fast near infinity. Moreover, assume that

$$D(Z) \sim \begin{cases} (1 + Z)^{\alpha+1/2} & (|Z + 1| \ll 1), \\ C_p(1 - Z)^{\beta+1/2} & (|Z - 1| \ll 1). \end{cases}$$

Then we consider the class of polynomials satisfying

$$P_Nq_N(x) \sim \begin{cases} (2\pi)^{-1/2}[D^{-1}(Z^{-1}(x))^{2/3}Z^N(x) + D^{-1}(Z(x))Z^{-N}(x)] \quad (|x \pm 1| \gg N^{-2/3}), & \text{(A10)} \\ N^{\alpha+1/2}(-1)^N(N\sqrt{1-X^2})^{-\alpha}J_\alpha(N\sqrt{1-x^2}) \quad (|x+1| \ll N^{-2/3}), & \text{(A11)} \\ N^{\beta+1/2}C_p(N\sqrt{1-x^2})^{-\beta}J_\beta(N\sqrt{1-x^2}) \quad (|x-1| \ll N^{-2/3}). & \text{(A12)} \end{cases}$$

This assumption is satisfied by all the orthogonal polynomials with weight function $a(x)$:

$$a(x) = (1 + x)^\alpha (1 - x)^\beta b(x) \tag{A13}$$

where $b(x)$ is a nonnegative polynomial on $[-1, 1]$, and in particular for the polynomials given by (A3)–(A6).

For this review we omit now the (very sophisticated) asymptotic expansions that lead to the error estimates. In Fig. 10 the error of the pseudospectral Chebyshev approximation to $u_t + u_x = 0$ for data with l smooth derivatives is given. Observe the different behavior in different regions in the x, t-plane, a behavior that cannot easily be obtained by the analysis based on energy methods.

REFERENCES

[1] F. BAER [1964], *Integration with the spectral vorticity equation,* J. Atmos. Sci., 21, pp. 260–276.

[2] F. BAER AND G. W. PLATZMAN [1961], *A procedure for numerical integration of the spectral vorticity equation,* J. Meteor., 18, pp. 393–410.

[3] S. H. BOKHARI, M. Y. HUSSAINI AND S. A. ORSZAG [1982], *Fast orthogonal derivations on* STAR, Comput. Math. Appl., 8, pp. 367–377.

[4] W. BOURKE [1972], *An efficient, one-level, primitive-equation spectral model,* Mon. Wea. Rev., 100, pp. 683–689.

[5] W. BOURKE [1974], *A multi-level spectral model.* I. *Formulation and hemispheric integrations,* Mon. Wea. Rev., 102, pp. 687–701.

[6] W. BOURKE, B. MCAVANEY, K. PURI AND R. THURLING [1977], *Global modelling of atmospheric flow by spectral methods,* in Methods in Computational Physics, Vol. 17, General Circulation Models of the Atmosphere, J. Chang, ed., Academic Press, New York, pp. 267–324.

[7] W. BOURKE, K. PURI AND R. THURLING [1974], *Numerical prediction for the Southern Hemisphere via the spectral method,* GARP WGNE Report No. 7, pp. 22–42.

[8] M. E. BRACHET AND S. A. ORSZAG [1983], *Secondary instability of free shear flows,* J. Fluid Mech., to appear.

[9] B. L. BUZBEE, F. W. DORR, J. A. GEORGE AND G. H. GOLUB [1971], *The direct solution of the discrete Poisson equation on irregular regions,* SIAM J. Numer. Anal., 8, pp. 722–736.

[10] B. P. BYRNAK [1975], *An operational spectral semi-implicit one-layer model,* The Danish Meteorological Institute, Meddelelser Nr. 25.

[11] C. CANUTO, H. FUJII AND A. QUARTERONI [1983], *Approximation of symmetry breaking bifurcations for the Rayleigh convection problem,* SIAM J. Numer. Anal., 20, pp. 873–884.

[12] C. CANUTO, Y. MADAY AND A. QUARTERONI [1982], *Analysis for the combined finite element and Fourier interpolation,* Numer. Math., 39, pp. 205–220.

[13] C. CANUTO AND A. QUARTERONI [1980], *Propriétés d'approximation dans les espaces de Sobolev de systèmes de polynômes orthogonaux,* C. R. Acad. Sci. Paris, 290, Ser. A, pp. 925–928.

[14] ——— [1981a], *Spectral and pseudo-spectral methods for parabolic problems with nonperiodic boundary conditions,* Calcolo, 18, pp. 197–218.

[15] ——— [1981b], *Spectral methods for hyperbolic equations,* Rend. Sem. Mat. Univers. Politecn. Tornio, 39, pp. 21–31.

[16] ——— [1982a], *Approximation results for orthogonal polynomials in Sobolev spaces,* Math. Comp., 38, pp. 67–86.

[17] ——— [1982b], *Error estimates for spectral and pseudo-spectral approximations of hyperbolic equations,* SIAM J. Numer. Anal., 19, pp. 629–642.

[19] G. COMTE-BELLOT [1965], *Ecoulement turbulent entre deux parois paralles,* PUBL Sci. Tech. M. Min. Air., 419 Paria.

[20] J. W. COOLEY AND J. W. TUKEY [1965], *An algorithm for the calculation of complex Fourier series,* Math. Comp., 19, pp. 297–301.

[21] P. CORNILLE [1982], *A pseudospectral scheme for the numerical calculation of shocks,* J. Comput. Phys., 47, pp. 146–159.

[22] J. CURRY, J. HERRING, J. LONCARIC AND S. A. ORSZAG [1983], *Order and disorder in two- and three-dimensional Bénard convection,* J. Fluid Mech., to appear.

[23] R. DALEY, I. SIMMONDS AND J. HENDERSON [1974], *A Galerkin approach to short-term forecasting in the Northern Hemisphere,* GARP WGNE Report No. 7, pp. 43–45.

[24] P. J. DAVIS AND P. RABINOWITZ [1975], *Methods of Numerical Integration,* Academic Press, New York.

[25] J. W. DEARDORFF [1970a], *A numerical study of three-dimensional turbulent channel flow at large Reynolds numbers,* J. Fluid Mech., 41, pp. 453–480.

[26] ——— [1973], *The use of subgrid transport equations in a three-dimensional model of atmospheric turbulence,* J. Fluids Engrg., pp. 429–438.

[27] ——— [1970], *A three-dimensional numerical investigation of the idealized planetary boundary layer,* Geophys. Fluid Dynamics, 1, pp. 377–410.

[28] M. DEVILLE, P. HALDENWANG AND G. LABROSSE [1981], *Comparison of time integration (finite difference and spectral) for the nonlinear Burgers equation,* Proc. IV GAMM Conference, Vieweg Verlag, R. F. A.

[29] S. A. ORSZAG, M. ISRAELI AND M. O. DEVILLE [1983], *Splitting methods for incompressible flow problems,* submitted to J. Comput. Phys.

[30] E. DORON, A. HOLLINGSWORTH, B. J. HOSKINS AND A. J. SIMMONS [1974], *A comparison of grid-point and spectral methods in a meteorological problem,* Quart. J. Roy. Met. Soc., 100, pp. 371–383.

[31] E. ELIASEN AND B. MACHENHAUER [1974], *On the spectral representation of the vertical variation of the meteorological fields in numerical integrations of a primitive equations model,* GARP WGNE Report No. 7, pp. 83–93.

[32] E. ELIASEN, B. MACHENHAUER AND E. RASMUSSEN [1970], *On a numerical method for integration of the hydrodynamical equations with a spectral representation of the horizonal fields,* Report No. 2, Institut for Theoretical Meteorology, University of Copenhagen.

[33] H. W. ELLSAESSER [1966], *Evaluation of spectral versus grid methods of hemispheric numerical weather prediction,* J. Appl. Meteor., 5, pp. 246–262.

[34] W. J. FEIEREISEN, W. C. REYNOLDS AND J. H. FERZIGER [1981], *Numerical simulation of a compressible homogeneous, turbulent shear flow,* Report No. TF-13, Dept of Mechanical Engineering, Stanford Univ., Stanford, CA.

[35] T. W. FLATTERY [1970], *Spectral models for global analysis and forecasting,* Air Weather Service Technical Report 242.

[36] B. FORNBERG [1975], *On a Fourier method for the integration of hyperbolic equations,* SIAM J. Numer. Anal., 12, pp. 509–528.

[37] ——— [1977], *A numerical study of two-dimensional turbulence,* J. Comput. Phys., 25, pp. 1–31.

[38] D. FUNARO [1983a], *Analysis of the DuFort–Frankel method for linear differential systems,* SIAM J. Numer. Anal., to appear.

[39] ——— [1983b], *Error estimates for spectral approximations of linear hyperbolic equations over an hypercube,* Calcolo, to appear.

[40] B. GALERKIN [1915], *Rods and plates: Series occurring in various questions concerning the elastic equilibrium of rods and plates,* Vestnik Inzhenerov, 19, pp. 897–908.

[41] J. GAZDAG [1976], *Time-differencing schemes and transform methods,* J. Comput. Phys., 20, pp. 196–207.

[42] I. GOLDHIRSH AND S. A. ORSZAG [1983], *Two-dimensional convection in a rigid box,* to appear.

[43] T. GORDON AND W. STERN [1974], *Spectral modelling at* GFDL, GARP WGNE Report No. 7, pp. 46–82.

[44] D. GOTTLIEB [1981], *The stability of pseudospectral Chebyshev methods,* Math. Comput., 36, pp. 107–118.

[45] D. GOTTLIEB, M. GUNZBURGER AND E. TURKEL [1982], *On numerical boundary treatment for hyperbolic systems,* SIAM J. Numer. Anal., 19, pp. 671–697.

[46] D. GOTTLIEB AND B. GUSTAFSSON [1976], *Generalized DuFort–Frankel methods for parabolic initial-boundary value problems,* SIAM J. Numer. Anal., 13, pp. 129–144.

[47] D. GOTTLIEB AND L. LUSTMAN [1981], *The DuFort–Frankel Chebyshev method for parabolic initial boundary value problems,* ICASE Report No 81–42; Comput. & Fluids, to appear.

[48] ——— [1982], *The spectrum of the Chebyshev collocation operator for the heat equation,* ICASE Report No. 82–12. SIAM J. Numer. Anal., 20 (1983), pp. 909–921.

[49] D. GOTTLIEB, L. LUSTMAN AND S. A. ORSZAG [1981], *Spectral calculations of one-dimensional inviscid compressible flow,* SIAM J. Sci. Statist. Comput., 2, pp. 296–310.

[50] D. GOTTLIEB AND S. A. ORSZAG [1977], *Numerical Analysis of Spectral Methods: Theory and Applications,* CBMS Regional Conference Series in Applied Mathematics 26, Society for Industrial and Applied Mathematics, Philadelphia.

[51] D. GOTTLIEB, S. A. ORSZAG AND E. TURKEL [1981], *Stability of pseudospectral and finite difference methods for variable coefficient problems,* Math. Comp., 37, pp. 293–305.

[52] D. GOTTLIEB AND E. TURKEL [1980], *On time discretization for spectral methods,* Stud. Appl. Math., 63, pp. 67–86.

[53] A. HAGEMAN AND D. M. YOUNG, eds. [1981], *Applied Iterative Methods,* Academic Press, New York.

[54] G. J. HALTINER AND R. T. WILLIAMS [1980], *Numerical Prediction and Dynamical Meteorology,* 2nd ed., John Wiley, New York.

[55] A. HARTEN [1983a], *On the symmetric form of systems of conservation laws with entropy,* J. Comput. Phys., 49, pp. 151–164.

[56] ——— [1983b], *High resolution schemes for hyperbolic conservation laws,* J. Comput. Phys., 49, pp. 357–393.

[57] T. HERBERT [1983], *Secondary instability of plane channel flow to subharmonic three-dimensional disturbances,* Phys. Fluids, 26, pp. 871–874.

[58] J. R. HERRING, S. A. ORSZAG, R. H. KRAICHNAN AND D. G. FOX [1974], *Decay of two-dimensional homogeneous turbulence,* J. Fluid Mech., 66, pp. 417–444.

[59] R. S. HIRSH AND T. D. TAYLOR [1982], *Recent developments at* APL *in the use of pseudo-spectral techniques,* Proc. EUROMECH Conference on Spectral Methods, Nice, France.

[60] R. S. HIRSH, T. D. TAYLOR, M. M. NADWORNY AND J. L. KERR [1982], *Techniques for efficient implementation of pseudo-spectral methods and comparison with finite difference solutions of the Navier-Stokes equations,* 8th International Conference on Numerical Methods in Fluid Dynamics Proceedings, Lectures Notes in Physics 170, E. Krause, ed., Springer-Verlag, New York.

[61] T. L. HOLST [1979], *A fast, conservative algorithm for solving the transonic full-potential equation,* AIAA Paper 79-1456.

[62] B. J. HOSKINS AND A. J. SIMMONS [1974], *The development of spectral methods in the U. K. University of Atmospheric Modelling Group,* GARP WGNE Report No. 7, pp. 94–99.

[63] M. Y. HUSSAINI, D. A. KOPRIVA, M. D. SALAS AND T. A. ZANG [1983], *Spectral methods for Euler equations,* AIAA-83-1942-CP Proc. Sixth AIAA Computational Fluid Dynamics Conference, Danvers, MA.

[64] M. Y. HUSSAINI, M. D. SALAS AND T. A. ZANG [1983], *Spectral methods for inviscid, compressible flows,* in Advances in Computational Transonics, W. G. Habshi, ed., Pineridge Press, Swansea, UK.

[65] G. JACQUES [1976], *Numerical experiments with a two-layer pseudospectral model of general circulation,* M.Sc. Thesis, McGill Univ., Montreal.

[66] A. JAMESON [1979], *Acceleration of transonic potential flow calculations on arbitrary meshes by the multiple grid method,* AIAA Paper 79-1458.

[67] A. KASAHARA [1978], *Further studies on a spectral model of the global barotropic primitive equations with Hough harmonic expansions,* J. Atmos. Sci., 35, pp. 2043–2051.

[68] P. S. KLEBANOFF, K. D. TIDSTROM AND L. M. SARGENT [1962], *The three-dimensional nature of boundary layer instability,* J. Fluid Mech., 12, pp. 1–34.

[69] L. KLEISER [1982a], *Numerische Simulationen zum laminar-turbulenten Umschlagsprozeß der ebenen Poiseuille-Strömung,* Ph.D. thesis, Universität Karlsruhe, Kernforschungszentrum Karlsruhe, KLK 3271.

[70] ——— [1982b], *Spectral simulations of laminar-turbulent transition in plane Poiseuille flow and comparison with experiments,* 8th International Conference on Numerical Methods in Fluid Dynamics Proceedings, Lecture Notes in Physics 170, E. Krause, ed., Springer-Verlag, New York, pp. 280–285.

[71] L. KLEISER AND U. SCHUMANN [1980], *Treatment of incompressibility and boundary conditions in* 3-D *numerical spectral simulations of plane channel flows,* Proc. 3rd GAMM Conference on Numerical Methods in Fluid Mechanics, E. H. Hirschel, ed., Vieweg, Braunschweig, pp. 165–173.

[72] S. J. KLINE, W. C. REYNOLDS, F. A. SCHRAUB AND D. W. RUNSTADLER [1967], *The structure of turbulent boundary layers,* J. Fluid Mech., 30, pp. 741–773.

[73] L. S. KOVASZNAY, H. KOMODA AND B. R. VASUDEVA [1962], *Detailed flow field in transition,* Proc. 1962 Heat Transfer and Fluid Mechanics Institute, pp. 1–26.

[74] H.-O. KREISS AND J. OLIGER [1972], *Comparison of accurate methods for the integration of hyperbolic equations,* Tellus, 24, pp. 199.

[75] ——— [1979], *Stability of the Fourier method,* SIAM J. Numer. Anal., 16, pp. 421–433.

[76] S. KUBOTA [1959], *Surface spherical harmonic representations of the system of equations for analysis,* Pap. Meteorol, Geophys., 10, pp. 145–166.

[77] S. KUBOTA, M. HIROSE, Y. KIKUCHI AND Y. KURIHARA [1961], *Barotropic forecasting with the use of surface spherical harmonic representation,* Pap. Meteorol. Geophys. 12, pp. 199–215.

[78] J. LAMBIOTTE, S. BOKHARI, M. Y. HUSSAINI AND S. A. ORSZAG [1982], *Navier–Stokes solutions on the Cyber-*203 *by a pseudospectral technique,* 10th IMACS World Congress on System Simulation and Scientific Computation, Montreal.

[79] P. D. LAX [1978], *Accuracy and resolution in the computation of solutions of linear and nonlinear equations,* in Recent Advances in Numerical Analysis, Proc. Symposium, Mathematical Research Center, University of Wisconsin, Academic Press, New York, pp. 107–117.

[80] A. LEONARD AND A. WRAY [1982], *New numerical method for the simulation of three-dimensional flow in a pipe,* 8th International Conference on Numerical Methods in Fluid Dynamics Proceedings, Lecture Notes in Physics 170, E. Krause, ed., Springer-Verlag, pp. 335–342.

[81] E. N. LORENZ [1960], *Maximum simplification of the dynamic equations,* Tellus, 12, pp. 243–254.

[82] M. MACARAEG [1983], *Numerical solution of the axisymmetric flow in a heated rotating spherical shell,* Ph.D. thesis, Atmospheric Sciences Division, UTSI.

[83] B. MACHENHAUER [1973], *On the use of the spectral method in numerical integrations of atmospheric models,* Proc. Symposium on Difference and Spectral Methods for Atmosphere and Ocean Dynamics Problems, USSR Academy of Sciences, Siberian Branch, Novosibirsk.

[84] B. MACHENHAUER AND R. DALEY [1972], *A baroclinic primitive equation model with a spectral representation in three dimensions,* Institute of Theoretical Meteorology, University of Copenhagen, Report No. 4.

[85] ——— [1974], *Hemispheric spectral model,* GARP Publications Series No. 14, pp. 226–251.

[86] B. MACHENHAUER AND E. RASMUSSEN [1972], *On the integration of the spectral hydrodynamical equations by a transform method,* Institute for Theoretical Meteorology, University of Copenhagen, Report No. 3.

[87] A. MAJDA, J. MCDONOUGH AND S. OSHER [1978], *The Fourier method for nonsmooth initial data,* Math. Comput., 32, pp. 1041–1081.

[88] Y. MADAY [1981], *Sur quelques propriétes des approximations par des méthodes spectrales dans les espaces de Sobolev à poids, applications à la résolution de problèmes non linéaires,* Thése de troisième cycle, University of Paris VI.

[89] Y. MADAY AND A. QUARTERONI [1981], *Legendre and Chebyshev spectral approximations of Burgers' equation,* Numer. Math., 37, pp. 321–332.

[90] ——— [1982a], *Approximation of Burgers' equation by pseudospectral methods,* RAIRO Anal. Numer., 16, pp. 375–404.

[91] ——— [1982b], *Spectral and pseudospectral approximations of Navier–Stokes equations,* SIAM J. Numer. Anal., 19, pp. 761–780.

[92] N. N. MANSOUR, P. MOIN, W. C. REYNOLDS AND J. H. FERZIGER [1979], *Improved methods for large eddy simulations of turbulence,* in Turbulent Shear Flows I, F. Durst, B. E. Launder, F. W. Schmidt, J. H. Whitelaw, eds., Springer-Verlag, New York, pp. 386–401.

[93] P. S. MARCUS [1981], *Effects of truncation on modal representations of thermal convection,* J. Fluid Mech., 103, pp. 241–255.

[94] ——— [1983], *Simulation of Taylor-Couette flow* - I, II, submitted to J. Fluid Mech.

[95] P. S. MARCUS, S. A. ORSZAG AND A. T. PATERA [1982], *Simulation of cylindrical Couette flow,* 8th International Conference on Numerical Methods in Fluid Dynamics Proceedings, Lecture Notes in Physics 170, E. Krause, ed., Springer-Verlag, New York, pp. 371–376.

[96] J. B. MCLAUGHLIN AND S. A. ORSZAG [1982], *Transition from periodic to chaotic thermal convection,* J. Fluid Mech., 122, pp. 123–142.

[97] B. MERCIER [1981], *Analyse numérique des méthodes spectrales,* Note CEA-N-2278, Commissariat a l'Energie Atomique, Centre d'Etudes de Limeil, 94190 Villeneuve-Saint Georges, France.

[98] ——— [1982], *Stabilité et convergence des méthodes spectrales polynômiales: application à l'équation d'advection,* RAIRO Anal. Numer, 16, pp. 67–100.

[99] P. E. MERILEES [1966], *Harmonic representation applied to large scale atmospheric waves,* AMRG Publications in Meteorology, 83, McGill Univ., Montreal.

[100] ——— [1968], *The equations of motion in spectral form,* J. Atmos. Sci., 25, pp. 736–743.

[101] ——— [1972], *Truncation error in a spectral model,* Atmosphere, 10, pp. 1–9.

[102] ——— [1973], *An alternative scheme for the summation of a series of spherical harmonics,* J. Appl. Meteor., 132, pp. 224–227.

[103] ——— [1974], *Numerical experiments with the pseudospectral method in spherical coordinates,* Atmosphere, 12, pp. 77–96.

[104] P. E. MERILEES, P. DUCHARME AND G. JACQUES [1977], *Experiments with a polar filter and a one-dimensional semi-implicit algorithm,* Atmosphere, 15, pp. 19–32.

[105] B. METIVET AND Y. MORCHOISNE [1982a], *Resolution of unsteady Navier-Stokes equations by multi-domain spectral method,* Proc. EUROMECH Conference on Spectral Methods, Nice, France.

[106] ——— [1982b], *Multi-domain spectral technique for viscous flow calculation,* Proc. IV GAMM Conference on Numerical Methods in Fluid Mechanics, H. Viviand, ed., Vieweg, pp. 207–219.

[107] P. MOIN AND J. KIM [1980], *On the numerical solution of time-dependent viscous incompressible fluid flows involving solid boundaries,* J. Comput. Phys., 35, pp. 381–392.

[108] P. MOIN, W. C. REYNOLDS AND J. H. FERZIGER [1978], *Large eddy simulation of incompressible turbulent channel flow,* Report TF-12, Dept. of Mechanical Engineering, Stanford Univ., Stanford, CA.

[109] Y. MORCHOISNE [1979], *Resolution of Navier–Stokes equations by a space-time pseudospectral method,* La recherche aerospatiale, No. 1979-5, pp. 293–306 (English translation).

[110] ——— [1981a], *Pseudo-spectral space-time calculations of incompressible viscous flows,* AIAA 19th Aerospace Sciences Meeting, St. Louis, AIAA Paper No 81-0109.

[111] ——— [1981b], *Pseudo-spectral methods for homogeneous or inhomogeneous flows,* Proc. 3rd Turbulent Shear Flow, Symposium, Davis, CA, Chapter 19, pp. 1–5.

[112] E. M. MURMAN AND J. D. COLE [1971], *Calculation of plane steady transonic flow,* AIAA J., 9, pp. 114–121.

[113] S. A. ORSZAG [1970], *Transform method for calculation of vector coupled sums: Application to the spectral form of the vorticity equation,* J. Atmos. Sci., 27, pp. 890–895.

[114] ——— [1971a], *Numerical simulations of incompressible flows within simple boundaries: accuracy,* J. Fluid Mech., 49, pp. 75–112.

[115] ——— [1971b], *Accurate solution of the Orr–Sommerfeld stability equation,* J. Fluid Mech., 50, pp. 689–703.

[116] ——— [1972], *Comparison of pseudospectral and spectral approximations,* Stud. Appl. Math., 51, pp. 253–259.

[117] ——— [1974], *Fourier series on spheres,* Mon. Wea. Rev., 102, pp. 56–75.

[118] ——— [1976], *Turbulence and transition: A progress report,* Fifth International Conference on Numerical Fluid Dynamics Proceedings, A. I. vander Vooren and Zandbergen, eds. Lecture Notes in Physics 59, Springer-Verlag, New York, 39–51.

[119] ——— [1980], *Spectral methods for problems in complex geometries,* J. Comput. Phys., 37, pp. 70–92.

[120] S. A. ORSZAG AND D. GOTTLIEB [1980], *High resolution spectral calculations of inviscid compressible flows,* in Approximation Methods for Navier–Stokes Problems, Lecture Notes in Mathematics 771, Springer-Verlag, New York, pp. 381–398.

[121] S. A. ORSZAG AND M. ISRAELI [1974], *Numerical simulation of viscous incompressible flows,* Annual Review of Fluid Mechanics, 6, Annual Reviews, Inc., Palo Alto, CA, pp. 281–318.

[122] S. A. ORSZAG AND L. C. KELLS [1980], *Transition to turbulence in plane Poiseuille flow and plane Couette flow,* J. Fluid Mech., 96, pp. 159–205.

[123] S. A. ORSZAG AND A. T. PATERA [1980], *Subcritical transition to turbulence in plane channel flows,* Phys. Rev. Lett., 45, pp. 989–993.

[124] ——— [1981a], *Subcritical transition to turbulence in planar shear flows,* in Transition and Turbulence, R. E. Meyer, ed., Academic Press, New York, pp. 127–146.

[125] ——— [1981b], *Calculation of Von Karman's constant for turbulent channel flow,* Phys. Rev. Lett., 47, pp. 832–835.

[126] S. A. ORSZAG AND A. T. PATERA [1983], *Secondary instability of wall-bounded shear flows,* J. Fluid Mech., 128, pp. 347–385.

[127] S. A. ORSZAG AND G. S. PATERSON, JR. [1972a], *Numerical simulation of turbulence*, Statiscal Models and Turbulence, Springer-Verlag, New York, pp. 127–147.

[128] ——— [1972b], *Numerical simulation of three dimensional homogeneous isotropic turbulence*, Phys. Rev. Lett., 28, pp. 76–79.

[129] J. E. PASCIAK [1980], *Spectral and pseudo-spectral methods for advection equations*, Math. Comp., 35, pp. 1081–1092.

[130] ——— [1982], *Spectral methods for a nonlinear initial value problem involving pseudo-differential operators*, SIAM J. Numer. Anal., 19, pp. 142–154.

[131] A. T. PATERA AND S. A. ORSZAG [1980], *Transition and turbulence in plane channel flows*, Proc. 7th International Conference on Numerical Methods in Fluid Dynamics, R. W. MacCormack and W. C. Reynolds, eds., Springer-Verlag, New York, pp. 329–335.

[133] G. W. PLATZMAN [1960], *The spectral form of the vorticity equation*, J. Meteor., 17, pp. 635–644.

[134] ——— [1961], *An approximation to the product of discrete functions*, J. Meteor., 18, pp. 31–37.

[135] A. QUARTERONI [1982], *Theoretical and computational aspects of spectral methods*, in Computing Methods in Applied Sciences and Engineering, V. R. Glowinski and J. L. Lions, eds., North-Holland, Amsterdam, pp. 325–345.

[136] J. J. RILEY AND R. W. METCALFE [1980a], *Direct numerical simulation of a perturbed, turbulent mixing layer*, AIAA 18th Aerospace Sciences Meeting, Pasadena, CA, Paper No. AIAA-80-0274.

[137] ——— [1980b], *Direct numerical simulations of the turbulent wake of an axisymetric body*, Turbulent Shear Flows II, Bradbury, et al., eds., Springer-Verlag, New York. Also available as Flow Research Report 135, (March 1979) and NASA CR-152282.

[138] J. J. RILEY, R. W. METCALFE AND M. A. WEISSMAN [1981], *Direct numerical simulation of homogeneous turbulence in density-stratified fluids*, Flow Research Report 183.

[139] T. J. RIVLIN [1974], *The Chebyshev Polynomials*, John Wiley, New York.

[140] A. J. ROBERT [1966], *The integration of a low order spectral form of the primitive meteorological equations*, J. Meteor. Soc. Japan, 44, pp. 237–244.

[141] R. S. ROGALLO [1977], *An ILLIAC program for the numerical simulation of homogeneous, incompressible turbulence*, NASA TM-73203.

[142] L. SAKELL [1982], *Solution to the Euler equation of motion, pseudospectral techniques*, 10th IMACS World Congress on System, Simulation and Scientific Computing.

[143] M. D. SALAS, T. A. ZANG AND M. Y. HUSSAINI [1982], *Shock-fitted Euler solutions to shock-vortex interactions*, 8th International Conference on Numerical Methods in Fluid Dynamics Proceedings, Lecture Notes in Physics 170, E. Krause, ed., Springer-Verlag, New York, pp. 461–467.

[144] U. SCHUMANN [1975], *Subgrid scale model for finite difference simulations of turbulent flows in plane channels and annuli*, J. Comput. Phys., 18, pp. 376–404.

[145] ——— [1980], *Fast elliptic solvers and their application in fluid dynamics*, in Computational Fluid Dynamics, W. Kollman, ed., Hemisphere, Washington, pp. 401–430.

[146] U. SCHUMANN, G. GRÖTZBACH AND L. KLEISER [1980], *Direct numerical simulation of turbulence*, in Prediction Methods for Turbulent Flows. W. Kollman, ed., Hemisphere, Washington, pp. 123–258.

[147] J. SELA [1980], *Spectral modelling at the National Meteorological Center*, Mon. Wea. Rev., 108, pp. 1279–1292.

[148] E. D. SIGGIA AND A. ZIPPELIUS [1981], *Dynamics of defects in Rayleigh Bernard convection*, Phys. Rev. A, 24, pp. 1036–1049.

[149] I. SILBERMAN [1954], *Planetary waves in the atmosphere*, J. Meteor., 11, pp. 27–34.

[150] I. SIMMONDS [1975], *The spectral representation of moisture*, J. Appl. Meteor., 14, pp. 175–179.

[151] A. J. SIMMONS AND B. J. HOSKINS [1975], *A comparison of spectral and finite-difference simulations of a growing baroclinic wave*, Quart. J. Roy. Meteor. Soc., 101, pp. 551–565.

[152] J. SMAGORINSKY [1963], *General circulation experiments with the primitive equations*, Mon. Wea. Rev. 91, pp. 99–164.

[153] A. J. SROKOWSKI AND S. A. ORSZAG [1977], *Mass flow requirements for* LFC *wing suction*, AIAA Paper 77-1222.

[154] C. L. STREETT [1983], *A spectral method for the solution of transonic potential flow about an arbitrary airfoil*, AIAA-83-1949-CP Proc. Sixth AIAA Computational Fluid Dynamics Conference, Danvers, MA.

[155] C. L. STREETT, T. A. ZANG AND M. Y. HUSSAINI [1983], *Spectral multigrid methods with applications to transonic potential flow*, to appear.

[156] J. T. STUART [1965], *Hydrodynamic stability*, Appl. Mech. Rev., 18, pp. 523–531.

[157] E. TADMOR [1983], *Skew-selfadjoint form for systems of conservation laws*, J. Math. Anal. Appl., to appear.

[158] ——— [1983], *Finite difference, spectral and Galerkin methods for time-dependent problems*, ICASE Report No 83-22, Hampden, VA, and NASA CR #172149.

[159] I. TANI [1969], *Boundary layer transition*, in Annual Review of Fluid Mechanics, 1, Annual Reviews, Inc., CA, pp. 169–196.

[160] T. D. TAYLOR, R. S. HIRSH AND M. M. NADWORNY [1981], *Comparison of* FFT, *direct inversion and conjugate gradient methods for solution of flow equations by pseudospectral methods*, Proc. 4th GAMM Conference on Numerical Methods, Paris, France; submitted to Comput. Fluids.

[161] T. D. TAYLOR, R. B. MYERS AND J. H. ALBERT [1981], *Pseudospectral calculations of shock waves, rarefaction waves and contact surfaces*, Comput. Fluids, 9, pp. 469–473.

[162] A. TOOMRE, D. O. GOUGH AND E. A. SPIEGEL [1977], *Numerical solutions of single-mode convection equations*, J. Fluid Mech., 79, pp. 1–31.

[163] E. TURKEL [1980], *Numerical methods for large-scale time-dependent partial differential equations*, in Computational Fluid Dynamics, 2, Von Karman Institute, Hemisphere Washington, pp. 127–262.

[164] Y. S. WONG, T. A. ZANG AND M. Y. HUSSAINI [1983], *Efficient iterative techniques for the solution of spectral equations*, submitted to J. Comput. Phys.

[165] A. WRAY AND M. Y. HUSSAINI [1980], *Numerical experiments in boundary-layer stability*, AIAA Paper 80-0275, Aerospace Sciences Meeting.

[166] T. A. ZANG AND M. Y. HUSSAINI [1980], *Mixed spectral/finite difference approximations for slightly viscous flows*, Lecture Notes in Physics 141, Springer-Verlag, New York, pp. 461–466.

[167] T. A. ZANG D. A. KOPRIVA AND M. Y. HUSSAINI [1983], *Pseudospectral calculation of shock turbulence interactions*, Proc. Third International Conference on Numerical Methods in Laminar and Turbulent Flow, C. Taylor, ed., Pineridge Press, Swansea, UK.

[168] T. A. ZANG, Y. S. WONG AND M. Y. HUSSAINI [1982], *Spectral multigrid methods for elliptic equations*, J. Comput. Phys., 48, pp. 485–501.

[169] ——— [1983], *Spectral multigrid methods for elliptic equations* II, J. Comput. Phys., to appear.

VARIATIONAL METHODS IN THE THEORETICAL ANALYSIS OF SPECTRAL APPROXIMATIONS

CLAUDIO CANUTO* AND ALFIO QUARTERONI*

Abstract. Stability and optimal rates of convergence for Chebyshev pseudo-spectral approximations are given. We consider elliptic problems with nonconstant coefficients and Dirichlet or Neumann-type boundary conditions. They are set into a variational form which satisfies the inf-sup conditions of Nečas-Babuška.

1. Introduction and Preliminaries. In the last decade, spectral methods have been enjoying a remarkable success in practical computation. This led a growing number of mathematicians to analyze those methods from the theoretical point of view. The aim is to provide spectral methods with a rigorous justification obtaining in particular their properties of stability and convergence.

However, only a few applications have a theoretical justification through the results achieved until now in this direction. The problems which have been attacked with greater success by the theoretical analysis are those with periodic boundary conditions. For the nonperiodic problems the most interesting Chebyshev methods have been studied mainly in the case of constant higher order coefficients and Dirichlet boundary conditions. The mathematical investigation seems actually more difficult than in other numerical methods, such as for instance in finite elements.

Among the techniques that have been used we recall: i) the analysis of the asymptotic behavior of the Fourier coefficients of the solution; ii) energy estimates; iii) the analysis of the differential equation satisfied by the approximate solution; iv) the investigation of the Green's function for the discrete problem.

An important and powerful tool to derive energy estimates and convergence results in numerical analysis is

* Istituto di Analisi Numerica del C.N.R. and Istituto di Matematica Applicata dell'Università di Pavia - 27100 Pavia, Italy.

the variational principle. We mean here the search for minima for quadratic functionals as well as for saddle points for bilinear forms. This enlarges considerably the range of applications of the variational principle.

The abstract framework in which the differential problem is set is as follows.

$$(1.1) \quad \text{Find } u \in W : \quad B(u,v) = f(v) \quad \forall v \in V,$$

where $B(w,v)$ is a bilinear form on the function spaces W and V, and f is a continuous linear functional on V.

The use of variational principles is crucial in the finite element method, to derive stability and convergence, even to suggest new discretizations of the original problem. As for spectral methods, the most often used such as pseudo-spectral (collocation) methods, are not originally formulated in a variational way. However, the possibility of recognizing the solution of a pseudo-spectral problem as the solution of a variational problem too has been proved. This allows one to carry out a very accurate analysis of these methods (|10|, |18|, |13|, |5|, |4|).

It is our belief that using the variational principle is advantageous in the numerical analysis of spectral methods. In this paper, we try to help the reader think the same way by investigating some significant problems. We would like to emphasize that in spectral methods it is not always possible to assume $W=V$ in (1.1). Moreover, even if this choice is allowed, the bilinear form $B(w,v)$ may not be positive definite, so that energy estimates may not be obtained by setting $v=u$ in (1.1). Sect. 2 and 3 give two different examples of these situations. However, the stability and the convergence of the spectral scheme are assured, provided the form $B(w,v)$ fulfills some properties of continuity and coerciveness, as stated in the next two theorems.

<u>Theorem 1.1</u> - <u>Let</u> W <u>and</u> V <u>be two real Hilbert spaces with norm</u> $\|\cdot\|_W$ <u>and</u> $\|\cdot\|_V$ <u>respectively.</u>
<u>Let</u> $B(w,v)$ <u>be a bilinear form on</u> $W \times V$ <u>such that:</u>

$$(1.2) \quad \forall w \in W, \ \forall v \in V, \ |B(w,v)| \le C_1 \|w\|_W \|v\|_V \quad (C_1 < \infty),$$

$$(1.3) \quad \forall w \in W, \quad \sup_{v \in V} \frac{|B(w,v)|}{\|v\|_V} \ge C_2 \|w\|_W \quad (C_2 > 0),$$

$$(1.4) \quad \forall v \in V, \ v \neq 0 \ \sup_{w \in W} |B(w,v)| > 0.$$

Let $f \in V'$ (the dual space of V). Then the problem (1.1) has a unique solution, which satisfies:

$$(1.5) \quad \| u \|_W \leq C_2^{-1} \| f \|_{V'} . \square$$

Remark 1.1 - The previous theorem is due to J.Nečas (|17|, Thm. 3.1). We stated it in the form of |1| (Thm. 5.2.1).

The conditions (1.3)-(1.4) are called the coerciveness conditions.

The condition (1.4) may be equivalently replaced by the more restrictive one:

$$(1.6) \quad \forall v \in V, \quad \sup_{w \in W} \frac{|B(w,v)|}{\| w \|_W} \geq C_3 \| v \|_V \quad (C_3 > 0).$$

Eventually we note that, if W and V have finite dimension , the condition (1.4) is fulfilled iff dimW = dimV.□

Now we assume that the problem (1.1) is approximated by a sequence of finite dimensional problems:

$$(1.7) \quad \text{Find } u_N \in W_N \; : \; B_N(u_N, v) = f_N(v) \;\; \forall v \in V_N,$$

where $B_N(w,v)$ is a bilinear form on the subspaces $W_N \subseteq W$ and $V_N \subseteq V$, and f_N is a continuous linear functional on V_N.

The next theorem provides an abstract estimate of the error between the solutions of (1.1) and (1.7). Its proof is classical; we give it for the reader's convenience . In the case $B_N = B$ and $f_N = f$, it was given in |1|, Thm. 6.2.1

Theorem 1.2 - Assume that the form B(w,v) on WxV satisfies the hypotheses of Thm. 1.1. Moreover, assume that the form $B_N(w,v)$ on $W_N x V_N$ satisfies the same hypotheses with constants $C_1(N)$, $C_2(N)$ which could depend on N. Let $u \in W$ and $u_N \in W_N$ be the solutions of (1.1) and (1.7) respectively. Then:

$$(1.8) \quad \| u - u_N \|_W \leq \inf_{w \in W_N} \{ (1 + \frac{C_1}{C_2(N)}) \| u - w \|_W + \frac{1}{C_2(N)} \sup_{v \in V_N} [\frac{|B(w,v) - B_N(w,v)|}{\| v \|_V} + \frac{|f(v) - f_N(v)|}{\| v \|_V}] \}.$$

Proof - For any $w \in W_N$ and $v \in V_N$, we have by (1.1) and (1.7)

$$B_N(u_N-w,v) = B(u-w,v)+B(w,v)-B_N(w,v)+f_N(v)-f(v).$$

The coerciveness of B_N and the continuity of B yield

$$C_2(N)\ \|u_N-w\|_W \le C_1\ \|u-w\|_W + \sup_{v\in V_N}\left\{\frac{|B(w,v)-B_N(w,v)|}{\|v\|_V} + \frac{|f(v)-f_N(v)|}{\|v\|_V}\right\}.$$

Then the result follows from the triangular inequality.□

In spectral methods of polynomial type (Chebyshev, Jacobi, Legendre), weighted L^2 norms of functions and their derivatives are the natural norms in which stability and convergence are established. This leads to the introduction of the spaces H^s_ρ of the functions having such norms finite.

Furthermore, the right hand side of (1.8) can be bounded by using several operators of orthogonal projection and of interpolation (at the collocation nodes). So, the asymptotic behavior of the error is strictly related to approximation properties of such operators in the norms of H^s_ρ. Hence we recall here some definitions and results which are fundamental in our analysis.

Set $\Omega=(-1,1) \subseteq \mathbb{R}$, and let $\rho=\rho(x)$ be a weight function on Ω. In the sequel three different weights will intervene: the Chebyshev weight $\omega(x)=(1-x^2)^{-\frac{1}{2}}$, the Jacobi weight $\eta(x)=\omega^{-1}(x)$, and the weight $\rho\equiv 1$. We define

$$L^2_\rho = \{\ \phi : \Omega \to \mathbb{R} \mid \phi \text{ is measurable and } \|\phi\|_{0,\rho} < \infty\}$$

with $\|\phi\|_{0,\rho}=(\phi,\phi)^{\frac{1}{2}}_\rho$ and $(\phi,\psi)_\rho=\int_\Omega \phi(x)\psi(x)\rho(x)dx$.

For any integer $s\ge 0$ we set

$$H^s_\rho = \{\phi \in L^2_\rho \mid D^k\phi \in L^2_\rho,\ 0\le k\le s\} \quad (D^k=d^k/dx^k),$$

endowed with the norm $\|\phi\|^2_{s,\rho} = \sum_{k=0}^{s} \|D^k\phi\|^2_{0,\rho}$.

The spaces H^s_ρ of non integer order $s>0$ can be defined by the theory of interpolation of spaces (see, e.g., |2|). Since the spaces H^s_ρ, $s>0$ non-integer, are needed only in a few proofs, the reader unfamiliar with such spaces may refer our results to the case s integer. $H^1_{o,\rho}$ will eventually denote the subspace of H^1_ρ of the functions vanishing on the boundary.

Let $\mathbb{P}_N$ be the space of algebraic polynomials of degree $\leq N$, and set $S_N=\{p\in \mathbb{P}_N \mid p(-1)=p(1)=0\}$. We define P_N as the orthogonal projection operator upon S_N with respect to the inner product of L^2_ω (i.e. $P_N\phi$ is the truncation of order N of the Chebyshev series of ϕ).

The following approximation result holds (|3|, Thm. 2.2)

$$(1.9) \quad \forall\phi\in H^s_\omega \cap H^1_{o,\omega},\ \|\phi-P_N\phi\|_{r,\omega}\leq CN^{e(r,s)}\|\phi\|_{s,\omega},\ 0\leq r\leq s,$$

where $e(r,s)$ is $2r-1/2-s$ if $r>1$ and $3r/2-s$ if $r\leq 1$. Hereafter C denotes a generic positive constant independent of N.

Let now $x_j=\cos \frac{j\pi}{N}$, $0\leq j\leq N$, be the Chebyshev nodes, and for any continuous function ϕ let $P_c\phi$ denote the polynomial of $\mathbb{P}_N$ interpolating ϕ at the nodes x_j. Then (|3|, Thm. 3.1)

$$(1.10) \quad \forall\phi\in H^s_\omega,\ s>\tfrac{1}{2},\quad \|\phi-P_c\phi\|_{r,\omega}\leq CN^{2r-s}\|\phi\|_{s,\omega},\ 0\leq r\leq s.$$

Denoting by ω_j the weights of the Chebyshev-Gauss-Lobatto quadrature formula ($\omega_o=\omega_N=\pi/2N$, $\omega_j=\pi/N$ $1\leq j\leq N-1$), we define a "discrete" inner product and norm as follows:

$$(1.11) \quad (\phi,\psi)_{N,\omega}=\sum_{j=0}^{N}\phi(x_j)\psi(x_j)\omega_j,\quad \|\phi\|_{N,\omega}=(\phi,\phi)^{\frac{1}{2}}_{N,\omega}.$$

Then

$$(1.12) \quad \begin{cases} (\phi,\psi)_{N,\omega}=(\phi,\psi)_\omega & \text{if } \phi\psi\in \mathbb{P}_{2N-1}, \text{ and} \\ C'\|\phi\|_\omega \leq \|\phi\|_{N,\omega} \leq C''\|\phi\|_\omega & \text{if } \phi\in \mathbb{P}_N (|3|, \text{Lemma } 3.2). \end{cases}$$

The use of the operators P_N and P_c in the numerical analysis of spectral methods is well known (see, e.g., |9|). The exponent of N^{-1} in (1.9) and (1.10) is the largest possible (see |3|, Sect. 2). However, better approximation properties can be obtained by different projection operators. Namely, let Π_N^s denote the orthogonal projection operator upon S_N with respect to the inner product of H_ω^s. Then (|14|, Thm.1.6, and |12|, Ch.IV,Thm.3)

$$(1.13)\quad \forall\phi\in H_\omega^s\cap H_{o,\omega}^1,\ \|\phi-\Pi_N^s\phi\|_{r,\omega}\le CN^{r-s}\|\phi\|_{s,\omega}\ ,\quad 0\le r\le s.$$

Different applications of such estimates have been given in |4|,|5|,|14|,|15|,|16|. A readable survey is given in |19|.

In this paper we deal with the Chebyshev approximation of two second order elliptic problems, one with Dirichlet boundary conditions and the other with Neumann (or mixed) boundary conditions. The Dirichlet problem is the following:

$$(1.14)\qquad -(\alpha u_x)_x + (\beta u)_x + \gamma u = f \text{ in } \Omega;\ u(-1)=u(1)=0,$$

with α, β and γ depending on x.

After integration by parts, problem (1.14) is set into the variational formulation (1.1) with the same spaces $V=W=H_{o,\omega}^1$. However, it is easily seen that the choice v=u cannot help in proving stability. Actually, $u_x(u\omega)_x$ can be strictly negative somewhere in Ω, hence for a suitable $\alpha=\alpha(x)>0$ the integral $\int_\Omega \alpha u_x(u\omega)_x dx$ can be negative. This forces us to choose a function v different from u, and a variational formulation like (1.1) will ensure the stability. The idea we are going to follow is to define a function v in such a way that $\int_\Omega \alpha u_x(v\omega)_x dx=\int_\Omega \alpha(u_x)^2\omega dx$. A similar choice is possible for the collocation problem too.

The error estimate we prove in Sect.2 for such a problem is essentially of the following type:

$$\|u-u_N\|_{1,\omega}\le CN^{1-s}\|u\|_{s,\omega}\ .$$

Next we consider the following Neumann problem:

$$(1.15)\qquad -u_{xx} + \beta u_x + \gamma u = f \ \text{ in } \Omega;\ u_x(-1)=u_x(1)=0,$$

with β and γ depending on x. Also general boundary conditions of mixed type can be treated (see Remark 3.1). For such problems, integration by parts is not allowed, since u is not zero at the boundary and the weight is singular there. Nevertheless , it is possible to formulate the problem in a variational form in which the spaces W and V do not coincide (roughly speaking, functions in W are more regular than those in V). We prove asymptotic stability for a Chebyshev pseudo-spectral approximation, and an error estimate of the kind:

$$\|u-u_N\|_{2,\eta} \le CN^{2-s} \|u\|_{s,\omega} .$$

All the results presented here are derived under minimal regularity assumptions on the coefficients (essentially, they must be Hölder continuous of exponent >1/2). Hence our analysis applies to physical problems with rough coefficients. We shall use the spaces $C^{k,\nu}$ of functions with k-th derivative Hölder continuous in Ω with exponent ν, provided with the norm

$$\|\phi\|_{C^{k,\nu}} = \sup_{x\in\Omega,\, 0\le i\le k} |D^i\phi(x)| + \sup_{x,y\in\Omega} \frac{|D^k\phi(x)-D^k\phi(y)|}{|x-y|^\nu} .$$

L^∞ will denote the space of (essentially) bounded functions in Ω. Eventually, if V is a Banach space, V' denotes its topological dual, and < , > the duality pair.

The extension of these results to several spatial dimensions can be done at the expense of some further technical difficulties only.

Some sharper estimates which improve the rates of convergence of $u-u_N$ in the L^2 norm will be presented elsewhere.

The analysis of the evolution problems corresponding to problems (1.14) and (1.15) has been carried out in some special cases only. In the general case, it presents more difficulties, and it is now under investigation.

2. The Dirichlet boundary value problem. Throughout this section we set $\|u\| = \|u\|_{1,\omega}$ for the sake of simplicity. We are interested in the analysis of Chebyshev spectral methods for the problem (1.14). We assume α, β and $\gamma \in L^\infty$ with $\alpha(x) \ge \alpha_0 > 0$ in Ω. Further assumptions on α, β and γ will be given later.

A bilinear form $B(w,v)$ on $H^1_{o,\omega} \times H^1_{o,\omega}$ defined by

$$(2.1) \qquad B(w,v) = \int_\Omega \left[\alpha w_x \left(v_x + \frac{\omega_x}{\omega} v\right) + (\beta w)_x v + \gamma w v\right] \omega dx$$

is associated to the operator in (1.14). We represent $B(w,v)$ through the form $B(w,v)=a(w,v)+b(w,v)$ where

$$(2.2) \qquad a(w,v) = \int_\Omega \alpha w_x \left(v_x + \frac{\omega_x}{\omega} v\right) \omega dx$$

is the bilinear form associated to the principal part of the operator (1.14). If $u \in H^1_{o,\omega}$ is the solution of (1.14) for a given $f \in (H^1_{o,\omega})'$, then u is the solution of the variational problem

$$(2.3) \qquad B(u,v) = \langle f,v \rangle \qquad \forall v \in H^1_{o,\omega}$$

and vice versa. This follows by integration by parts, using |5|, Lemma 1.1.

The Chebyshev pseudo-spectral approximation of problem (1.14) is defined as follows:

$$(2.4) \qquad \begin{cases} u_N \in S_N = \{p \in \mathbb{P}_N \mid p(-1)=p(1)=0\} \\ -[P_c(\alpha u_{N,x})]_x + [P_c(\beta u_N)]_x + \gamma u_N = f \quad \text{at } x=x_j, \quad j=1,\dots,N-1. \end{cases}$$

This problem is equivalent to the variational formulation

$$(2.5) \qquad B_N(u_N,v) = (f,v)_{N,\omega} \qquad \forall v \in S_N,$$

where $B_N(w,v)=a_N(w,v)+b_N(w,v)$ on $S_N \times S_N$, with

$$(2.6) \qquad a_N(w,v) = \int_\Omega P_c(\alpha w_x)\left(v_x + \frac{\omega_x}{\omega} v\right)\omega dx \ ,$$

$$(2.7) \qquad b_N(w,v) = \int_\Omega [P_c(\beta w)]_x v \omega dx + (\gamma w,v)_{N,\omega}.$$

Problem (2.3) and its approximation (2.5) are set into the abstract form of Thm. 1.1.

Throughout this section we shall assume that problem (2.3) is well posed in $H^1_{o,\omega}$, i.e.,

$$(2.8) \qquad \begin{cases} \text{the form } B(w,v) \text{ satisfies the hypotheses of Theorem 1.1} \\ \text{with } W = V = H^1_{o,\omega}. \end{cases}$$

This assumption is fulfilled in particular in the case $\beta \equiv \gamma \equiv 0$, as is seen in the next theorem. The techniques used in its proof will be extended in a natural way to cover the analysis of problem (2.5).

Theorem 2.1 - There exist two constants $C_1>0$, $C_2>0$ such that the form $a(w,v)$ on $H^1_{0,\omega}\times H^1_{0,\omega}$ satisfies the hypotheses of Theorem 1.1.

Proof - The form a is continuous on $H^1_{0,\omega}\times H^1_{0,\omega}$ since $\|\frac{\omega_x}{\omega} v\| \leq C\|v\|$ (|5|, Lemma 1.1).

In order to check condition (1.3) we need the following result.

Lemma 2.1 - Let $w\in H^1_{0,\omega}$. The differential problem

$$(2.9)\qquad v_x + \frac{\omega_x}{\omega} v = w_x \quad , \quad v\in H^1_{0,\omega}$$

has a unique solution iff $w\in W_0=\{w\in H^1_{0,\omega} | \int_\Omega w_x \omega dx=0\}$. Moreover, the operator $Tw=v$ is a topological isomorphism from W_0 onto $Z=\{v\in H^1_{0,\omega} | \int_\Omega \omega_x \omega^{-1} v dx=0\}$.

Proof - If $v\in H^1_{0,\omega}$ satisfies (2.9), then $\int_\Omega w_x \omega dx=\int_\Omega (v\omega)_x dx=0$ since $v\omega$ is continuous on $\bar\Omega$ and vanishes at $x=\pm 1$. Conversely, assume $w\in W_0$. The general solution of eq. (2.9) has the form

$$(2.10)\qquad v(x)=\omega^{-1}(x)(C+\int_{-1}^{x} w_t \omega dt), \quad C\in\mathbb{R}.$$

If $C=0$, $v\in H^1_{0,\omega}$. Actually, owing to (2.9) it is enough to check that $\chi=\frac{\omega_x}{\omega} v\in L^2_\omega$. Since χ is continuous in Ω, we can localize the analysis in a neighbourhood of the endpoints; $w\in W_0$ implies $\chi(x)= x\omega(x)\int_{-1}^{x} w_t\omega dt=-x\omega(x)\int_x^1 w_t \omega dt$, hence we get $\|\chi\|_{0,\omega}\leq C\|w_x\|_{0,\omega}$ by applying Lemma 10.1 of |11| with $\alpha=1/4$, $g(t)=u_t t^{-\frac{1}{2}}$ near $t=0$. If $C\neq 0$ then $v\notin H^1_\omega$, as $\omega^{-1}\notin H^1_\omega$. Finally, if $v\in Z$ then $v_x+\omega_x\omega^{-1}v$ is in L^2_ω and it has zero average on Ω, so it is the derivative of a function $w\in H^1_{0,\omega}$. This accomplishes the proof of the lemma.□

Let us consider now the symmetric bilinear form $((w,v))_\alpha=\int_\Omega \alpha w_x v_x \omega dx$. It defines an equivalent inner product on $H^1_{0,\omega}$. By the previous lemma we get

$$(2.11)\qquad a(u,Tw)=((u,w))_\alpha \quad \forall u\in H^1_{0,\omega} \quad , \quad \forall w\in W_0,$$

hence there exists a suitable constant $C>0$ such that

(2.12) $a(w,Tw)\geq C\|w\|\|Tw\|$, $\forall w\in W_o$,

(2.13) $a(T^{-1}z,z)\geq C\|z\|\|T^{-1}z\|$, $\forall z\in Z$.

This implies that the form $a(w,v)$ (resp. $a^t(v,w)\equiv a(w,v)$) satisfies the hypotheses of Theorem 1.1 for the spaces $W=W_o$ and $V=Z$ (resp. $W=Z$ and $V=W_o$), which are closed subspaces of $H^1_{o,\omega}$ of codimension one. We define $\phi\in H^1_{o,\omega}$ by the relations

(2.14) $\|\phi\|=1$, $((\phi,w))_\alpha=0$ $\forall w\in W_o$,

(which is equivalent to $a(\phi,v)=0$, $\forall v\in Z$), and $\psi\in H^1_{o,\omega}$ by the relations

(2.15) $\|\psi\| = 1$, $((\psi,v))_\alpha = 0$ $\forall v\in Z$.

We set $C_o=a(\phi,\psi)$ and we note that $C_o\neq 0$. Indeed, if $C_o=0$, then $a(\phi,v)=0$ $\forall v\in H^1_{o,\omega}$, hence $-(a\phi_x)_x\equiv 0$ in the sense of distributions for $\phi\in H^1_{o,\omega}\subset H^1_o$. This implies $\phi\equiv 0$, which contradicts (2.14). Moreover we can assume $C_o>0$, so that we have trivially

(2.16) $a(\phi,\psi) = C_o\|\phi\|\|\psi\|$.

Finally, by the properties of a^t previously proved, we can define $z\in Z$ as the solution of

(2.17) $a^t(z,w) = -a(w,\psi)$, $\forall w\in W_o$.

One has $\|z\|\leq C\|w\|\lesssim C$.

Let now w be any function in $H^1_{o,\omega}$. It can be uniquely written as $w=w_o+\lambda\phi$, for $w_o\in W_o$ and $\lambda\in\mathbb{R}$. We set

(2.18) $v = Tw_o + \lambda(\psi+z)$.

By the previous estimates, one has $\|v\|\leq C\|w\|$. Moreover,

$$a(w,v) = a(w_o,Tw_o) + \lambda a(w_o,\psi+z) + \lambda a(\phi,Tw_o)$$

$$+ \lambda^2 a(\phi,\psi) + \lambda^2 a(\phi,z) \geq C\|w_o\|^2 + \lambda^2 C_o \geq C'\|w\|^2 .$$

Since $a(w_o,\psi+z) = 0$ by (2.17), and $a(\phi,Tw_o) = a(\phi,z) = 0$ by (2.14), the proof of (1.3) is complete.

In order to prove (1. 4), let $v = v_o + \lambda\psi$ for $v_o \in Z$, $\lambda \in \mathbb{R}$, and define

(2.19) $w = T^{-1}(v_o - \lambda z) + \lambda\phi$.

Then, as before $a(w,v) \geq C'\|w\|^2 > 0$ if $v \neq 0$. This completes the proof of the theorem.□

Now we want to prove the uniform coerciveness of the discrete forms $a_N(u,v)$. To this end we shall need a regularity result for the orthogonal complement to the subspace W_o. Furthermore, we shall give an estimate which will be used to evaluate the distance between the forms $a(u,v)$ and $a_N(u,v)$ for polynomial u and v.

From now on, we shall assume

(2.20) $\alpha \in C^{o,\nu}$ for a given $\nu > \frac{1}{2}$.

This is a very mild regularity requirement on the coefficient of the second order elliptic operator; however (2.20) will be enough to obtain the properties of stability and convergence of the method. So the present analysis applies to the approximation of physical problems with non smooth data.

Lemma 2.2 - The function ϕ defined in (2.14) belongs to H^θ_ω with $\theta = \nu + 3/4$.

Proof - For fixed $w \in W_o$, define $v_o(x) = \int_{-1}^{x} w_t \omega dt$. Then $v_o \in \tilde{V} = \{v \in H^1_{o,\omega^{-1}} \mid \int_\Omega v_x \omega^{-1} dx = 0\}$, and ϕ satisfies $\int_\Omega \alpha\phi_x v_{o,x} dx = 0$, $\forall v_o \in \tilde{V}$. Let $\tilde{v} \in H^1_{o,\omega^{-1}}$ be a function such that $\int_\Omega \tilde{v}\omega^{-1} dx = 1$. Then any $v \in H^1_{o,\omega^{-1}}$ can be written as $v = v_o + \mu\tilde{v}$, where $v_o \in \tilde{V}$ and $\mu = \int_\Omega v_x \omega^{-1} dx$. It follows that

$$\int_\Omega \alpha\phi_x v_x dx = -K\int_\Omega v\omega^{-1}_x dx, \quad \forall v \in H^1_{o,\omega^{-1}},$$

where $K = \int_\Omega \alpha\phi_x \tilde{v}_x dx$. Thus $(\alpha\phi_x)_x = Kx\omega \in L^p$ for any $p<2$.

Owing to this regularity and to the assumption (2.20), one has $\phi \in H^{1+\nu}$ (see Tartar(|20|, Thm.4) - see also Campanato (|6|, Thm.16) in a slightly different functional framework, though).
We get the final result by Theorem 4.1 of the Appendix.□

Lemma 2.3 - Assume that $\alpha \in C^{k,\nu}$ with $k+\nu > \frac{1}{2}$. Then

(2.21) $\|\alpha p-P_c(\alpha p)\|_{o,\omega} \le CN^{-t}\|\alpha\|_{C^{k,\nu}}\|p\|_{t,\omega}$, $0\le t\le s\equiv k+\nu$, $\forall p\in \mathbb{P}_N$

Proof - If $g(x)$ is a function on $(-1,1)$, we set $\tilde g(\theta) = g(\cos\theta)$. $\tilde P_c\tilde g$ will denote the trigonometric interpolant of $\tilde g$ of degree N at the points $\theta_j = j\pi/(2N+1)$, $j = 0,\dots,2N$. Then (|3|, Thm.1.2)

$$\|\alpha p-P_c(\alpha p)\|_{o,\omega} = C\|\widetilde{\alpha p} - \tilde P_c(\widetilde{\alpha p})\|_{L^2(0,2\pi)} \le C\, N^{-s}\|\widetilde{\alpha p}\|_{H^s(0,2\pi)} .$$

By using the inequality

$$\|\widetilde{\alpha p}\|_{H^s(0,2\pi)} \le C\|\tilde\alpha\|_{C^{k,\nu}}\|\tilde p\|_{H^s(0,2\pi)} ,$$

the Fourier inverse inequality (see |3|, Proposition 1.1), and going back to α and p in Ω, we get the result. □

Theorem 2.2 - Assume $\alpha\in C^{o,\nu}$ for a suitable $\nu>1/2$. There exists $N_o>0$ such that for any integer $N\ge N_o$ the form $a_N(w,v)$ on $S_N\times S_N$ satisfies the hypotheses of Theorem 1.1 , with two constants C_1 and C_2 independent of N.

Proof- For any $w\in S_N$ and $v\in S_N$, one has

$$a_N(w,v) = a(w,v) - \int_\Omega (\alpha w_x - P_c(\alpha w_x))(v\omega)_x dx.$$

By the continuity of $a(w,v)$ and by the estimate (2.21) with $p = w_x$ and $t = 0$, we obtain the continuity of $a_N(w,v)$.

We prove now condition(1.3). To this end, we use the property that the operator T defined in Lemma 2.1 preserves the polynomials. Indeed setting $W_N = W_o\cap S_N$ and $Z_N = Z\cap S_N$, we have

(2.22) T is an isomorphism of W_N onto Z_N.

In order to check (2.22), let $w\in W_N$. Then $v=Tw$ is given by (2.10) with C=0. After finding an antiderivative for $w_t\omega$ we get $v(x)=p(x)+\mu\omega^{-1}(x)\arcsin(x)$, with $p\in S_N$ and $\mu\in\mathbb{R}$. Since $\omega^{-1}(x)\arcsin(x)$ does not belong to H^1_ω, then necessarily $\mu=0$ and $v\in S_N$. This gives (2.22).

Using the property (2.22) and reasoning as in the proof of Theorem 2.1, one can verify that $a_N(w,v)$ is uniformly coercive on $W_N\times Z_N$. In order to prove the theorem, we introduce the orthogonal complements of W_N and of Z_N. To this end we define

(2.23) $\phi_N \in S_N$: $((\phi-\phi_N, w))_\alpha = 0 \quad \forall w \in S_N$

and

(2.24) $\psi_N \in S_N$: $((\psi-\psi_N, v))_\alpha = 0 \quad \forall v \in S_N$.

By (2.14) and (2.15), ϕ_N is orthogonal to W_N and ψ_N is orthogonal to Z_N with respect to the inner product $((.,.))_\alpha$ in S_N. Furthermore, the density of H^σ_ω in H^1_ω for all $\sigma>1$, and the estimate (1.13), allows one to deduce that

(2.25) $\phi_N \to \phi$ and $\psi_N \to \psi$ in H^1_ω, as $N \to \infty$.

Furthermore, since ϕ is more regular by Lemma 2.2, it is possible to estimate the order of convergence of ϕ_N to ϕ by using (1.13) once more, namely

(2.26) $\|\phi-\phi_N\| \le CN^{1-\theta} \|\phi\|_{\theta,\omega}$.

Finally, let $z_N \in Z_N$ be the solution of the problem

(2.27) $a_N(w, z_N) = -a_N(w, \psi_N) \quad \forall w \in W_N$.

By the uniform coerciveness and the continuity of $a_N(w,v)$, we deduce that $\|z_N\| \le C\|\psi_N\| \le C'$.

Let $w = w_N + \lambda\phi_N$ be an arbitrary element in S_N, with $w_N \in W_N$ and $\lambda \in \mathbb{R}$. As in (2.19) we set

(2.28) $v = Tw_N + \lambda(\psi_N + z_N) \in S_N$.

By Lemma 2.1 and by (2.25), $\|v\| \le C\ \|w\|$, with C>0 independent of N. Hence (1.3) is verified if we show that $a_N(w,v) \ge C\ \|w\|^2$. We have by (2.27) and (2.33)

$$\begin{aligned} a_N(w,v) &= a_N(w_N, Tw_N) + \lambda a_N(w_N, \psi_N + z_N) + \lambda a_N(\phi_N, Tw_N) + \\ (2.29) \qquad & \lambda^2 a_N(\phi_N, \psi_N + z_N) = a_N(w_N, Tw_N) + \lambda^2 a(\phi,\psi) + \\ & \lambda\,[\,a_N(\phi_N, v) - a(\phi_N, v)\,] + \lambda^2 [a(\phi_N, \psi_N) - a(\phi,\psi)]. \end{aligned}$$

The two terms in square brackets vanish as $N \to \infty$. For the first term the following estimate holds:

$$(2.30)\qquad |a_N(\phi_N,v)-a(\phi_N,v)| \le CN^{1-\theta}\|\phi\|_{\theta,\omega}\ \|v\| \qquad \forall v\in H^1_{o,\omega}.$$

In order to prove (2.30), denote by $\tilde{\phi}_N$ any element in S_N. Then, by the continuity of the forms $a(w,v)$ and $a_N(w,v)$ and by Lemma 2.3, we get:

$$|a_N(\phi_N,v)-a(\phi_N,v)|\le|a_N(\phi_N-\tilde{\phi}_N,v)|+|a_N(\tilde{\phi}_N,v)-a(\tilde{\phi}_N,v)|+$$
$$|a(\tilde{\phi}_N-\phi_N,v)|\le C\{\|\phi_N-\tilde{\phi}_N\|+N^{1-\theta}\|\alpha\|_{C^{o,\nu}}\|\tilde{\phi}_N\|_{\theta,\omega}\}\|v\|.$$

Choosing as $\tilde{\phi}_N$ the orthogonal projection of ϕ on S_N with respect to the norm of H^θ_ω, from (1.13) it follows:

$$(2.31)\qquad \|\tilde{\phi}_N\|_{\theta,\omega}\le\|\phi\|_{\theta,\omega},\qquad \|\phi-\tilde{\phi}_N\|\le CN^{1-\theta}\|\phi\|_{\theta,\omega}.$$

Thanks to (2.26) we obtain (2.30).

On the other hand, (2.25) and the continuity of the form $a(w,v)$ give

$$(2.32)\qquad a(\phi_N,\psi_N)\to a(\phi,\psi)\qquad \text{as } N\to\infty.$$

By the coerciveness of a_N on $W_N x Z_N$ and by (2.16), it follows that

$$a_N(w_N,Tw_N)+\lambda^2 a(\phi,\psi)\ge C\ \|w_N\|^2+C_o|\lambda|^2\ge C'\ \|w\|^2.$$

Hence, recalling (2.30) and (2.32), there exists $N_o>0$ such that, for any $N\ge N_o$ one has $a_N(w,v)\ge(C'/2)\cdot\|w\|^2$. This proves (1.3).

At last, we note that $a_N(w,v)$ satisfies (1.4) (see the Remark 1.1). Hence, the Theorem is completely proved. □

Our next purpose will be to derive the analogue of Theorem 2.2 for the whole form $B_N(w,v)$. To this end, we shall consider $B_N(w,v)$ as a perturbation of the uniformly coercive form $a_N(w,v)$ through the lower order form $b_N(w,v)$. The following lemma provides a convergence result which will be used in the analysis.

<u>Lemma 2.4 - Let</u> $v\in H^1_{o,\omega}$ <u>be given, and let</u> $v_N\in S_N$ <u>be the solution of the problem</u>

$$(2.33)\qquad a_N(w,v_N)=a(w,v)\qquad \forall w\in S_N.$$

Then, if ν is defined by (2.20) it follows that

(2.34) $$\|v-v_N\|_{o,\omega} \le CN^{-\nu}\|v\| .$$

Proof - First of all, note that $\|v_N\| \le C\|v\|$ by the previous theorem and the continuity of the form $a(w,v)$. We prove (2.34) using a classical Aubin-Nitsche argument (see, e.g., |7|, pg. 136). One has:

$$\|v-v_N\|_{o,\omega} = \sup_{\zeta \in L^2_\omega} \frac{(\zeta, v-v_N)_\omega}{\|\zeta\|_{o,\omega}} = \sup_{\zeta \in L^2_\omega} \frac{a(\phi, v-v_N)}{\|\zeta\|_{o,\omega}} .$$

Here, for a given $\zeta \in L^2_\omega$, $\phi \in H^1_{o,\omega}$ is the solution of the variational problem $a(\phi,w)=(\zeta,w)_\omega$ $\forall\, w \in H^1_{o,\omega}$, that is it satisfies: $-(\alpha\phi_x)_x = \zeta$ in Ω.

By the Tartar's result (|20|, Thm.4), one gets $\phi \in H^{1+\nu}$ with $\|\phi\|_{1+\nu,\omega} \le C\|\zeta\|_{o,\omega}$.

If $\phi_N \in S_N$ denotes the orthogonal projection of ϕ on S_N with respect to the norm of $H^{1+\nu}_\omega$, we have by (2.33) that

$$\begin{aligned} a(\phi,v-v_N) &= a(\phi-\phi_N, v-v_N) + a(\phi_N, v-v_N) \\ &= a(\phi-\phi_N, v-v_N) - \int_\Omega [\alpha\phi_{N,x} - P_c(\alpha\phi_{N,x})](v_N\omega)_x dx . \end{aligned}$$

This, (1.2), (1.13) and Lemma 2.2 give (2.34). □

In the next lemma, an abstract estimate of the error on the lower order terms of the operator in (1.14) is given. It will be used in proving both the coerciveness of the form $B_N(w,v)$ and the convergence of the approximation (2.4).

Lemma 2.5 - Let w', w'', v', v'' be arbitrary elements in S_N. Then:

$$|b(w',v') - b_N(w'',v'')| \le C\|w'\|\,\|v'-v''\|_{o,\omega} + C\|v''\|\,\|w'-w''\|_{o,\omega}$$

(2.35) $$+ C\|v''\|\,(\|\beta w'' - P_c(\beta w'')\|_{o,\omega} + \|\gamma w'' - P_c(\gamma w'')\|_{o,\omega}$$

$$+ \|\gamma w'' - P_{N-1}(\gamma w'')\|_{o,\omega}) .$$

Proof - We have:

$$b(w',v')-b_N(w'',v'')=[b(w',v')-b(w'',v'')]+$$
$$\int_\Omega [\beta w''-P_c(\beta w'')](v''\omega)_x dx+\{(\gamma w'',v'')_\omega-(\gamma w'',v'')_{N,\omega}\}.$$

The bracketed term and the integral can be bounded by the Cauchy-Schwarz inequality. As for the last term, we apply the abstract estimate (see |5|, proof of Lemma 3.2):

$$(2.36)\qquad |(g,v'')_\omega-(g,v'')_{N,\omega}|\le C\,\|v''\|\,(\|g-P_c g\|_{o,\omega}+\|g-P_{N-1}g\|_{o,\omega})$$

□

We can finally prove the uniform coerciveness of the form $B_N(w,v)$.

Theorem 2.3 - Assume that (2.8) and (2.20) hold. Moreover, suppose that $\beta,\gamma\in C^{o,\nu}$ for $\nu>\frac{1}{2}$. There exists $N_1>0$ such that for any integer $N\ge N_1$ the form $B_N(w,v)$ on $S_N \times S_N$ satisfies the hypotheses of Theorem 1.1, with constants C_1 and C_2 independent of N.

Proof - Owing to Remark 1.1, it is enough to check that (1.2) and (1.3) hold.

Continuity (1.2) for $B_N(w,v)$ is an easy consequence of the continuity of $a_N(w,v)$ (see Theorem 2.2) and of $b_N(w,v)$, via Lemma 2.5.

We check now (1.3). To this end let $w\in S_N$ be fixed. Due to (2.8) there exists $v\in H^1_{o,\omega}$ such that

$$B(w,v)=a(w,v)+b(w,v)\ge \bar{C}_2\,\|w\|\,\|v\|$$

for a suitable constant $\bar{C}_2>0$ independent of N. Define $v_N\in S_N$ by means of (2.33). Then

$$B_N(w,v_N)=B(w,v)+b_N(w,v_N)-b(w,v).$$

By Lemma 2.5 with $w'=w''=w$, $v'=v_N$ and $v''=v$, by (2.34) and the estimates (1.9) and (1.10), we have that

$$|b_N(w,v_N)-b(w,v)|\le\frac{\bar{C}_2}{2}\,\|w\|\,\|v\|$$

for N large enough. Then (1.3) holds. □

As a first consequence of Theorems 2.3 and 1.1 we get the following result.

<u>Corollary 2.1</u> - <u>The pseudo-spectral problem</u> (2.4) <u>has a unique solution if N is sufficiently large.</u>
<u>Furthermore, there exists a constant</u> C <u>independent of N such that the following stability estimate holds:</u>

$$(2.37)\qquad \|u_N\|_{1,\omega} \le C\, \|f\|_{C^{0,0}}. \ \square$$

The last result of this section is a convergence estimate for the solution of problem (2.4).

<u>Theorem 2.4</u> - <u>Assume</u> $\alpha,\beta,\gamma \in C^{k,\nu}$ <u>with</u> $k+\nu>\frac{1}{2}$, <u>and</u> $f \in H^{\sigma}_{\omega}$ <u>with</u> $\sigma>\frac{1}{2}$. <u>Under the hypothesis</u> (2.8), <u>let</u> $u \in H^1_{0,\omega} \cap H^s_{\omega}$ <u>be the solution of</u> (2.3), <u>and for any</u> $N \ge N_1$ <u>let</u> $u_N \in S_N$ <u>be the solution of</u> (2.4). <u>Then, provided</u> $k+\nu>s-1$,

$$(2.38)\qquad \|u-u_N\|_{1,\omega} \le C' N^{1-s}\, \|u\|_{s,\omega} + C''\, N^{-\sigma}\, \|f\|_{\sigma,\omega},$$

<u>with</u> C' <u>depending on</u> α,β <u>and</u> γ.

<u>Proof</u> - We apply the abstract results of Theorem 1.2; w is now the orthogonal projection of u onto S_N with respect to the norm of H^s_{ω}. Then, using (1.13) it follows that

$$(2.39)\qquad \|u-w\| \le C N^{1-s}\, \|u\|_{s,\omega}.$$

From (2.21) with $p=w_x$ and from (2.35) with $w'=w''=w$ and $v'=v''=v$, using also the error estimates (1.9) and (1.10), we derive that

$$(2.40)\qquad \sup_{v \in S_N} \frac{|B(w,v)-B_N(w,v)|}{\|v\|} \le C' N^{1-s}\, \|w\|_{s,\omega}.$$

Furthermore, using (1.9),(1.10) once more, and (2.36) with g=f and v"=v, we have that

$$\sup_{v \in S_N} \frac{|(f,v)-(f,v)_{N,\omega}|}{\|v\|} \le C'' N^{-\sigma} \|f\|_{\sigma,\omega} . \tag{2.41}$$

Owing to (1.8), the estimate (2.38) is now deduced from (2.39),(2.40) and (2.41). □

3. The Neumann boundary value problem. In this section we apply the abstract Theorems 1.1 and 1.2 to the analysis of a Chebyshev pseudo-spectral approximation of problem (1.15). For simplicity, we shall assume that the coefficients β and γ be infinitely differentiable in $\bar{\Omega}$, but the same analysis can be carried out under the weaker assumption that $\beta,\gamma \in C^{k,\nu}$ for some k,ν satisfying $k+\nu>\frac{1}{2}$.

The function spaces W and V, and the bilinear form B(w,v) used for setting problem (1.15) in the framework of Theorem 1.1 will be naturally suggested by its Chebyshev approximation. It is defined as follows:

$$\begin{cases} u_N \in S_N = \{p \in \mathbb{P}_N \mid p_x(-1)=p_x(1)=0\}, \\ -u_{N,xx} + P_c(\beta u_{N,x}) + \gamma u_N = f \quad \text{at } x=x_j \;, \; 1\le j\le N-1. \end{cases} \tag{3.1}$$

We denote by $(\cdot,\cdot)_{N,\eta}$ the "discrete" Jacobi inner product, which is defined as follows:

$$(\phi,\psi)_{N,\eta} = (\phi,(1-x^2)\psi)_{N,\omega} \tag{3.2}$$

(η being the Jacobi weight, i.e., $\eta(x)=(1-x^2)\omega(x)$).

By multiplying the j-th equation in (3.1) by $v(x_j)(1-x_j^2)\omega_j$ and summing up over j, we get that (3.1) is equivalent to

$$B_N(u_N,v) = (f,v)_{N,\eta} \qquad \forall v \in \mathbb{P}_{N-2}, \tag{3.3}$$

where

$$B_N(w,v) = (-w_{xx} + P_c(\beta w_x) + \gamma w, v)_{N,\eta} \tag{3.4}$$

is a bilinear form defined on $S_N \times \mathbb{P}_{N-2}$.

In (3.1), equation (1.15) is not collocated at the boundary points, since the Neumann conditions are imposed there. Therefore, a discrete inner product independent of the boundary values is needed in the variational formulation (3.3). The simplest way to fulfill this requirement is to modify the Chebyshev inner product as done in (3.2). This guarantees a convenient polynomial precision, as well.

The Jacobi discrete inner product is an approximation of the inner product of L^2_η, in the sense that they coincide if $wv \in \mathbb{P}_{N-3}$. In analogy with (3.3) and (3.4), this suggests the following form for problem (1.15). Define

$$(3.5) \qquad \tilde{H}^2_\eta = \{ w \in H^2_\eta \mid w_x(-1) = w_x(1) = 0 \} ,$$

provided with the norm of H^2_η.

(We note that $H^2_\eta \subset C^1(\bar{\Omega})$, hence $w_x(\pm 1)$ makes sense.)

Then, assuming $u \in \tilde{H}^2_\eta$, problem (1.15) is equivalent to

$$(3.6) \qquad B(u,v) = (f,v)_\eta \qquad \forall v \in L^2_\eta ,$$

where

$$(3.7) \qquad B(w,v) = \int_\Omega (-w_{xx} + \beta w_x + \gamma w) v \eta dx$$

is a continuous bilinear form on $\tilde{H}^2_\eta \times L^2_\eta$.

We shall assume from now on that the Neumann problem (1.15) is coercive on H^1, that is

$$(3.8) \qquad \int_\Omega ((w_x)^2 + \beta w_x w + \gamma w^2) dx \geq C \, \|w\|_1^2 \qquad \forall w \in H^1 .$$

This hypothesis assures that problem (1.15) is well posed in H^1. (Instead of (3.8), we could ask something weaker, i.e., that problem (1.15) with $f=0$ admits the trivial solution only). As a consequence of (3.8), we will prove in the next lemma that (1.15) is well posed also in $\tilde{H}^2_\eta$.

Lemma 3.1 - For any $g \in L^2_\eta$, there exists a unique $w \in \tilde{H}^2_\eta$ such that

$$(3.9) \qquad \mathscr{B} w \equiv -w_{xx} + \beta w_x + \gamma w = g \quad \text{in } \Omega ,$$

$$(3.10) \qquad \|w\|_{2,\eta} \leq C \, \|g\|_{o,\eta} .$$

Proof - We prove this result with the help of the theory of interpolation of spaces. To this end, we note first that if $g \in (H^1)'$ (the dual space of H^1), then by (3.8) and the Lax-Milgram lemma there exists a unique solution $w \in H^1$ with $\|w\|_1 \le C \|g\|'_1$. On the other hand, if $g \in L^2$, then by (3.9) w is in H^2 and $\|w\|_2 \le C\|g\|_o$. Hence, if $g \in (H^s)'$, for $0<s<1$, then $w \in H^{1+s}$ and $\|w\|_{1+s} \le C \|g\|'_s$ (|2|, Thm. 4.1.2 and Corollary 4.5.2). Using Theorem 4.2 of the Appendix, we get $L^2_\eta \subset (H^{1/4})'$ with continuous injection, thus $w \in H^{7/4}$. Then, from (3.9) we deduce that $w_{xx} \in L^2_\eta$, that is $w \in H^2_\eta$ and (3.10) holds. □

Corollary 3.1 - The bilinear form $B(w,v)$ on $\tilde{H}^2_\eta x L^2$ satisfies the hypotheses of Theorem 1.1.

Proof - Take $v=\mathscr{B}w$ in (1.3), $w=\mathscr{B}^{-1}v$ in (1.4) and use (3.10) □

Now we can prove the uniform coerciveness of the form $B_N(w,v)$ on $S_N x\, \mathbb{P}_{N-2}$. The strategy is to use compactness arguments to control the error between $B_N(w,v)$ and $B(w,v)$, which is due only to the lower order terms.

Theorem 3.1 - There exists $N_o>0$ such that for any $N \ge N_o$ the form B_N on $S_N x\, \mathbb{P}_{N-2}$ satisfies the hypotheses of Theorem 1.1 with constants C_1 and C_2 independent of N.

Proof - Owing to the precision of the Jacobi discrete inner product we have

$$(3.11) \qquad B_N(w,v)=(-w_{xx},v)_\eta+(\beta w_x+\gamma w,v)_{N,\eta}.$$

We prove (1.2). The first term in (3.11) can be bounded by the Cauchy-Schwarz inequality. We shall deal only with the term $(\beta w_x,v)_{N,\eta}$; the same procedure applies to the remaining term. By the Cauchy-Schwarz inequality we get:

$$|(\beta w_x),v)_{N,\eta}| \le C \|\beta w_x\|_{N,\eta} \|v\|_{o,\eta},$$

where $\|\phi\|^2_{N,\eta} = (\phi,\phi)_{N,\eta}$. On the other hand, since $1-x^2 \le 1$ in Ω, we have by (1.12)

$$\|\beta w_x\|_{N,\eta} \le \|\beta w_x\|_{N,\omega} \le C \|P_c(\beta w_x)\|_{o,\omega}.$$

Finally, by Lemma 2.3 and Thms. 4.1 and 4.2 of the Appendix,

$$\|P_c(\beta w_x)\|_{o,\omega} \le \|\beta w_x\|_{o,\omega} + \|\beta w_x - P_c(\beta w_x)\|_{o,\omega}$$

$$\le C \|w\|_{1,\omega} \le C' \|w\|_{2,\eta} .$$

We check now (1.3). Given $w \in S_N$, set $v=-w_{xx}+P_{N-2}(\beta w_x)+P_{N-2}(\gamma w) \in \mathbb{P}_{N-2}$. Then $\|v\|_{o,\eta} \le C \|w\|_{2,\eta}$.

Define the error E by the relation:

$$E = B_N(w,v) - (\mathscr{B}w, \mathscr{B}w)_\eta .$$

Let I denote the identity operator (i.e., Iz=z for any function z). Using repeatedly the Cauchy-Schwarz inequality, the relations (1.12) and the bound $1-x^2 \le 1$ in Ω, some lengthy but not difficult calculations lead to:

$$|E| \le C \|w\|_{2,\eta} \cdot \{ \|(I-P_{N-1})\beta w_x\|_{o,\omega} + \|(I-P_c)\beta w_x\|_{o,\omega} +$$

$$\|(I-P_{N-1})\gamma w\|_{o,\omega} + \|(I-P_c)\gamma w\|_{o,\omega} \}.$$

Using the estimates (1.9),(1.10) and Thms. 4.1 and 4.2 of the Appendix, it follows that

$$|E| \le C' N^{-1/2} \|w\|_{2,\eta} .$$

Hence, applying Lemma 3.1 gives

$$B_N(w,v) \ge (C - C' N^{-1/2}) \|w\|^2_{2,\eta}$$

so that (1.3) holds if N is large enough. Finally, (1.4) holds since $\dim S_N = \dim \mathbb{P}_{N-2}$ (see Remark 1.1).□

As far as convergence is concerned, we can prove the following result.

Theorem 3.2 - Let $f \in H^\sigma_\omega$ be given, with $\sigma > 1/2$. Let $u \in \tilde{H}^2_\eta$ be the solution of (3.6), and for $N \ge N_o$, let $u_N \in S_N$ be the solution of (3.3). Then

(3.12) $$\|u-u_N\|_{2,\eta} \le C N^{-\sigma} \{ \|u\|_{\sigma+2,\omega} + \|f\|_{\sigma,\omega} \},$$

with C depending on β and γ.

Proof - Note that, by classical regularity theorems, $u \in H^{\sigma+2}_\omega$

if $f \in H^{\sigma}_{\omega}$. We apply the abstract Theorem 1.2. We define $w \in S_N$ through the conditions: $w_x = \Pi^1_{N-1} u_x$, $w(-1)=u(-1)$. Then, using (1.13) we have:

$$(3.13) \qquad \|u-w\|_{2,\omega} \leq CN^{-\sigma} \|u\|_{\sigma+2,\omega}.$$

The supremum in (1.8) is bounded by:

$$\frac{C}{\|v\|_{o,\eta}} \{ |(\beta w_x + \gamma w, v)_\eta - (\beta w_x + \gamma w, v)_{N,\eta}| + |(f,v)_\eta - (f,v)_{N,\eta}| \} .$$

We consider only the term $E = (\beta w_x, v)_\eta - (\beta w_x, v)_{N,\eta}$ (the others can be dealt with similarly).

Working as in the previous proof, and using (1.9),(1.10) and (1.13), we get:

$$|E| = |((I-P_{N-1})\beta w_x, v)_\eta + ((P_{N-1}-P_c)\beta w_x, v)_{N,\eta}|$$

$$\leq C \|v\|_{o,\eta} \{ \|(I-P_{N-1})\beta w_x\|_{o,\omega} + \|(I-P_c)\beta w_x\|_{o,\omega} \}$$

$$\leq C \|v\|_{o,\eta} \{ \|(I-P_{N-1})\beta(I-\Pi^1_{N-1})u_x\|_{o,\omega} +$$

$$\|(I-P_c)\beta(I-\Pi^1_{N-1})u_x\|_{o,\omega} + \|(I-P_{N-1})\beta u_x\|_{o,\omega} +$$

$$\|(I-P_c)\beta u_x\|_{o,\omega} \} \leq C \|v\|_{o,\eta} \{ (1+N^{-1}) \|\beta(I-\Pi^1_{N-1}u_x\|_{1,\omega} +$$

$$2N^{-\sigma} \|\beta u_x\|_{\sigma+2,\omega} \} \leq C \|v\|_{o,\eta} N^{-\sigma} \|u\|_{\sigma+2,\omega} . \square$$

Remark 3.1 - We have considered in this section the Neumann boundary conditions for simplicity of exposition only. However, the same results can be established by the same proofs for the case of mixed boundary conditions:

$$\lambda_- u_x(-1) + \mu_- u(-1) = \lambda_+ u_x(1) + \mu_+ u(1) = 0,$$

provided the coefficients $\lambda_-, \mu_-, \lambda_+, \mu_+$ could guarantee the well posedness of the associated boundary value problem. In particular, this occurs for the mixed Dirichlet-Neumann boundary conditions $u_x(-1) = u(1) = 0$.

Recently, Gottlieb and Lustman |8| showed that the eigenvalues for the Chebyshev approximations of such problems are all real, negative and distinct. □

4. Appendix. Here we give (for the reader's convenience) some inclusion properties among the families of spaces H^s_ω, H^s and H^s_η .

Theorem 4.1 - For any real $s \geq 1/4$, $H^s \subset H^{s-1/4}_\omega$ with continuous injection.

Proof - The main tool is Thm. 11.2 of |11|. According to it, $u \to \rho^{-s} u$ maps H^s into L^2 with continuity if $0 \leq s < 1/2$ and $\rho = \rho(x)$ behaves like the distance of x from the boundary of Ω. Taking $\rho = \omega^{-2}$ we get $H^{1/4} \subset L^2_\omega$. Applying this result to the k-th derivative of u ($k>0$ integer), we get $H^{k+1/4} \subset H^k_\omega$. We conclude the proof by an interpolation argument. □

Theorem 4.2 - For any real $s \geq 0$, $H^s_\eta \subset H^{s-1/4}$ with continuous injection.

Proof - The inclusion $H^{1/4} \subset L^2_\omega \subset L^2$ gives by duality $L^2 \subset L^2_\eta \subset H^{-1/4}$, where H^{-s} for $0 \leq s < \frac{1}{2}$ denotes the dual space of H^s. Reasoning on the k-th derivative ($k>0$ integer) yields $H^k_\eta \subset H^{k-1/4}$. Once more we conclude with an interpolation argument. □

REFERENCES

|1| I.BABUŠKA,A.K.AZIZ, Survey Lectures on the Mathematical Foundations of the Finite Element Method, in The Mathematical Foundations of the Finite Element Method with Applications to Partial Differential Equations (A.K.Aziz, Editor), pp.3-359, Academic Press (New York), 1972.

|2| J.BERGH,J.LÖFSTRÖM, Interpolation Spaces. An Introduction, Springer-Verlag (Berlin-Heidelberg -New York), 1976.

|3| C.CANUTO,A.QUARTERONI, Approximation results for orthogonal polynomials in Sobolev spaces, Math.Comput., 38 (1982), pp.67-86.

|4| C.CANUTO,A.QUARTERONI, Error estimates for spectral and pseudo-spectral approximations of hyperbolic equations, SIAM J. Numer. Anal., 19 (1982), pp.629-642.

|5| C.CANUTO,A.QUARTERONI, Spectral and pseudo-spectral methods for parabolic problems with nonperiodic boundary conditions, Calcolo, 18 (1981), pp.197-217.

|6| S.CAMPANATO, Equazioni ellittiche del II° ordine e spazi $L^{(2,\lambda)}$, Annali di Mat.Pura e Appl., 69 (1965), pp.321-382.

|7| P.G.CIARLET, The Finite Element Method for Elliptic Problems, North-Holland (Amsterdam), 1978.

|8| D.GOTTLIEB, L.LUSTMAN, ICASE Report No. 82-12.

|9| D.GOTTLIEB,S.A.ORSZAG, Numerical Analysis of Spectral Methods: Theory and Applications, CBMS Regional Conference Series in Applied Mathematics 26, SIAM (Philadelphia), 1977.

|10| H.O.KREISS,J.OLIGER, Stability of the Fourier method, SIAM J. Numer. Anal., 16(1979), pp.421-433.

|11| J.L.LIONS,E.MAGENES, Non Homogeneous Boundary Value Problems and Applications, vol.1, Springer-Verlag (Berlin-Heidelberg-New York), 1972.

|12| Y.MADAY, Thèse de III Cycle, University of Paris VI, 1981.

|13| Y.MADAY,A.QUARTERONI, Spectral and pseudo-spectral approximations of Navier-Stokes equations, SIAM J. Numer. Anal., 19 (1982), pp. 761-780.

|14| Y.MADAY,A.QUARTERONI, Legendre and Chebyshev spectral approximations of Burgers' equation, Numer. Math., 37 (1981), pp.321-332.

|15| Y.MADAY,A.QUARTERONI, Approximation of Burgers' equation by pseudo-spectral methods, to appear on R.A.I.R.O. An. Num.

|16| B.MERCIER, Stabilité et convergence des méthodes spéctrales polynômiales: application à l'équation d'advection, R.A.I.R.O. An. Num., 16(1982), pp.67-100.

|17| J.NEČAS, Sur une méthode pour resoudre les équations aux dérivées partielles du type elliptique, voisine de la variationnelle, Ann.Sc.Norm.Sup.Pisa, 16 (1962), pp.305-326.

|18| J.E.PASCIAK, Spectral and pseudo-spectral methods for advection equations, Math.Comput., 35(1980), pp.1081-1092.

|19| A.QUARTERONI, Theoretical and computational aspects of spectral methods, to appear on the Proceedings of the Fifth International Conference on Computing Methods in Applied Sciences and Engineering, Versailles 1981, to be published by North-Holland (Amsterdam).

|20| L.TARTAR, Théorème d'interpolation nonlinéaire et applications, C.R.Acad.Sc.Paris, Sér.A, v.270, 1970, pp.1729-1731.

SPECTRAL METHODS FOR TWO-DIMENSIONAL SHOCKS

DAVID GOTTLIEB*, LIVIU LUSTMAN**, AND CRAIG STREETT#

Abstract. Chebyshev collocation methods are applied to two-dimensional gasdynamic problems with shocks, namely inviscid flow (Euler equations) and transonic flow (full potential equation). Boundary conditions, numerical dissipation and iteration techniques are considered, and computational results are presented and compared with finite-difference results.

1. Introduction. The first numerical experiments in spectral calculations of discontinuous flows have been published fairly recently. Viscous cases have been presented in [12], where a mixed spectral/finite difference approximation is employed. For inviscid flow -- the Euler equations in a shock tube problem -- a purely spectral method was applied [4], and the following properties reported:

(1) The spectral (Chebyshev collocation) methods are conservative, a necessary condition for correct shock speed.

(2) The shock position is sharply defined within one mesh interval, rather than having a smeared transition over a few intervals, as is typical in finite differences calculations.

(3) The Gibbs phenomenon, inherent in discontinuous data, can be corrected to produce excellent numerical values. In fact, spectral accuracy is obtained despite the apparent lack of smoothness.

---

*Senior Staff Scientist, Institute for Computer Applications in Science and Engineering, NASA Langley Research Center, Hampton, VA 23665. Work supported under NASA Contract Nos. NAS1-15810 and NAS1-16394 and Air Force Office of Scientific Research Contract No. 83-0039.

**Senior Scientist, Systems and Applied Sciences Corporation, NASA Langley Research Center, Hampton, VA 23665. Work supported under NASA Contract No. 16572.

#Aerospace Technician, Transonic Aerodynamics Division, Theoretical Aerodynamics Branch, NASA Langley Research Center, Hampton, VA 23665

(4) In order to achieve these results one needs special appropriate boundary conditions; also some form of dissipation, e.g., smoothing high modes, is needed (see [8]).

In view of the successful treatment of one-dimensional problems the extension to several dimensions seemed the natural next step. In the sequel we present a summary of our work on two physically relevant problems with shocks in plane gasdynamics.

The first is the immediate extension of the shock tube problem, namely, two-dimensional compressible inviscid flow (Euler equations). In this case, we found that most of the conclusions drawn from the one-dimensional problem still apply. We are again able to accurately capture shocks, even though these are not aligned with the numerical grid employed. The main difference is in the more involved boundary treatment and also in the fact that the two-dimensional flow computations appear to need stronger dissipation for stability.

The second problem we treated is transonic potential flow. This is probably one of the best behaved nonlinear problems, and in its spectral treatment we have advanced remarkably fast. Not only have we successfully solved typical simple test cases, but we are near a production code for general airfoils with lift.

We believe that these results show that spectral methods are a valid approach to certain problems in computational fluid dynamics. Of course, much additional basic research is still to be done.

2. Euler Equations. We shall write the equations in full conservation form, using the conserved quantities -- density, momentum, and energy -- as our variables. This is, of course, needed in order to capture shocks correctly. Thus, our basic equations are:

(2.1)

$$\frac{\partial}{\partial t}\begin{bmatrix}\rho\\ m\\ n\\ e\end{bmatrix}+\frac{\partial}{\partial x}\begin{bmatrix}m\\ \frac{\gamma-1}{\rho}\left(\rho e-\frac{n^2}{2}\right)+\frac{3-\gamma}{2}\frac{m^2}{\rho}\\ \frac{mn}{\rho}\\ \frac{m}{\rho^2}\left(\gamma\rho e-\frac{\gamma-1}{2}\left(m^2+n^2\right)\right)\end{bmatrix}+\frac{\partial}{\partial y}\begin{bmatrix}n\\ \frac{mn}{\rho}\\ \frac{\gamma-1}{2}\left(\rho e-\frac{m^2}{2}\right)+\frac{3-\gamma}{2}\frac{n^2}{\rho}\\ \frac{n}{\rho^2}\left(\gamma\rho e-(\gamma-1)\left(m^2+n^2\right)\right)\end{bmatrix}=0$$

with ρ the density, e the energy and m,n the horizontal and vertical momenta; γ is the specific heat ratio.

We shall consider a very simple situation: a shock impinging on a solid flat plate and being reflected. The flow is assumed to be supersonic everywhere, which is the case if the initial Mach number is high enough. Figure 1 presents a diagram of the flow, together with some of the boundary conditions, and we summarize below the (constant) values of the various physical quantities in the three regions shown.

supersonic inflow:

ρ,m,n,e prescribed

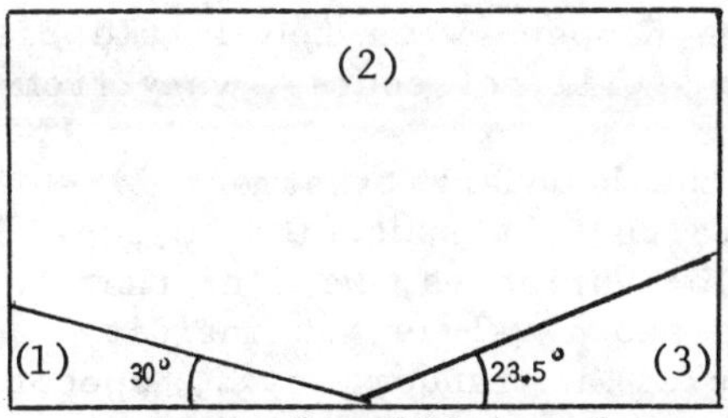

supersonic outflow:

no data prescribed

physical solid boundary
$n = 0$

Figure 1.

Table I

	Region		
	1	2	3
Density	1.0000	1.8621	3.1441
Horizontal velocity	1.0000	0.8843	0.7972
Vertical velocity	0.0000	-0.2005	0.0000
Pressure	0.0794	0.1951	0.4132
Energy	0.6984	1.2532	2.0322
Horizontal momentum	1.0000	1.6466	2.5066
Vertical momentum	0.0000	-0.3373	0.0000
Mach number	3.0000	2.3673	1.8587
Speed of sound	0.3333	0.3830	0.4289

The domain is covered by a Chebyshev grid in x and y; the equations are then solved by modified Euler unsplit time stepping:

$$u_t + f_x + g_y = 0, \tag{2.2}$$

$$\begin{aligned} u^* &= u^n - \frac{\Delta t}{2}(f_x+g_y)^n, \\ u^{n+1} &= u^n - \Delta t(f_x^*+g_y^*). \end{aligned} \tag{2.3}$$

The derivatives are computed spectrally, and f and g are obtained from u by collocation.

In addition to this straightforward procedure, one must impose numerical boundary conditions and provide some dissipation, as we know from theory [8] and one dimensional examples. The main area of investigation was to find (by trial and error) which are the correct boundary conditions and how much dissipation is needed. We must emphasize that these are crucial questions for a spectral code; under wrong boundary conditions the calculations become explosively unstable. The same happens when too little dissipation is present. The "correct amounts" will, of course, vary from problem to problem.

Let us address the boundary treatment first. In Figure 1, it is clear that some data must be specified on top. By the standard count of inflow and outflow variables, we find that three quantities can be given, based on the known values "at infinity" in region 2. However, they cannot be picked at random, e.g., specify ρ,m,n and let e evolve. After experimenting with this arbitrary assignment, with obviously unstable results, we proceeded to the usual characteristic extrapolation. This proved so satisfactory that we did not try other boundary conditions on top.

Let us detail the extrapolation procedure. As a first step, one must write the characteristic equations in the normal direction, for a linearized system. In our case, this means ignoring the $\frac{\partial}{\partial x}$ terms in (2.1) and producing (see, e.g. [1]):

$$\text{(2.4)}\qquad \begin{aligned} &\text{I.} && \left(\frac{\partial}{\partial t} + v_0 \frac{\partial}{\partial y}\right)\left(\frac{\rho}{\rho_0} - \frac{p}{\rho_0 c_0{}^2}\right) = 0 \\ &\text{II.} && \left(\frac{\partial}{\partial t} + v_0 \frac{\partial}{\partial y}\right)\left(\frac{u}{c_0}\right) = 0 \\ &\text{III.} && \left(\frac{\partial}{\partial t} + (v_0+c_0) \frac{\partial}{\partial y}\right)\left(\frac{p}{\rho_0 c_0{}^2} + \frac{v}{c_0}\right) = 0 \\ &\text{IV.} && \frac{\partial}{\partial t} + (v_0-c_0) \frac{\partial}{\partial y}\Big)\left(\frac{p}{\rho_0 c_0{}^2} - \frac{v}{c_0}\right) = 0 \,. \end{aligned}$$

Here p stands for pressure, c for local speed of sound; the values v_0, c_0, ρ_0 belong to the state about which we linearize.

Equations (2.4) define four characteristic quantities: $\frac{\rho}{\rho_0} - \frac{p}{\rho_0 c_0{}^2}$, $\frac{u}{c_0}$, $\frac{p}{\rho_0 c_0{}^2} \pm \frac{v}{c_0}$ as functions of our computational (conserved) variables ρ,m,n,e. We shall identify them by the corresponding roman numerals in (2.4), and extrapolate using the basic rule: an inflow variable is determined by boundary data; an outflow variable is determined by the numerical method.

On top, $v < 0 < v + c$; therefore, I, II and IV are inflow variables, while III is an outflow variable. The extrapolation is:

$$\begin{aligned} I_{\text{corrected}} &= I_{\text{exact}} \\ II_{\text{corrected}} &= II_{\text{exact}} \\ III_{\text{corrected}} &= III_{\text{computed}} \\ IV_{\text{corrected}} &= IV_{\text{exact}} \end{aligned} \tag{2.5}$$

In the formulas above, the subscript "corrected" means the value after extrapolation; "exact" is computed using known data of region 2, and "computed" means the value resulting from the time step. What we have is a set of equations for $(\rho,m,n,e)^{\text{top}}_{\text{corrected}}$, to be solved after applying (2.3).

There remains, of course, the question of the linearization values ρ_0, c_0, v_0. Since data are available on the top, we took these to be the exact values for region 2. As we supplied so much exact information, it is not surprising that the top boundary never caused any more trouble.

Consider now the boundary correction at the rigid wall. The equations (2.4) still apply, but the situation is different. Since $v - c < v = 0 < v + c$, there is one inflow quantity III, one outflow quantity IV, and two characteristic quantities I and II. The only boundary data that we have is $v = 0$; thus we are forced to treat the characteristic quantities as outflow - as the theory allows - and set them, after correction, to their "computed" values:

$$\begin{aligned} I_{\text{corrected}} &= I_{\text{computed}} \\ II_{\text{corrected}} &= II_{\text{computed}} \\ v_{\text{corrected}} &= 0 \\ IV_{\text{corrected}} &= IV_{\text{computed}} \end{aligned} \tag{2.6}$$

Formulas (2.6) have the same validity as (2.5) but, unfortunately, not the same numerical effectiveness. First, the linearizing values ρ_0, c_0 are not known ($v_0 = 0$, obviously), and we simply set them to their values at the former time step. In this form, equations (2.6) still produced instability in some cases.

Since this proved disappointing, we tried the standard finite difference boundary condition:

$$\frac{\partial}{\partial y}\begin{Bmatrix} \rho \\ m \\ e \end{Bmatrix} = 0, \qquad n = 0 \qquad \text{at the wall.} \tag{2.7}$$

This can be justified by symmetry arguments, and can be implemented either in a finite difference fashion:

$$(2.8)\qquad (\rho,m,e)\big|_{y=0} = (\rho,m,e)\big|_{y=h}$$
$$n\big|_{y=h} = -n\big|_{y=h}$$

or in a spectral fashion:

$$(2.9)\qquad (\text{Spectral derivative at } y=0)\begin{Bmatrix}\rho\\ m\\ e\end{Bmatrix} = 0$$
$$n\big|_{y=0} = 0$$

Equations (2.9) simply mean that the value of the function at $y = 0$ should be replaced by some quantity, such that the new interpolation polynomial has a zero derivative at $y = 0$. This is done by considering the explicit coefficients $D_{y=0,\,y_k}$ defined for the function $\phi(y)$ by:

$$(2.10)\qquad \frac{d\phi}{dy}\Big|_{y=0} = \sum_{y_k = \text{Chebyshev points}} D_{y=0,\,y_k}\,\phi(y_k)$$

Again some cases thus treated were divergent so we abandoned (2.7) as a boundary condition. As an afterthought, one may generalize that finite difference boundary treatment of derivatives will be unsafe in a spectral method because of its strong global influence.

Finally, we hit upon a combination of conserved and characteristic quantities that gave stable results. It is produced by a restatement of (2.4) in terms of the conserved quantities, substitution of $n = 0$ and recomputation of characteristic vectors.

For its final form there are three "characteristic equations" as follows:

$$\text{I}^0.\quad \frac{d}{dt}(u_0\rho - m) = 0$$

$$(2.11)\qquad \text{II}^0.\quad \frac{d}{dt}(s_0^2\rho - e) = 0$$

$$\text{IV}^0.\quad \left(\frac{\partial}{\partial t} + c_0\frac{\partial}{\partial y}\right)\left(\frac{\gamma-1}{2}u_0^2\rho + (1-\gamma)mn_0 + nc_0 + (\gamma-1)e\right) = 0,$$

and the correction is effected by:

$$(2.12)\qquad (\text{I}^0,\ \text{II}^0,\ \text{IV}^0)_{\text{corrected}} = (\text{I}^0,\ \text{II}^0,\ \text{IV}^0)_{\text{computed}}$$
$$n_{\text{corrected}} = 0.$$

Solving (2.12) for the new quantities $\rho_{\text{corrected}}$, $m_{\text{corrected}}$, $e_{\text{corrected}}$, we get:

$$\text{(2.13)} \qquad \begin{aligned} \rho_{\text{corrected}} &= \left(\rho - \frac{n}{c_0}\right)\text{computed} \\ m_{\text{corrected}} &= \left(m - \frac{u_0 n}{c_0}\right)\text{computed} \\ e_{\text{corrected}} &= \left(e - \frac{s_0^2 n}{c_0}\right)\text{computed}. \end{aligned}$$

The quantity s_0^2 is a linearization value, defined by:

$$s_0^2 = \frac{c_0^2}{\gamma-1} + \frac{1}{2}u_0^2.$$

The linearization is done about the values at the previous time step.

Another boundary condition that was tested was based on the one-dimensional Riemann invariants. Despite the fact that there are no two-dimensional analogs, we used their expressions as if there were a tube in the y direction, stopped at $y = 0$. Explicitly:

$$\text{(2.14)} \qquad \begin{aligned} &\text{i.} \quad \left(\frac{\partial}{\partial t} + v\frac{\partial}{\partial y}\right)\left(\frac{p}{\rho^\gamma}\right) = 0 \\ &\text{ii.} \quad \left(\frac{\partial}{\partial t} + v\frac{\partial}{\partial y}\right)u = 0 \\ &\text{iii.} \quad \left(\frac{\partial}{\partial t} + (v+c)\frac{\partial}{\partial y}\right)\left(v + \frac{2c}{\gamma-1}\right) = 0 \\ &\text{iv.} \quad \left(\frac{\partial}{\partial t} + (v-c)\frac{\partial}{\partial y}\right)\left(v - \frac{2c}{\gamma-1}\right) = 0. \end{aligned}$$

The extrapolation is accomplished by:

$$\text{(2.15)} \qquad \begin{aligned} (\text{i, ii, iv})_{\text{corrected}} &= (\text{i, ii, iv})_{\text{computed}} \\ v_{\text{corrected}} &= 0. \end{aligned}$$

The equations above are solved without needing any linearization values; the resulting scheme is stable, but converges to less accurate values than (2.12).

At this point we come to the question of dissipation. In the one-dimensional case smoothing high modes was all that was needed to stabilize computations, even for strong shocks (weak shocks remained stable with no dissipation at all). In the flat plate problem there is a similar behavior, modified by geometry. Namely, consider the following subcases of the shock reflection problem:

a. Shock leaving the plate (wedge shock)

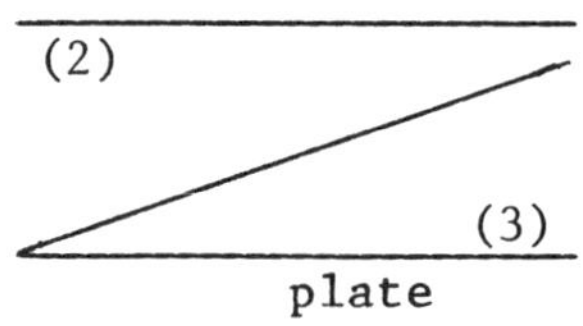

If we start with initial data taken from region 2 all over the domain, in time region 3 will develop, with the shock in the correct location. This is the most tolerant problem; it is stable even without high mode smoothing, and will allow (2.6), (2.9), (2.12), or (2.15) as a boundary correction. However, (2.9) produces large wiggles in the converged solution.

b. Shock approaching, but not hitting the plate

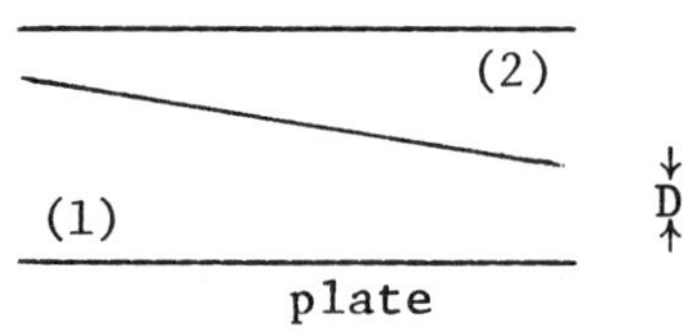

This problem will run without mode smoothing, but only if the interval D is large enough. If the shock comes near the plate, the problem becomes unstable if high modes are not reduced.

Finally, if we want to solve the full shock reflection problem, we find that mode smoothing is necessary, but no longer sufficient for stability. Typically, out of some continuous initial data the shock will evolve, hit the plate, start to reflect and then abruptly the computation diverges. Interestingly enough, even in these divergent runs the shock position is correct. Thus, we decided on an additional diffusion mechanism, based on this sharp shock capture property. After a time long enough for the shock to appear and hit the plate, we started smoothing the solution by applying a one-sided Shumann filter every 20, 50, or 100 time steps. This, of course, involves locating the shock, in order not to smooth across it and smear the data.

This procedure introduced sufficient dissipation to stabilize the computations. A typical result is shown in Figure 2. Note that even this very coarse 9×9 grid is effective in capturing oblique shocks. The "initial guess" respects boundary data at the inflow, but is otherwise smooth, and all shocks and variables must evolve to the correct positions and values. By refining the mesh, better

EXACT SOLN FOR COMPARISON
RHO

1.86	1.86	1.86	1.86	1.86	1.86	1.86	1.86	1.86
1.86	1.86	1.86	1.86	1.86	1.86	1.86	1.86	1.86
1.86	1.86	1.86	1.86	1.86	1.86	1.86	1.86	1.86
1.86	1.86	1.86	1.86	1.86	1.86	1.86	1.86	1.86
1.00	1.00	1.86	1.86	1.86	1.86	1.86	1.86	1.86
1.00	1.00	1.00	1.00	1.86	1.86	1.86	1.86	1.86
1.00	1.00	1.00	1.00	1.86	1.86	1.86	3.14	3.14
1.00	1.00	1.00	1.00	1.00	1.86	3.14	3.14	3.14
1.00	1.00	1.00	1.00	1.00	3.14	3.14	3.14	3.14

INITIAL GUESS
RHO

1.86	1.86	1.86	1.86	1.86	1.86	1.86	1.86	1.86
1.86	1.81	1.78	1.76	1.75	1.75	1.74	1.74	1.74
1.86	1.74	1.68	1.65	1.64	1.63	1.63	1.63	1.63
1.86	1.61	1.55	1.53	1.53	1.53	1.52	1.52	1.52
1.00	1.29	1.37	1.41	1.42	1.43	1.43	1.42	1.42
1.00	1.16	1.26	1.31	1.34	1.35	1.35	1.35	1.35
1.00	1.11	1.20	1.25	1.28	1.29	1.30	1.29	1.29
1.00	1.09	1.17	1.22	1.25	1.27	1.27	1.27	1.27
1.00	1.09	1.17	1.22	1.25	1.27	1.27	1.27	1.27

AT TIME = 7.700
RHO

1.86	1.86	1.86	1.86	1.87	1.87	1.89	1.89	1.88
1.86	1.86	1.86	1.87	1.88	1.88	1.89	1.89	1.89
1.86	1.85	1.87	1.88	1.88	1.89	1.90	1.91	1.91
1.86	1.92	1.87	1.86	1.89	1.91	1.92	1.93	1.93
1.00	.98	1.80	1.97	1.92	1.91	1.94	1.94	1.92
1.00	.97	1.08	1.05	1.81	1.94	1.92	1.99	2.03
1.00	1.00	1.02	.96	1.85	1.98	1.95	2.89	3.08
1.00	1.01	1.00	1.03	1.12	2.03	3.00	3.03	3.01
1.00	1.02	.98	1.08	.98	2.75	2.98	3.05	3.10

Figure 2. Shock reflection problem. Shocks do not appear straight because of unequal interval of Chebyshev mesh.

converged solutions may be obtained. As in one dimension, the final converged solutions still exhibit oscillations, due to the Gibbs phenomenon. These can be corrected by a final cosmetic filter, possibly using the spectral shock locator [4]. Some fine tuning is needed, and we did not research further into this problem, being interested mainly in basic stability questions.

In summary, spectral methods applied to Euler equations in two dimensions exhibit conservation and fine shock resolution, but also show a strong dependence on numerical boundary treatment and numerical dissipation. These last two subjects are still not well understood, and form a wide research field, both for computational experiments and for theoretical work.

3. Transonic Flow. Consider the flow about a symmetric wing. The equation to be solved is:

$$(\rho\Phi_x)_x + (\rho\Phi_y)_y = 0 \tag{3.1}$$

where Φ is the velocity potential.

$$u = \Phi_x \qquad v = \Phi_y \tag{3.2}$$

and the density ρ is determined from Bernoulli's law:

$$\rho = \left[1 - \frac{\gamma-1}{2} M_\infty^2(\Phi_x^2 + \Phi_y^2 - 1)\right]^{\frac{1}{\gamma-1}}. \tag{3.3}$$

Here γ is the specific heat ratio and M_∞ is the free stream Mach number.

At infinity, the potential has to satisfy the condition:

$$\Phi \sim x \rightarrow u \sim 1$$

pertaining to uniform flow with (normalized) horizontal speed 1.

The wing is located on the x axis and its shape is given by:

$$y = \tau f(x) \qquad 0 \leqslant x \leqslant 1 \tag{3.5}$$

with τ a "small" parameter. For simplicity, we use a parabolic airfoil

$$f(x) = x(1-x) \tag{3.6}$$

and correspondingly:

$$0.5\tau = \text{thickness ratio}. \tag{3.7}$$

We impose the boundary condition at the surface:

(3.8) $$\frac{\partial\Phi}{\partial y} = \tau \frac{df}{dx}\frac{\partial\Phi}{\partial x} \quad \text{at} \quad y = 0,\ 0 \leq x \leq 1$$

which approximates, to the first order in τ, the exact condition:

(3.9) $$\frac{\partial\Phi}{\partial n} = 0 \quad \text{on the body} \quad y = \tau f(x)$$

Because of symmetry,

(3.10) $$\frac{\partial\Phi}{\partial y} = 0$$

on the rest of the x axis.

The outstanding property of equation (3.1) -- from a formal mathematical point of view -- is that it changes type from elliptic to hyperbolic as the local speed goes from subsonic to supersonic. We shall describe briefly the qualitative behavior of the flows it represents.

As long as the Mach number M_∞ is small enough, the flow over the wing is symmetrical and smooth; the discontinuities of $\frac{\partial\Phi}{\partial y}$ at $x = 0,1$ do not propagate into the field. The speed is either subsonic everywhere, or a small supersonic pocket may develop over the wing. As M_∞ increases over a critical value (depending on the thickness ratio for a given wing shape) a shock wave develops near the trailing edge, across which the speed reduces abruptly from supersonic to subsonic.

Thus the physical situation is no longer symmetric, although there is symmetry in the differential equation and boundary conditions. There is a symmetrical solution to our problem, having two shocks, but one of them is an unphysical rarefaction shock near the leading edge. This is clearly a case of non-uniqueness of weak solutions, where an "entropy" condition must be employed to select the relevant one.

It is seen, therefore, that an entropy inequality -- or equivalently a desymmetrizing procedure -- must be enforced, in addition to equations (3.1, 3.4, 3.8, 3.10). Numerically, this is implemented by using upwind derivatives (the Cole-Murman scheme [9]) or a modified density approach [5].

Equation (3.1) will be solved by an iterative approach, which may be thought as advancing in artificial, nonphysical time towards a steady state. A simple-minded evolution equation, apparently of diffusion type, is:

(3.11) $$\Phi_t = (\rho\Phi_x)_x + (\rho\Phi_y)_y .$$

There is, however, a better approach based on:

$$\alpha\Phi_t + \beta\Phi_{xt} = (\rho\Phi_x)_x + (\rho\Phi_y)_y. \tag{3.12}$$

The coefficients α and β are variable; α is nonzero at subsonic points, β nonzero at supersonic points. Equation (3.12) preserves a finite signal speed of propagation, as needed in hyperbolic cases [7].

Let us discuss now the spectral implementation of the procedure described above. Obviously, we have first to discretize our problem. Because of the singularities at the leading and trailing edges, we decided to split the computational domain into three subdomains as shown in Figure 3, and use in each of them a Chebyshev grid in x and y. Iterations may be performed separately in the three subdomains, after which the values of Φ are redefined on the interfaces $x = 0,1$ in such a way that Φ and the spectral derivative Φ_x remain continuous. We were able to prove that this procedure is stable [3], and numerical work amply supports this theoretical result.

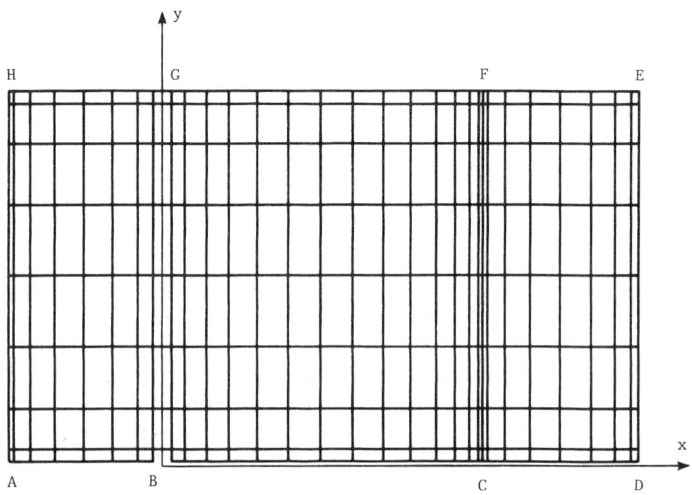

Figure 3. Three Chebyshev meshes. BC represents the wing.

The desymmetrizing technique used was the artificial density method. As spectral computation of derivatives is non-local, we could not employ the Cole-Murman approach. The modified density $\tilde{\rho}$ is obtained by biasing the density towards upstream values in supersonic regions only. The formulas used are:

$$\tilde{\rho} = \rho - \mu \overleftarrow{\delta}\rho \qquad \text{(1st order)} \tag{3.13}$$

$$\tilde{\rho} = \rho - \mu(\overleftarrow{\delta}\rho - \varepsilon\overleftarrow{\delta}E^{-1}\rho) \qquad \text{(2nd order)}. \tag{3.14}$$

In these expressions the finite difference operators δ, E are applied in the main flow direction, i.e. with increasing x. The switches μ and ε are computed as:

$$\mu = \max\left(0,\ 1 - \frac{1}{M^2}\right) \tag{3.15}$$

$$\varepsilon = \max\left(0,\ 1 - \frac{|\delta^0 \rho|}{K}\right) \tag{3.16}$$

with M the local Mach number and K of the order of the density jump across the shock. The second order density, after proper adjustment of the parameter K, produces significantly better results. Finally, we had to devise adequate iteration techniques to advance towards our solution. For (3.11) we could advance explicitly, using a DuFort-Frankel algorithm. The equation:

$$\frac{\Phi_{ij}^{n+1} - \Phi_{ij}^{n-1}}{2\Delta t} = \left[(\rho\Phi_x)_x + (\rho\Phi_y)_y\right]_{ij}^n - \frac{\sigma\rho_{ij}^n}{(\Delta x_i)^2 + (\Delta y_j)^2}\left(\Phi_{ij}^{n+1} + \Phi_{ij}^{n-1} - 2\Phi_{ij}^n\right) \tag{3.17}$$

is used to calculate Φ_{ij}^{n+1} in terms of known quantities; the parameter Δt may be used for optimization and σ is a dimensionless quantity, about 3 or 4. This scheme is remarkably simple, unconditionally stable (although accuracy considerations set limitations on Δt) and amenable to some theoretical investigation, even for the difficult Chebyshev spectral space operators [2]. Unfortunately, it cannot treat truly transonic cases, becoming divergent soon after shock formation. While this is not surprising -- DuFort-Frankel schemes are tailored to purely parabolic problems, and (3.11) is not parabolic due to the dependence of ρ on $\nabla\Phi$ -- it has motivated us to look into implicit iteration techniques.

An implicit spectral procedure seems at first doomed to failure, since the spectral operators are full, ill-conditioned matrices, which cannot be conveniently inverted. The technique of approximate inverses [10] provides an elegant way to overcome this difficulty. Given the problem:

$$L_{sp} u = r \tag{3.18}$$

with L_{sp} the spectral operator, one finds an operator L which is easily invertible and near in some sense to L_{sp}; then one proceeds to define a sequence u^n by:

$$L(u^{n+1} - u^n) = \alpha_n (L_{sp} u^n - r) \tag{3.19}$$

which will converge to u, the solution of (3.18). Here the quantities α_n are suitable convergence factors or even preconditioning operators.

Evidently, the success of this scheme hinges on the choice of the operator L. It is a safe rule of thumb to take L as a finite difference operator, defined on the same grid as L_{sp}, and representing the same differential expression. In our case, L will be a five-point operator, since five points are sufficient to represent

$$(\rho\Phi_x)_x + (\rho\Phi_y)_y .$$

The operator may be further modified to mimic closely (3.12), say:

$$L(\Phi^{n+1} - \Phi^n) \sim (\alpha\Phi_t + \beta\Phi_{xt})^{n+1/2}\Delta t \tag{3.20}$$

for better convergence.

This will solve the problem of having L "near" L_{sp}. In order to invert L efficiently, one may use a dimensional split, operating separately in the x and y directions (Douglas-Gunn ADI). Another possibility is an approximate L-U factorization of the five point operator, as described by Wong [11].

Thus we reach the final form of our transonic algorithm:

a. Compute Φ_x, Φ_y.

b. Correct Φ_y on the boundary, using $\Phi_y = \tau\Phi_x \frac{df}{dx}$ or $\Phi_y = 0$ respectively.

c. Compute ρ(or $\tilde{\rho}$) and the fluxes $\rho\Phi_x$, $\rho\Phi_y$.

d. Evaluate $(\rho\Phi_x)_x + (\rho\Phi_y)_y$.

e. Advance, either explicitly (DuFort-Frankel) or implicitly (Approximate Inverse). The approximate inverse will use, in defining the finite difference operators, ρ, μ, etc. as obtained in step (c).

f. Correct on subdomain interfaces, and impose Dirichlet conditions at infinity.

The derivatives in steps (a) and (d) are, of course, evaluated spectrally.

After experimenting with numerical schemes of this type, we reached the following conclusions:

1. The mesh needed for adequate resolution is significantly smaller than the corresponding finite difference mesh. Thus, the spectral precision contributes to efficiency by extracting information from a reduced data base.

2. Shock capture is satisfactory, although the modified density approach reduces spectral precision near shocks. A purely spectral desymmetrizing technique may be worth searching for.

3. DuFort-Frankel scheme, although inappropriate for shocks, resolves well smooth flows, converging in 100-200 (inexpensive) iterations.

4. The approximate inverse algorithms use substantially the same number of iterations to converge in the spectral case or in the purely finite difference case (where the residuals are computed by finite differences). Thus, there may be a tradeoff in CPU time

between the more complex spectral operator and the larger finite difference grid needed. Conceivably, existing codes may be upgraded by supplying a spectral residual, while leaving the iteration procedure unchanged.

Last, but not least, we present a comparison of a spectral code and a finite difference code. The relevant data are summarized below:

Table II

Spectral code: Douglas-Gunn approximate inverse
2nd order artificial density

Finite difference code: AF2X (see reference [6])

Parabolic airfoil, M_∞ = .85, 10% thickness ratio

	Mesh size	Number of Iterations	CPU time Cyber 175
Finite Differences	180 × 60	350	46 sec
Spectral	33 × 9	500	45 sec

As seen in Figure 4 the surface pressure results are virtually identical. It is also remarkable that, although the spectral code needed more iterations (also more complex iterations) the overall time is shorter, because of the reduced mesh employed. One should also recall that AF2X is a state of the art, highly optimized method.

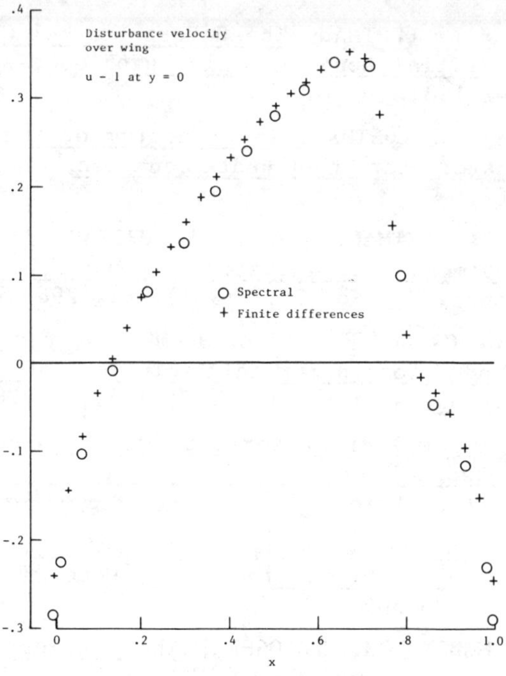

Figure 4.

In conclusion, we may certainly state that spectral methods have proved viable in transonics. More advanced calculations about general-shaped wings with lift have also been performed, and will be reported elsewhere.

4. Conclusions. Chebyshev collocation spectral methods have been applied to two-dimensional gasdynamic problems with shocks.

For the Euler equations we have demonstrated the ability of spectral methods to capture oblique shocks. Boundary condition implementation has been investigated, together with appropriate numerical dissipation. Further research is evidently needed in order to elucidate the role of dissipation and numerical boundary conditions in assuring stability.

For transonic flow, efficient codes based on an approximate inverse algorithm have been devised. Here is a case where the spectral method makes good its promise to reduce the overall computational effort by higher accuracy on smaller grids.

REFERENCES

[1] S. ABARBANEL and D. GOTTLIEB, Optimal time splitting for two- and three-dimensional Navier-Stokes equations with mixed derivatives, J. Comp. Phys., 41, No. 1, (1951), pp. 1-33.

[2] D. GOTTLIEB and L. LUSTMAN, The DuFort-Frankel Chebyshev method for parabolic initial boundary value problems, ICASE Report No. 81-42, December 1981.

[3] D. GOTTLIEB and L. LUSTMAN, The spectrum of the Chebyshev collocation operator for the heat equation, ICASE Report No. 82-12, May 1982.

[4] D. GOTTLIEB, L. LUSTMAN, and S. A. ORSZAG, Spectral calculations of one-dimensional inviscid compressible flows, SIAM J. Sci. Stat. Comput., 2, No. 3 , (1981), pp. 296-310.

[5] M. M. HAFEZ, J. C. SOUTH, and E. M. MURMAN, Artificial compressibility methods for numerical solution of transonic full potential equation, AIAA Paper No. 78-1148, (1978).

[6] T. L. HOLST, An implicit algorithm for the conservative transonic full potential equation using an arbitrary mesh, AIAA Paper No. 78-1113, (1978).

[7] A. JAMESON, Iterative solution of transonic flows over airfoils and wings, including flows at Mach 1, Comm. Pure Appl. Math., 27, (1974), pp. 283-309.

[8] A. MAJDA, J. MCDONOUGH, S. OSHER, The Fourier method for nonsmooth initial data, Math. Comp. 32, (1978), pp. 1041-1081.

[9] E. M. MURMAN and J. D. COLE, Calculation of plane steady transonic flow, AIAA J., 9, No. 1, (1971), pp. 114-121.

[10] S. A. ORSZAG, Spectral methods for problems in complex geometries, J. Comp. Phys., 37, (1980), pp. 70-92.

[11] Y. S. WONG, Iterative methods for problems in numerical analysis, D. Phil. Thesis, Oxford University, 1978.

[12] T. A. ZANG and M. Y. HUSSAINI, Mixed spectral/finite differences approximation for slightly viscous flows, ICASE Report No. 80-15, June 1980.

EDDY RESOLVING MODELING OF OCEAN CIRCULATION PROCESSES USING SPECTRAL METHODS

D. B. HAIDVOGEL*

Abstract. Ocean eddies are turbulent mesoscale currents now known to represent a dynamically significant component of the global ocean circulation. Spectral modeling studies of oceanic eddies, and related theoretical research, have contributed substantially to our present understanding of mid-ocean dynamics and other eddy-related processes. Here, we briefly review how mid-ocean eddy currents arise from the competition between waves and turbulence on the rotating earth, and examine one important consequence of the energetic eddy field: turbulent dispersion and mixing in the ocean.

1. Introduction. During the past decade, physical oceanographers have directed a considerable effort to the observational exploration and theoretical modeling of mid-ocean mesoscale "eddies" - energetic time-dependent current systems dynamically equivalent to the synoptic-scale weather systems in the atmosphere. Eddies occur intermittently in time; however, they appear to be spatially ubiquitous, having been observed in all of the world's oceans. Generally, eddies are characterized by circulation speeds of 5 to 50 cm s^{-1}; space/time scales of tens to hundreds of kilometers and weeks to months; and vertical displacements of up to several hundred meters in the main thermocline (the mid-depth region of strong temperature gradient which separates the upper wind-driven gyre from the more sluggish abyssal waters). Figure 1 illustrates schematically the distribution and types of eddy variability typically encountered in the North Atlantic Ocean.

Recent intensive observational programs such as the Mid-Ocean Dynamics Experiment (MODE) have begun to map out the geographical variability in eddy properties and to examine low-order dynamical balances [1]. From these observations, a consistent picture emerges: Vigorous hydrodynamical instabilities in the narrow, but intense, oceanic current systems - e.g., the Gulf Stream and North Equatorial

*Woods Hole Oceanographic Institution, Woods Hole, MA 02543. Support was furnished by National Science Foundation (Grant No. OCE 81-9486) and Office of Naval Research (Contract No. N00014-C-0019, NR083-004).

Current systems in the North Atlantic - are associated with local finite-amplitude meander generation and eddy growth. From these generation sites eddy energy is communicated, by dynamic processes at present poorly understood, to the oceanic interior and/or the remainder of the ocean basin.

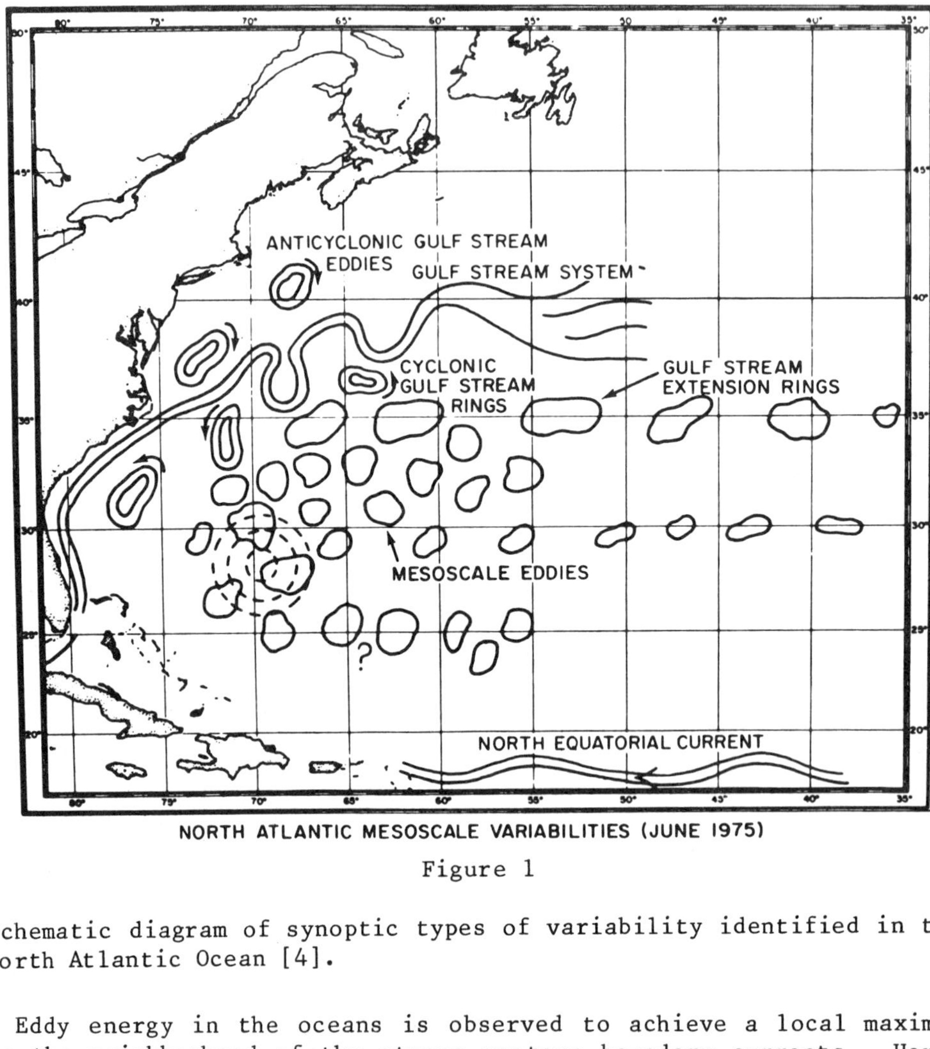

NORTH ATLANTIC MESOSCALE VARIABILITIES (JUNE 1975)

Figure 1

Schematic diagram of synoptic types of variability identified in the North Atlantic Ocean [4].

Eddy energy in the oceans is observed to achieve a local maximum in the neighborhood of the strong western boundary currents. Here, the mean and fluctuating components of motion co-exist at comparable amplitude (50-100 cm s^{-1}) in turbulent equilibrium. Although the associated eddy-mean field interactions are clearly the energy source for the fluctuations, the existence of the finite-amplitude eddy field also alters the time-mean circulation in fundamental ways. For instance, available observations and theoretical models both suggest that eddy-induction (vertical momentum transfer by the eddies) may be the principal driving mechanism for the observed deep mean circulation in the energetic boundary current regions [2].

In the mid-ocean, however, time-mean flows are extremely weak (1-5 cm s^{-1}) and spatially variable; eddies, where they exist, dominate the circulation. In such regions, presumed distant from active eddy generation sites, the eddy field is in approximate local (homogeneous) turbulent equilibrium with its environment. As discussed more fully below, this fact allows the mid-ocean eddy field to be treated using Fourier spectral techniques, independently of the global oceanic circulation in which it is embedded.

Understanding the eddies and their role in the general circulation has emerged as a central problem in ocean circulation dynamics. Moreover, it has become clear that eddies and eddy-related processes contribute importantly to a broad range of oceanic phenomena. These include: the large-scale global fluxes of momentum, heat and salt; the dispersal of passive scalars such as geochemicals and waste materials; and ocean-atmosphere coupling (climate). A comprehensive account of the current state of eddy science, including observational and theoretical summaries, has recently been produced [3].

2. Eddy-Resolving Models. A variety of high-resolution numerical models are being used to examine oceanic eddy phenomena [4]. Eddy-resolving general circulation models (EGCM's) are high-horizontal-resolution global (closed-basin) ocean models, driven by idealized mechanical and/or thermal forcing, in which both the large-scale circulation and mesoscale eddy field are explicitly resolved. In these models, mesoscale motions may spontaneously arise via internal processes as the equations are integrated in time. After reaching statistical equilibrium, long analysis records can be generated to yield reliable eddy and mean flow statistics. The results allow the investigation of the processes involved in the maintenance of the eddy field, and the characteristics of the interaction between eddies and the mean flow. Despite necessary idealization, EGCM's are beginning to reproduce many of the observed features of the ocean circulation including both its mean and eddy components [5].

Periodic (often called process) models and regional models resemble EGCM's in their use of fine horizontal resolution to achieve a simultaneous explicit representation of both the large-scale and mesoscale components of the oceanic circulation. They differ, however, in that periodic and regional models seek to determine the dynamical equilibrium characterizing the mesoscale variability in a local block of ocean, whose interconnection to the remainder of the oceanic gyre is either assumed statistically unimportant (via an assumption of horizontal homogeneity) or parameterized (in terms of some suitable boundary constraints on the local region). The resulting local dynamical models are used to examine the equilibrium properties of the mid-ocean eddy field as a function of local environmental processes and parameters such as turbulent energy level and space/time scales [6].

Spectral methods have contributed substantially to research in all three of these ocean modeling areas. In global domains, Fourier and Chebyshev polynomial approximation techniques have been applied to the problems of turbulent ocean flow above topography [7], the stability of oceanic jets [8], and the waves and mean circulation induced by time-dependent wind stresses [9]. The feasibility of, and requirements for, regional dynamical forecasting in the mid-ocean have also been examined, in part through the use of Chebyshev expansion techniques [10, 11].

Perhaps the most significant contribution of spectral methods to our present understanding of eddy dynamics and the ocean circulation has, however, been made in the application of periodic (homogeneous turbulence) models to the dynamics of the mid-ocean eddy field. Building upon many early simulations of damped two-dimensional turbulent flow (e.g., [12]), theoretical oceanographers have been able to assess the relative importance and dynamical significance of such competing geophysical effects as non-linear and bottom topographic interactions, planetary vorticity variations, stratification, local viscous effects, and forcing by the large-scale mean circulation. More recently, spectral models of mid-ocean turbulence have been used to examine a variety of applied problems involving global-scale scientific and resource management issues. Before discussing one such model application, we briefly review some of the known dynamic properties of the mid-ocean eddy field, placing particular emphasis on the effects of the earth's rotation. More detailed accounts, including discussions of the effects of stratification and topography, may be found in references [6] and [13].

3. <u>Mid-Ocean Turbulence</u>. To leading order, mesoscale motions in the mid-ocean are horizontally non-divergent, and in geostrophic and hydrostatic balance. In addition, for typical eddy length scales, geophysical variations in the vertical component of planetary vorticity can be approximated as a linear function of the north-south co-ordinate - the so-called "β-plane" approximation. Under these assumptions, and for convenience neglecting stratification and bottom topography, the resulting vorticity equation describing barotropic mid-ocean flow can be written [14]:

$$\frac{\partial}{\partial t} q + J(\psi, q) = D + F \tag{1}$$

where the total (or "potential") vorticity

$$q = \nabla^2\psi + f = \nabla^2\psi + (f_o + \beta y) \tag{2}$$

has contributions from the relative ($\zeta = \nabla^2\psi$) and planetary ($f = f_o + \beta y$) vorticities. Equation (1) requires that total vorticity be conserved following the horizontal fluid motion except for explicit non-conservative effects associated with dissipation (D) and forcing (F).

Consistent with the assumed spatial homogeneity of the mid-ocean eddy field, we require that the streamfunction $\psi(x,y,t)$ be periodic over the square domain $0 \leq x,y \leq 2\pi$. Accordingly, it is convenient to think of equation (1) as dimensionless - that is, as having been non-dimensionalized by $(x,y,t,\psi) \sim (L,L,L/U,UL)$. Here, L and U represent the dimensional size of the domain (divided by 2π) and the RMS velocity.

By temporarily suppressing latitudinal variations in planetary vorticity ($\beta = 0$), and all non-conservative effects ($D = F = 0$), equation (1) reduces to a statement of conservation of relative vorticity:

$$\frac{\partial \zeta}{\partial t} + J(\psi , \zeta) = 0. \tag{3}$$

Turbulent flows satisfying (3) conserve both total energy (E) and mean-square vorticity (enstrophy, Z). As is now well known, such two-dimensional flow must evolve so as to transfer kinetic energy to ever-larger scales of motion. Simultaneously, enstrophy (that is, mean-square vorticity) is transferred to finer spatial scales and wrapped up into ever finer filaments. Computer realizations of purely two-dimensional flow readily confirm these tendencies (e.g., Figure 1 in [6]).

Since two-dimensional turbulence has both energy and mean-square-vorticity as inviscid constants of motion, two inertial-ranges energy spectra may be defined [15]:

$$E_\varepsilon(\kappa) \sim \varepsilon^{2/3} \kappa^{-5/3} \tag{4a}$$

and

$$E_\rho(\kappa) \sim \rho^{2/3} \kappa^{-3} , \tag{4b}$$

where ε is the rate of cascade of kinetic energy (per unit mass), ρ is the rate of cascade of enstrophy, and κ is the non-directional wavenumber. (Logarithmic corrections are necessary to the second inertial range in order that the total rate of enstrophy transfer from all wavenumbers be independent of κ [16].) The (-5/3) range transfers energy to lower wavenumbers while the (-3) range cascades enstrophy to higher κ. For a nearly inviscid two-dimensional turbulent continuum flow in which energy is fed in at a constant rate to an intermediate band of wavenumbers near κ_o, the inertial range ideas suggest that a quasi-steady state develops in which $E(\kappa) = O(\kappa^{-5/3})$ for $\kappa \ll \kappa_o$ and $E(\kappa) = O(\kappa^{-3})$ for $\kappa \gg \kappa_o$, up to some viscous cutoff.

In general, solutions to the mid-ocean turbulence equation (1) must be sought numerically. In the Fourier spectral representation, we set:

$$(5) \qquad \psi(x,y,t) = \sum_{k=-(N/2)}^{(N/2)-1} \sum_{\ell=-(N/2)}^{(N/2)-1} A_{k\ell}(t)\, e^{i(kx + \ell y)}$$

where the truncation wavenumber $K = (N/2)$. By substitution of (5) into (1), the evolution equations for the spectral coefficients $A_{k\ell}(t)$ may be derived. These equations can then be advanced in time using standard time-stepping techniques - e.g., leapfrog with semi-implicit treatment of the diffusive terms. The non-linear product $J(\psi, q)$ may be computed so as to conserve energy and enstrophy, and without aliasing error, using the transform method [17]. A complete description of one model of this type is given in [18].

In numerical ocean simulations, it is now customary to incorporate dissipative effects in one of two ways: either as a bottom friction proportional to relative vorticity, or as a lateral friction of harmonic or biharmonic form. Often, these dissipative mechanisms are used in combination; therefore, in general:

$$(6) \qquad D = -\delta\nabla^2\psi + \nu\nabla^4\psi - \lambda\nabla^6\psi$$

where δ, ν and λ are poorly known turbulent diffusion co-efficients. To maintain a statistically steady eddy field in the presence of dissipation some energy input is necessary. Although a variety of forcing functions are possible, it is convenient to adopt the phase-coherent, constant energy injection mechanism

$$(7) \qquad F = -\gamma\psi/|\psi|^2 ,$$

typically applied over some narrow band of wavenumbers $\kappa_0 \leq \kappa = |k^2 + \ell^2|^{1/2} \leq \kappa_1$. Note, by formation of the energy equation from (1), that the forcing prescribed by (7) injects energy uniformly at rate γ into each forced wavenumber pair (k, ℓ).

Direct numerical verification of the inertial power-law ranges (4a,b) for inviscid two-dimensional flow have been sought in many studies. In contrast to the inertial range result that $E(\kappa) = 0(\kappa^{-3})$ for $\kappa > \kappa_1$, such numerical simulations yield two rather different asymptotic spectral states: quasi-equilibrium (quasi-inviscid) turbulence in which $E(\kappa) = 0(\kappa^{-1})$, and rapidly decaying (overdamped) turbulence in which $E(\kappa) = 0(\kappa^{-4})$. These asymptotic states appear to be achieved over a broad range of dissipational types and parameters [19]. Examples of energy spectra representative of the overdamped behavior are shown in Figure 2. The existence of pure inertial range behavior is therefore problematic, although a few extremely high resolution simulations have supported the existence of the κ^{-3} enstrophy-cascading range [20]. With commonly affordable levels of resolution ($N \leq 128$), however, it appears that only the quasi-equilibrium and overdamped states may be achieved in practice. Fortunately, the systematic tendencies for energy and enstrophy to be transferred towards lower and higher wavenumbers, respectively, are maintained in these power-law ranges.

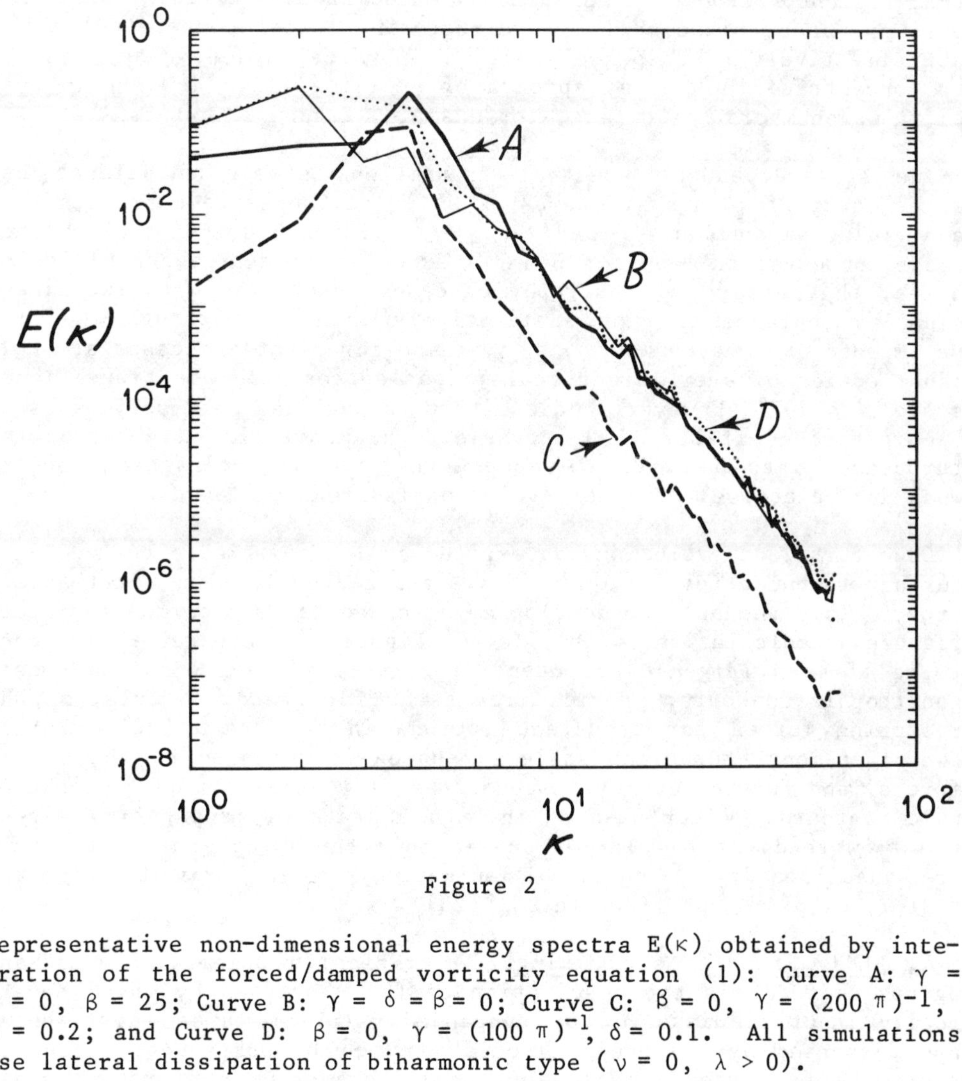

Figure 2

Representative non-dimensional energy spectra $E(\kappa)$ obtained by integration of the forced/damped vorticity equation (1): Curve A: $\gamma = \delta = 0$, $\beta = 25$; Curve B: $\gamma = \delta = \beta = 0$; Curve C: $\beta = 0$, $\gamma = (200\,\pi)^{-1}$, $\delta = 0.2$; and Curve D: $\beta = 0$, $\gamma = (100\,\pi)^{-1}$, $\delta = 0.1$. All simulations use lateral dissipation of biharmonic type ($\nu = 0$, $\lambda > 0$).

The introduction of competing dynamical mechanisms such as stratification, gradients of planetary vorticity, and topography is able to halt or to reverse the two-dimensional cascades of energy and enstrophy for some scales of motion. For example, with $\beta \neq 0$ Rossby (planetary) waves are possible and serve to redistribute energy by wave radiation. A competition may thus arise between turbulent and wave-like dynamics [21]. Associated with the restoring force provided by a large-scale planetary vorticity gradient is a scale-dependent boundary at which wave steepness is of order unity. This spectral boundary divides the energy spectrum into regions in which the mobility of energy is vastly different and at which, depending on the direction and speed of non-linear transfer within these spectral regions, turbulent energy transferred to this dividing scale may pile up. Thus, although the wave-restoring forces associated with β act effectively to disperse energy in physical space, they also act to concentrate the energy spectra of evolving fluid at certain specific wavenumbers. (Compare, for example, Figure 2: curves A and B.)

The scale-dependent spectral transition between turbulence and waves occurs for wave steepness of O(1) and therefore corresponds to a dividing wavenumber $\kappa_\beta \simeq (\beta/2U_{rms})^{1/2}$. Since β does not alter the gross conservation laws for E and Z except in the presence of boundaries, the energy and enstrophy cascades associated with two-dimensional turbulence are not qualitatively changed, although they may be impeded. For wavenumbers $\kappa >> \kappa_\beta$, these turbulent processes strongly favor eddies of ever larger scale; upon approaching the transitional scale $(\kappa_\beta)^{-1}$, however, eddies start propagating as Rossby waves as the restoring effects of β increase. When advection has turned the turbulence into a field of steep waves, mutual interactions become weak and selective. Further evolution is greatly slowed.

Once turbulent evolution has excited scales comparable to $(\kappa_\beta)^{-1}$, further turbulent evolution is characterized by growing zonal anisotropy. This tendency to develop zonal currents is a common end-state feature of calculations with β (e.g., Figure 3). More generally, contours of mean large-scale potential vorticity, the so-called quasi-geostrophic contours, provide free paths for steady circulation and restoring forces for transient motions in the ocean [22]. Purely two-dimensional turbulent scale expansion is defeated by these wave forces, and further evolution generates flow along the quasi-geostrophic contours. After passing through a state of propagating waves, the flow tends at the larger scales to alternating zonal jets which are quasi-steady. Such zonal anisotropy is a commonly observed feature of mid-ocean flows (e.g., [23]).

4. Eddy Stirring and Mixing. A critical problem facing oceanographers today is the description of the processes by which dynamically inactive dissolved or suspended materials are redistributed and dispersed by the ocean circulation. Such substances, often referred to as tracers, include naturally-occurring and anthropogenic chemical tracers (oxygen, tritium), biological nutrients and organisms (phytoplankton, fish larvae), and pollutants or waste materials either accidentally or intentionally released into the oceans (oil, nuclear waste). Particularly in the mid-ocean, the energetic meso-

scale eddy field is of probable importance to the dispersal of these tracers.

Crudely, but conveniently, eddy transport can be said to occur as either "stirring" or "mixing". The distinction may be visualized by considering the fate of a spot of red dye in the ocean. As the dye is redistributed by oceanic motions, it may be stirred into a number of undiluted red streaks or filaments of dye, separated by clear water. Alternately, it may be mixed, resulting in a pink cloud of reduced dye concentration. Stirring is thus transport without mixing.

The ability of eddies to transport dynamically passive (as well as active) substances in these ways has important scientific consequences in a range of fields including marine biology, geochemistry, and climate research [3]. In addition to these scientific applications, eddy stirring and mixing in the ocean relate directly to a variety of planetary-scale resource management issues. For instance, the advisability of, and preferred means for, chemical and nuclear waste disposal in the oceans depends importantly on knowledge of the anticipated eddy dispersal effects in the disposal regions.

Recent numerical studies using spectral models have begun to examine the stirring and mixing of a passive scalar in two-dimensional β-plane turbulence as a prototype for scalar dispersal within a single isopycnal layer in the mid-ocean. Briefly, specific realizations of a dynamically evolving mid-ocean velocity field are generated by direct integration of the forced/damped barotropic vorticity equation (1) over a spatially periodic domain and for representative mid-ocean values of the environmental parameters such as β and RMS velocity. An initial distribution of tracer (typically a Gaussian, or other localized spot) or a continuously-releasing tracer source is then initiated in mid-domain. The subsequent evolution of the tracer field is then followed by numerical integration of the advection-diffusion equation:

$$\frac{\partial\theta}{\partial t} + \vec{v} \cdot \nabla\theta = \eta\nabla^2\theta \tag{8}$$

where $\theta(x,y,t)$ represents the concentration of the tracer and $\vec{v}(x,y,t)$ is the dynamically simulated velocity field. Here, η is the "explicit diffusivity"; the associated diffusive term is assumed to properly parameterize mixing processes occurring on scales smaller than the grid scale. The spectral approximation techniques used to integrate (8) are identical to those applied to the vorticity equation.

It is well known that dispersal of a passive scalar will generally occur much more rapidly in turbulent than in laminar flow. For pure diffusion, the rate of dispersal is clearly related to the explicit

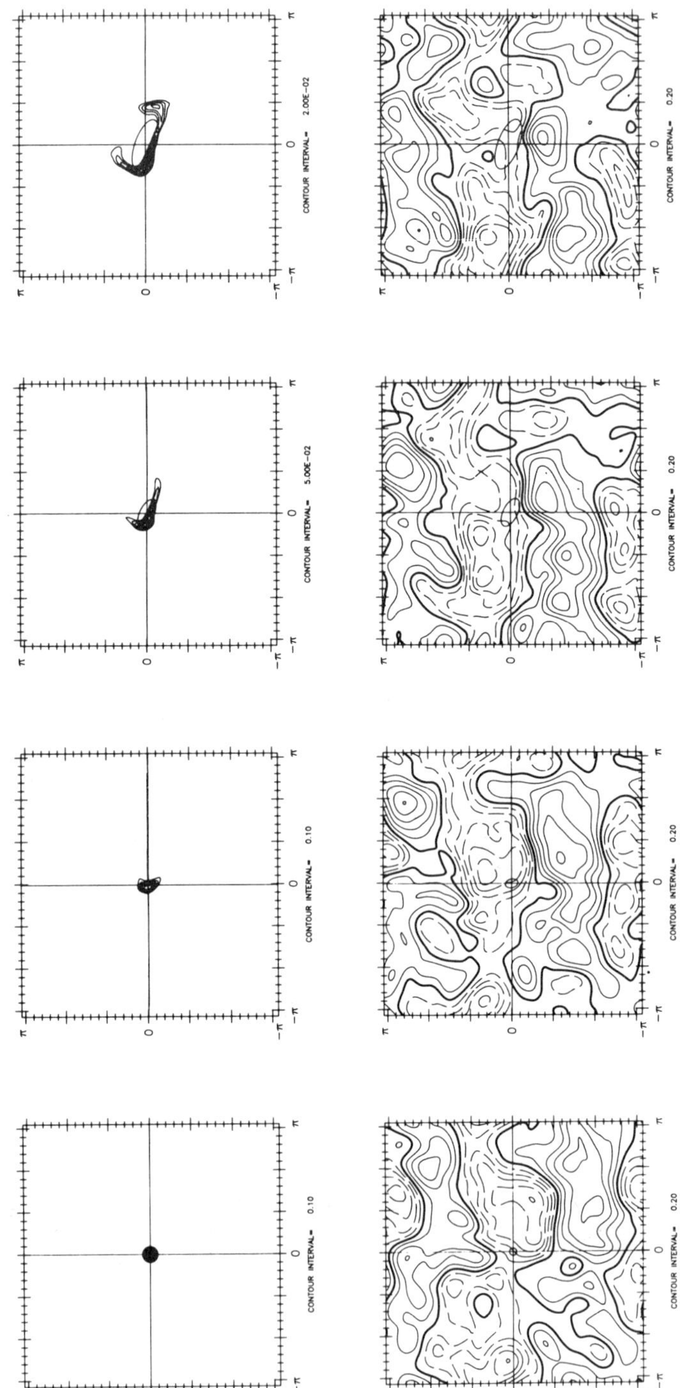
CONTOUR INTERVAL= 2.00E−02
CONTOUR INTERVAL= 5.00E−02
CONTOUR INTERVAL= 0.10
CONTOUR INTERVAL= 0.10
CONTOUR INTERVAL= 0.20
CONTOUR INTERVAL= 0.20
CONTOUR INTERVAL= 0.20
CONTOUR INTERVAL= 0.20

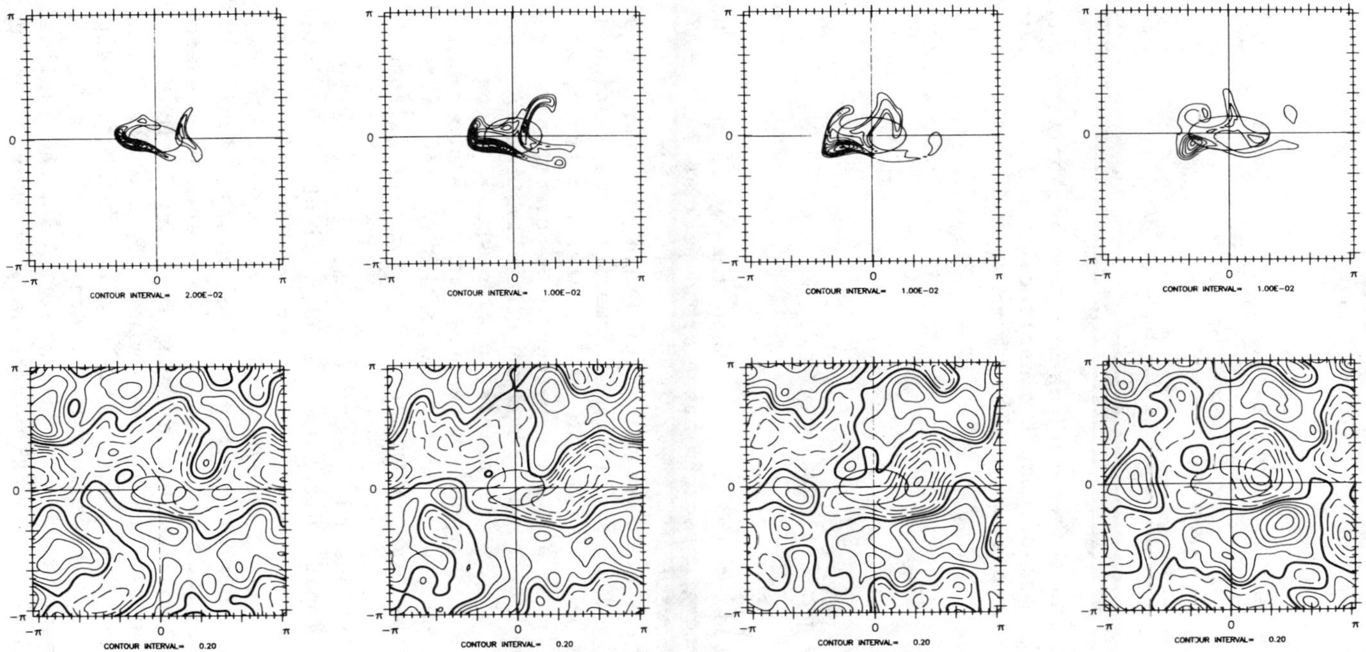

Figure 3

Simulated time series of contour maps showing the temporal evolution of the streamfunction (ψ, lower) and tracer fields (θ, upper) in a 1000-square kilometer region of the mid-ocean. The ψ field is generated from (1) using $\beta = 10$, $\gamma = 0.002$, $\delta = 0.1$, $\nu = 0.001$ and $\lambda = 0$. A small Gaussian spot of approximately 30 km in diameter, is initially released in the center of the simulated ocean. Subsequent maps, separated in time by about two weeks, show how the tracer is dispersed and fragmented by the simulated eddy field. Both the ψ and θ fields have been shifted to follow θ center of mass (indicated by the cross-hairs). The inscribed ellipse shows the second moment of the tracer distribution about the center of mass.

diffusivity, η. For the more general case of a tracer evolving in a turbulent fluid, it would be useful to parameterize the turbulent dispersal by an effective diffusivity, η_{eff}. In general, we expect η_{eff} to be substantially greater than η, the explicit diffusivity.

Consider the second moment of the tracer distribution about its center of mass:

$$(9) \qquad M_r = \frac{\iint r^2\, \theta(x,y)\, dxdy}{\iint \theta(x,y)\, dxdy}$$

where r is the radial distance from the center of mass. The time rate of change of this physical space moment clearly relates to the average spreading of material about the center of mass. It is therefore convenient to define:

$$(10) \qquad \eta_{eff} = 1/4 \frac{d}{dt}(M_r) \ .$$

Note that in the case of pure diffusion ($\vec{v} \to 0$), $M_r = 4\,\eta\, t$; hence, $\eta_{eff} \to \eta$ as $\vec{v} \to 0$.

Here energy is injected to the fluid within the wavenumber band $6 \leq \kappa \leq 8$; the resulting energy spectrum has a broad peak at $2 \leq \kappa \leq 7$, and a damped $[O(\kappa^{-4})]$ power-law at smaller scales. Figure 3 shows a representative physical space streamfunction field at intervals of 0.4 time units (about 15 days) for a single numerical realization of equation (1). This ψ field is in a statistically steady state, with kinetic energy constant to within a few percent. A corresponding set of maps that show the evolution of θ from an initially Gaussian spot with a "standard width" of 2 grid points is also shown. Note that contour intervals may change between successive maps. The cross-hairs in the figures indicate the position of the center of mass of the θ distribution. Also inscribed is the tracer's second moment ellipse.

Tracer dispersal simulations such as that summarized in Figure 3 clearly demonstrate the rapid sharpening of θ gradients associated with "stirring" by the explicitly resolved eddy velocity field. Because of the rapid transfer of tracer variance to small-scale features associated with advective stirring, diffusive loss ("mixing") of tracer variance is much more rapid than in the absence of the eddy field. By carrying out an ensemble of dispersal simulations, the ensemble-average spreading rate can be determined (Figure 4). For a short interval after the release of the tracer, the "effective diffusivity" - as measured by the time rate of change of the second moment of the tracer distribution - is little more than the explicit (subgridscale) diffusivity. It quickly increases, however, reaching an approximately constant value of O(1). In dimensional terms, the effective diffusivity is thus O(UL), as might be suggested by a turbulent mixing length argument. This result may also be anticipated theoretically using simple turbulence closure calculations [24].

A variety of simple models have been advanced to predict the spreading behavior of an instantaneous point release of tracer along an isopycnal in the mid-ocean (e.g., [25]). Typically, it is assumed that the mesoscale eddy field initially acts to strain out the tracer into ever larger and exponentially-growing "streaks" of constant width $L_S = (\eta/\gamma)^{1/2}$, where η is an explicit diffusivity associated with small-scale ocean mixing processes [26], and $\gamma = (u_x^2 + v_y^2)^{1/2}$ is the RMS rate of strain. The area occupied by the tracer is therefore also supposed to increase exponentially. Obviously, such a scenario is unlikely to continue forever; eventually, a well-mixed state must be approached. It is therefore important to determine how long appreciable streakiness of the spreading tracer path will persist.

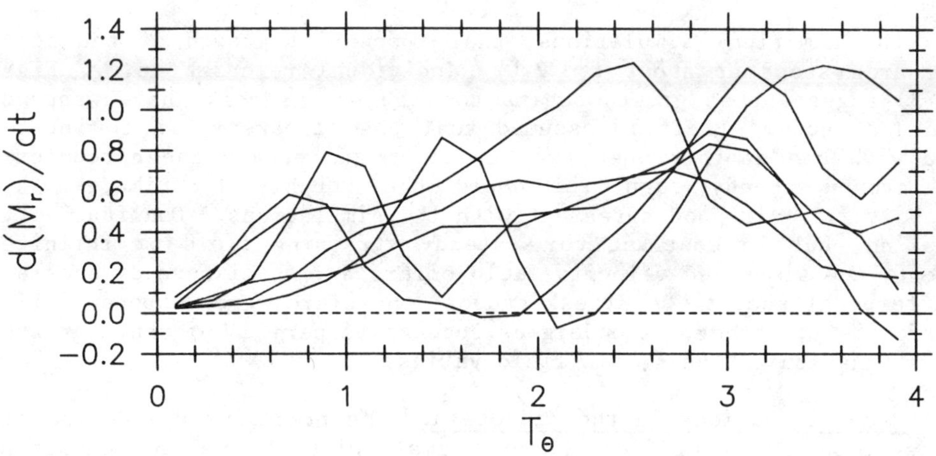

Figure 4

The time rate of change of the second moment M_r for six independent tracer dispersal simulations analagous to that shown in Figure 3. T_θ is the elapsed time since tracer release.

The "area occupied by the tracer" is a difficult concept to quantify in numerical simulations. For convenience, we follow a relative contour defined as 10 percent of the maximum concentration as a function of time. However, these "percentile contours" do not necessarily coincide with material line elements. Interpretation of the area occupied by these percentile contours is aided by a similarity solution for the advection/diffusion of a Gaussian blob in a pure straining field $\psi = \gamma xy/\sqrt{2}$ [27]. The similarity solution shows an O(1) adjustment time during which the streak is being formed and the blob is increasing linearly in area due to the action of the subgridscale diffusion. Once the streak has been formed, the "stretching" action of the straining field increases its length exponentially, while its width attempts to become thinner and thinner. However, the diffusivity η acts to spread the streak, keeping its width constant. Hence, the blob area grows exponentially after the adjustment period.

Based upon this simple theoretical model, we might expect the area of the 0.1 θ_{max} contour in the numerical model to grow exponentially with a time constant of $\gamma_{rms}/\sqrt{2}$. By contrast, on average, a blob in the numerical model increases in size at an exponential rate, virtually from t = 0 (Figure 5). The individual growth curves show that all six releases initially increase faster than the similarity result would suggest. The reason for this rapid growth is unclear. However, it may be that locally enhanced values of the rate of strain, in excess of the RMS value, initially dominate the dispersal process because of their shorter adjustment time.

In the numerical simulations, the exponential growth of the contour area stops at about t = 2.0. Individual releases (e.g., Figure 3) suggest that growth begins to slow when streaks have wrapped around an eddy. If it is assumed that the streaks begin to interfere with one another when the streak length equals the circumference around an eddy, then this should occur roughly at a time, t_{free}, of about 2.5 - in good agreement with the simulations. Qualitatively speaking, but in contrast to a steady straining field of infinite extent, the simulated velocity field offers a rich mixture of spatial and temporal scales. A streak cannot, therefore, grow exponentially indefinitely; rather, it undergoes successive periods of rapid growth and stagnation as the strain field varies.

5. Particle Motions in the Mid-Ocean. The preceding discussion of the evolving $\theta(x,y,t)$ field has not dealt with the related question of how individual tracer particles are dispersed. With suitable interpolation techniques, however, the Eulerian spectral models - which yield the equivalent of a time-series of velocity data on a fixed grid of points - can be used to predict Lagrangian circulation features such as particle trajectories and dispersal. For instance, within a turbulent flow evolving according to equation (1), we seed 80 Lagrangian particles, or "floats". The initial float placement - 5 clusters of 16 floats each - is shown in Figure 6. After launch, the trajectory of each float is determined by the kinematic requirement

(11) $$\frac{d\vec{x}_i}{dt} = \vec{u}_i(\vec{x}_i, t) \qquad 1 \le i \le 80$$

where $\vec{u}_i(x_i,t)$ is the simulated turbulent velocity field following the $\underline{i^{th}}$ float.

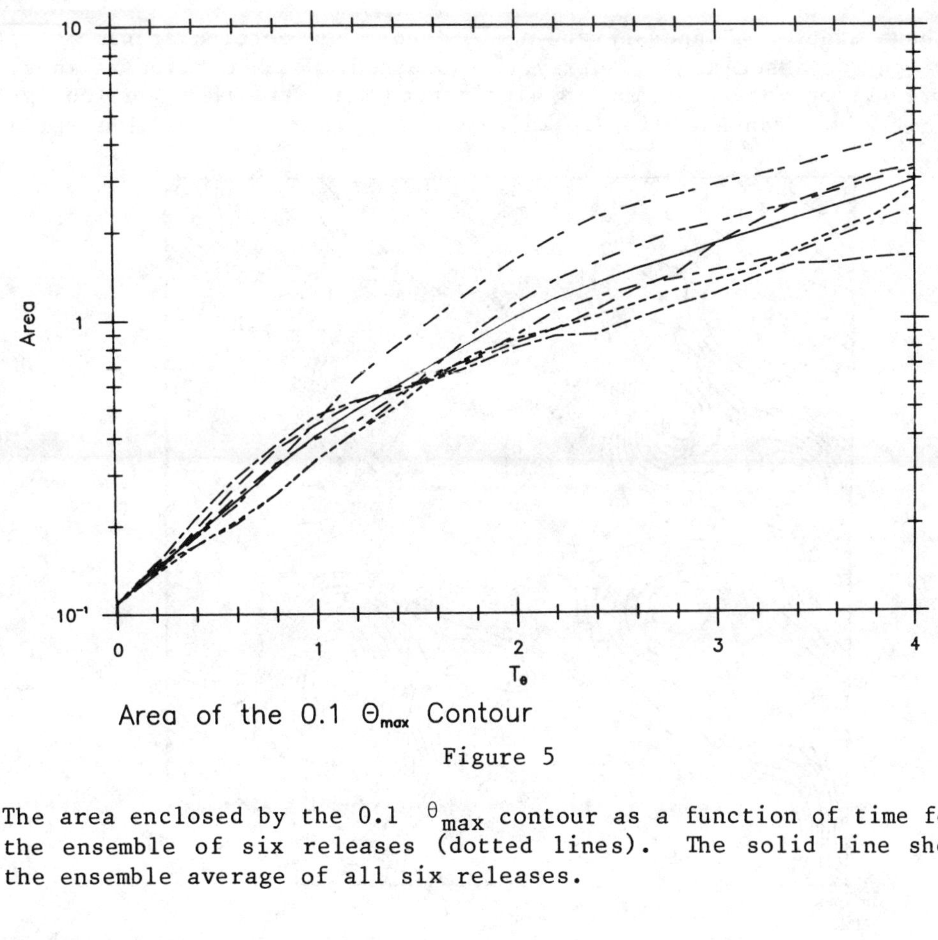

Area of the 0.1 Θ_{max} Contour

Figure 5

The area enclosed by the 0.1 θ_{max} contour as a function of time for the ensemble of six releases (dotted lines). The solid line shows the ensemble average of all six releases.

Equation (11) may be approximately solved using a Runge-Kutta, predictor-corrector, or other simple ODE integration technique. However, an important problem arises: Since the floats will in general lie between the grid-points of the Eulerian model, it is necessary to obtain the velocity values $\vec{u}_i(x_i,t)$ by interpolation from the fixed Eulerian grid. A variety of interpolation schemes, of varying efficiency and accuracy, can be used. Here we obtain the values of $\vec{u}_i(x_i,t)$ by direct summation of the Fourier series representation:

$$\vec{u}_i(\vec{x}_i, t^n) = [u_i(\vec{x}_i, t^n), v_i(\vec{x}_i, t^n)]$$

(12)

$$= \sum_{k=-K}^{K-1} \sum_{\ell=-K}^{K-1} (-i\ell, ik) A_{k\ell}(t^n) e^{i(kx+\ell y)}$$

(We have adopted a superscript 'n' to denote the particular time step at which the associated quantity is evaluated.) Although inefficient, the resulting interpolation is highly accurate, and allows us to ignore the importance of interpolation errors in the discussion which follows.

Figure 6

Initial streamfunction field and float placement for particle dispersal simulation in a (628 km)^2 mid-ocean domain. The ψ field is evolving according to (1) with $\beta = 2$, $\gamma = 0.0022$, $\delta = 0.1$, $\nu = 0.002$ and $\lambda = 0$.

Using this procedure, individual float tracks may be recovered from the time series of Eulerian velocity generated by the mid-ocean turbulence model. Figure 7 shows, in "spaghetti diagram" form, the trajectories of all 80 floats, followed for 230 days. We note that, while the Eulerian streamfunction field is periodic in space, the particle trajectories are not; they may be followed beyond the edges of the doubly periodic Eulerian domain.

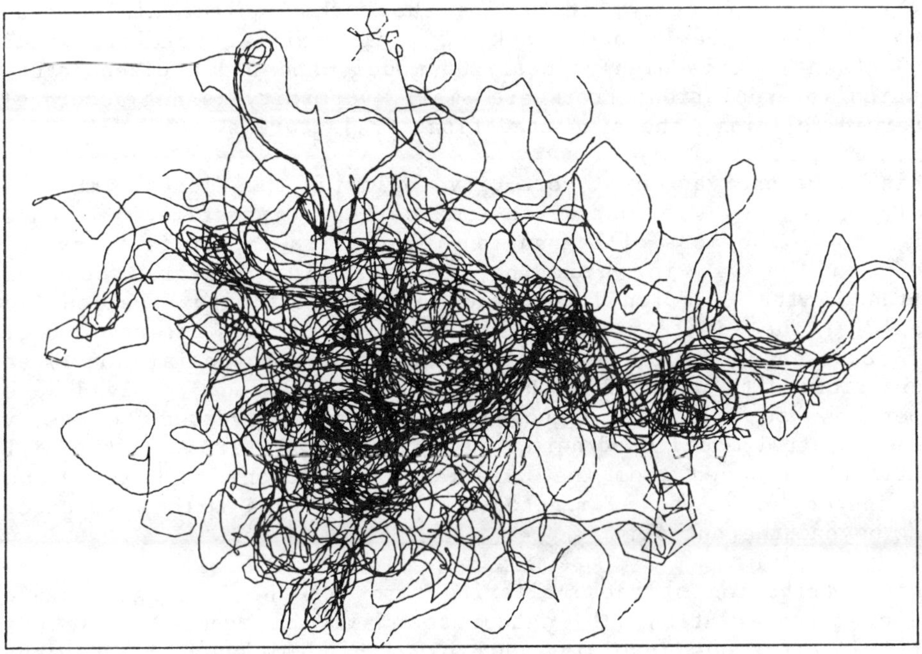

Figure 7

Cumulative float tracks ("spaghetti" diagram) for the 80 floats shown in Figure 6. Over a period of 230 days, the floats have escaped the initial domain ($0 \leq x,y \leq 628$ km). The edges of the inscribed perimeter are at x = −413, 1257 km and y = −109, 1036 km.

According to the continuum model, potential vorticity along any trajectory should change only as a result of the cumulative effects of the explicit non-conservative effects (dissipation and forcing) along that trajectory - i.e.,

$$\frac{d\, q_i}{dt} = D\,(\vec{x}_i, t) + F(\vec{x}_i, t) \, . \tag{13}$$

This Lagrangian statement of q-conservation can be integrated by the same procedure as that used for the kinematic equation (11), that is, by direct summation of the corresponding Fourier series. The result of integrating (13) is a dynamic "prediction" of the value of q following each float. A value of q_i $(\vec{x}_i, t^n)$ at the instantaneous position of the float may also be directly determined from the Eulerian vorticity spectrum, once the float track has been established.

By direct comparison of the q values predicted on the basis of vorticity conservation and those observed following the kinematically deduced trajectory, it can be shown that quite quickly after "launch", the floats lose track of their initial vorticity value [28]. Even in this highly idealized model, and under circumstances in which interpolation errors are small, vorticity is not accurately conserved following the simulated float trajectories.

This non-conservation of vorticity following the floats must ultimately be related, of course, to some sort of discretization/truncation error. The loss of Lagrangian information is, in fact, associated with the way in which the quadratic (non-linear) terms are treated in the Eulerian integration of the vorticity equation (1). Because of the finite horizontal resolution of the Eulerian model, here represented by the cut-off wavenumber K, potential non-linear interactions with wavenumbers higher than the maximum resolved wavenumber must be discarded, ignored, or otherwise treated improperly. In the spectral model used here, these "aliasing" terms are typically identically removed from the Eulerian calculation; that is, $J(\psi, q)$ is computed so as to eliminate any interactions with wavenumbers lying beyond the cut-off.

The contribution of these aliasing terms to the Lagrangian statement of q-conservation (13) cannot so easily be ignored. The quadratic interactions involving unresolved high wavenumber components are an integral part of the Lagrangian information inherent in the q-conservation statement. In truncating the Eulerian model at some upper cut-off wavenumber, some Lagrangian information is lost. Horizontal resolution is at the heart of the problem. Given sufficient resolution - i.e., a long enough tail to the energy spectrum, $E(\kappa)$ - the ignored aliasing terms would indeed be negligible in both the Eulerian <u>and</u> Lagrangian simulations.

The aliasing "error" terms can be thought of as an added non-conservative term in the Lagrangian balance of terms. Consequently, the Lagrangian statement of q-conservation for float trajectories derived by interpolation/integration from a Eulerian ocean model is not the continuum equation (13), but rather the modified balance

$$\frac{d\,q_i^n}{dt} = D(\vec{x}_i, t^n) + F(\vec{x}_i, t^n) + A(\vec{x}_i, t^n) \tag{14}$$

where $A(\vec{x}_i, t^n)$ are the aliasing error terms ignored in the computation of the Eulerian quadratic dynamics. These aliasing terms can be directly calculated; for the spectral simulation discussed here, the aliasing errors are the largest single term on the righthand side of (14). Identical comments hold for other conserved material properties such as a tracer.

In most applications, it is the statistical behavior of an ensemble of particles, rather than the detailed trajectory of each particle, which is of interest. It might therefore be hoped that, in the ensemble average, particle behavior is unaffected by the presence of A errors along each individual float track. However, such a result is probably unlikely. Even for an A having an unsystematic effect on the trajectories - analogous, perhaps, to a super-imposed random walk - the particle separation statistics must be influenced to some degree. (A random walk will quite effectively disperse particles.)

Compare, for example, two otherwise identical float dispersal simulations with different horizontal resolution ($N = 64$ and 128). An ensemble of 80 floats is released in each at identical space/time locations. The long-term spaghetti diagrams are shown in Figures 7 and 8. There is clearly a significant tendency for the low-resolution floats, in which the A errors are largest, to move apart more rapidly than in the high-resolution case. This visual impression is confirmed by direct calculation of the two-particle dispersion statistics for each of these cases. The particles disperse significantly more rapidly in the lower resolution simulations. Estimates of particle diffusivity are roughly three times larger in the lower resolution simulation.

It should be noted that the existence of the aliasing term, and its effect on Lagrangian conservation, may not be of particular importance for applications in which the dynamics of the particle motions are not of interest. [For instance, it is perfectly reasonable to compute particle dispersion (as done here) by interpolation from an Eulerian ocean simulation, as long as the resulting trajectories are not thought of as arising from a particular dynamical model.] Nonetheless, the rapid accumulation of Lagrangian data in the ocean (e.g., [23]) offers the future possibility of direct comparison between numerical simulations and observations of neutrally buoyant or surface-drifting floats. In such comparisons, it will be necessary to recognize the likely importance of these computational effects, which have no counterpart in the continuum ocean.

Acknowledgments. The author's application of spectral modeling techniques to ocean circulation processes is sponsored by the National Science Foundation (Grant No. OCE 81-9486) and the Office of Naval Research (Contract No. N00014-C-0019, NR083-004). Dr. J. C. McWilliams kindly provided the results summarized in Figure 2. Some of the computer simulations were conducted at the National Center for Atmospheric Research which is also supported by the National Science Foundation.

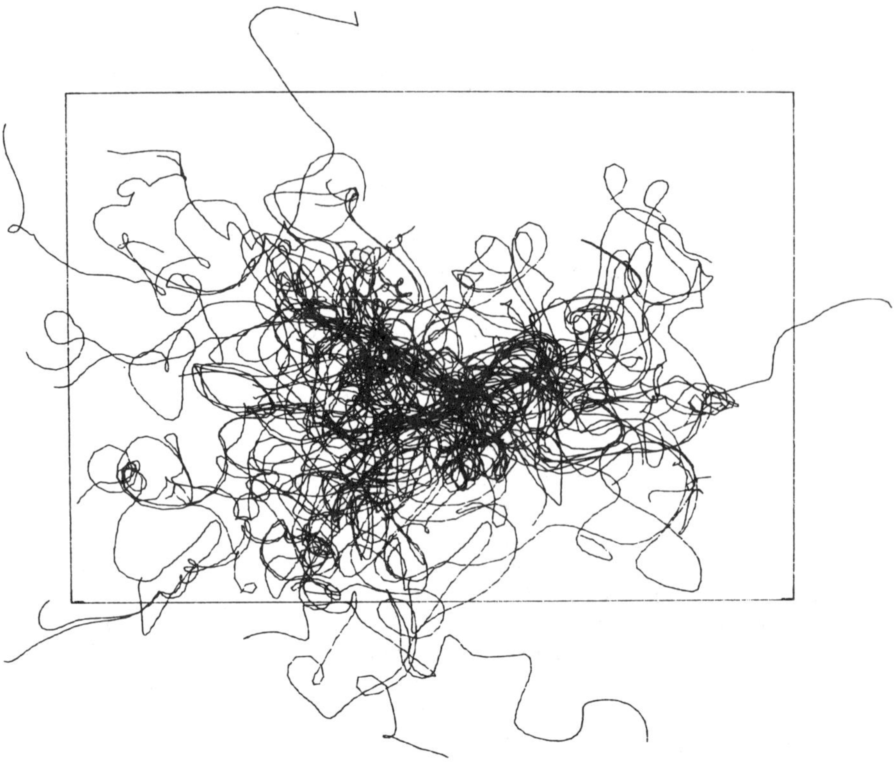

Figure 8

As in Figure 7, except with reduced horizontal resolution (N = 64). The perimeter drawn here is identical to that in Figure 7.

REFERENCES

[1] MODE Group, The mid-ocean dynamics experiment, Deep-Sea Res. 25, (1978), pp. 859-910.

[2] P. B. Rhines and W. R. Holland, A theoretical discussion of eddy-driven mean flows, Dyn. Atm. Oceans 3, (1979), pp. 289-325.

[3] A. R. Robinson, editor, Eddies in Marine Science, Springer-Verlag, New York, 1982.

[4] A. R. Robinson, D. E Harrison and D. B. Haidvogel, Mesoscale eddies and general ocean circulation models, Dyn. Atm. Oceans 3, (1979), pp. 143-180.

[5] W. J. Schmitz, Jr. and W. R. Holland, A preliminary comparison of selected eddy-resolving observations, J. Mar. Res. 40, (1982), pp. 75-117.

[6] D. B. Haidvogel, Periodic and regional models, In: Eddies in Marine Science, A. R. Robinson, editor, Springer-Verlag, New York, 1982.

[7] F. P. Bretherton and D. B. Haidvogel, Two-dimensional turbulence above topography, J. Fluid Mech. 78, (1976), pp. 129-154.

[8] D. B. Haidvogel, Quasigeostrophic regional and general circulation modeling: an efficient pseudospectral approximation technique, In: Computing Methods in Geophysical Mechanics, American Society of Mechanical Engineers AMD - Volume 25, New York, 1977.

[9] D. B. Haidvogel and P. B. Rhines, Waves and circulation driven by oscillatory winds in an idealized ocean basin, in preparation.

[10] D. B. Haidvogel, A. R. Robinson and E. E. Schulman, The accuracy, efficiency, and stability of three numerical models with application to open ocean problems, J. Comp. Phys. 34(1), (1980), pp. 1-53.

[11] A. R. Robinson and D. B. Haidvogel, Dynamical forecast experiments with a barotropic open ocean model, J. Phys. Oceanogr. 10, (1980), pp. 1909-1928.

[12] J. R. Herring, S. A. Orszag, R. H. Kraichnan and D. G. Fox, Decay of two-dimensional homogeneous turbulence, J. Fluid Mech. 66, (1974), pp. 417-443.

[13] P. B. Rhines, The dynamics of unsteady currents, In: The Sea, E. D. Goldberg, I. N. McCave, J. J. O'Brien and J. H. Steele (editors), Volume 6, Marine Modelling, Wiley, New York, (1977).

[14] G. Veronis, Dynamics of large-scale ocean circulation, In: Evolution of Physical Oceanography, B. A. Warren and C. Wunsch, editors, M.I.T. Press, Cambridge, MA, 1981.

[15] R. H. Kraichnan, Inertial ranges in two-dimensional turbulence, Phys. Fluids 3, (1967), pp. 1417-1421.

[16] R. H. Kraichnan, Inertial range transfer in two- and three-dimensional turbulence, J. Atm. Sci. 28, (1971), pp. 145-161.

[17] S. A. Orszag, Numerical simulation of incompressible flows within simple boundaries. I. Galerkin (Spectral) representations, Stud. App. Math. 50, (1971), pp. 293-327.

[18] M. K. Davey, Numerical Calculation of the Spectral of Representation of u • ∇ x for Doubly Periodic Flow, University of Washington, Department of Oceanography, Special Report No. 93, Seattle, WA, 1980.

[19] A. F. Bennett and D. B. Haidvogel, Low resolution numerical simulation of decaying two-dimensional turbulence, J. Atm. Sci., submitted for publication.

[20] S. A. Orszag, Turbulence and transition: a progress report, In: Proceedings of the Fifth International Conference on Numerical Methods in Fluid Dynamics, A. I. van de Vooren and P. J. Zandbergen, editors, Lecture Notes in Physics, No. 59, Springer-Verlag, New York, 1973.

[21] P. B. Rhines, Waves and turbulence on a β-plane, J. Fluid Mech. 69, (1975), pp.] 417-443.

[22] P. B. Rhines, Geostrophic turbulence, Ann. Rev. Fluid Mech. 11, (1979), pp. 401-441.

[23] J. F. Price, Particle dispersion in the western North Atlantic, in preparation.

[24] G. Holloway, Personal Communication.

[25] C.J.R. Garrett, On the streakiness of a dispersing tracer in two- and three-dimensional turbulence, J. Fluid Mech., submitted for publication.

[26] W. R. Young, P. B. Rhines and C.J.R. Garrett, Shear flow dispersion, internal waves and horizontal mixing in the ocean, J. Fluid Mech., submitted for publication.

[27] T. Keffer and D. B. Haidvogel, Numerical simulations of tracer streakiness, in preparation.

[28] D. B. Haidvogel, Particle motions in a turbulent ocean, in preparation.

ITERATIVE SPECTRAL METHODS AND SPECTRAL SOLUTIONS TO COMPRESSIBLE FLOWS

M. YOUSUFF HUSSAINI* AND THOMAS A. ZANG**

Abstract. A spectral multi-grid scheme is described which can solve pseudospectral discretizations of self-adjoint elliptic problems in $O(N \log N)$ operations. An iterative technique for efficiently implementing semi-implicit time-stepping for pseudospectral discretizations of Navier-Stokes equations is discussed. This approach can handle variable coefficient terms in an effective manner. Pseudospectral solutions of compressible flow problems are presented. These include one-dimensional problems and two-dimensional Euler solutions. Results are given both for shock-capturing approaches and for shock-fitting ones.

1. Introduction. Spectral methods have clear advantages provided that the discrete spectral equations can be solved efficiently and that the solution to the continuous problem is well behaved ([1], [2]). Efficient direct solutions of spectral equations are generally possible only for simple geometries and with explicit spectral equations, although a few constant coefficient cases can be solved cheaply by direct methods. In other circumstances iterative methods are necessary. The key approach of approximating the full spectral operator with a sparse finite difference one was developed by Orszag [3] and by Morchoisne [4]. The first half of this paper will describe some recent progress on iterative schemes for elliptic problems and on iterative solutions of semi-implicit, time-stepping procedures for Navier Stokes equations. The second part of the paper will describe some recent progress that has been made on the application of spectral methods to compressible flow problems with shock waves.

2. Spectral Multi-Grid Methods. The self-adjoint elliptic equation

$$-\nabla \cdot (a \nabla u) = f, \tag{2.1}$$

*Institute for Computer Applications in Science and Engineering, NASA Langley Research Center, Hampton, VA 23665, NASA Contracts No. NAS1-15810 and NAS1-16394.

**College of William and Mary, Williamsburg, VA 23185, NASA Grant NAG1-109.

where $u(\underline{x})$ is the solution, $f(\underline{x})$ the forcing and $a(\underline{x})$ the variable coefficient, arises in many contexts. In two or more dimensions direct "fast Poisson solvers" [5] are not applicable, even for the simplest discretizations and geometries. Perhaps the most efficient iterative schemes for finite difference and finite element discretizations of these problems employ multi-grid techniques ([6] - [8]). Theoretical estimates indicate that satisfactory convergence can be achieved in $O(N)$ arithmetic operations, where N is the total number of grid points. Zang, Wong, and Hussaini [9] have recently devised effective multi-grid procedures for the solution of pseudospectral discretizations of equation (2.1). A summary of that work follows along with some new developments.

For simplicity and clarity, the one-dimensional version of equation (2.1) with periodic boundary conditions on $[0,2\pi]$ will be used to explain the spectral multi-grid (SMG) approach. Write the Fourier pseudospectral discretization using N collocation, or grid, points as

$$LV = F \tag{2.2}$$

in obvious notation. Consider first a standard single-grid scheme employing Richardson relaxation

$$v \leftarrow v + \omega(F - Lv), \tag{2.3}$$

where v is the latest approximation to V and ω is a relaxation parameter. Label the real and positive eigenvalues of L as $\lambda_1, \lambda_2, \cdots, \lambda_N$ in order of increasing magnitude. The error at any stage, $v - V$, can be resolved into an expansion in the eigenvectors of L. Each iteration reduces the error component corresponding to λ_j to $\nu(\lambda_j)$ times its previous value, where

$$\nu(\lambda) = 1 - \omega\lambda. \tag{2.4}$$

The optimal choice of ω results from minimizing $|\nu(\lambda)|$ for $\lambda \in [\lambda_1, \lambda_N]$. This produces an optimal single-grid spectral radius

$$\rho = (\lambda_N - \lambda_1)/(\lambda_N + \lambda_1). \tag{2.5}$$

The efficiency of this single-grid approach is evident in the constant coefficient case $a(x) = 1$. The eigenvalues $\lambda_j = \lfloor (j/2) \rfloor^2$ and thus $\rho \cong 1 - 8/N^2$. (The fact that $\lambda_1 = 0$ is associated with the non-uniqueness of this special problem. One should really use λ_2 in place of λ_1 in equation (2.5) for this case.) This implies that $O(N^2)$ iterations are required to achieve convergence. Combined with the $O(N \log N)$ cost per iteration (due to the spectral evaluation of Lv) this produces a total cost of $O(N^3 \log N)$.

The spectral multi-grid approach can obtain the solution in $O(N \log N)$ operations since the number of iterations turns out to be independent of N. This is explained in what follows. Define a series of grids (or levels) for $k = 2,3,\cdots,K$, each consisting of

N_k uniformly spaced points, where $N_k = 2^k$. The solution to equation (2.2) is obtained by combining Richardson iterations on level K with Richardson iterations for related problems on the coarser levels $k < K$. Denote the relevant discrete problem at any level k by

(2.6) $$L_k V_k = F_k.$$

On the finest level K, $L_K = L$, $F_K = F$ and the solution $V_K = V$, the solution to equation (2.2). At any stage in the iterative solution process for equation (2.6), only an approximation v_k to the exact answer V_k is available. If this approximation is deemed adequate, then the approximation on the next-finer level $k+1$ is corrected via

(2.7) $$v_{k+1} \leftarrow v_{k+1} + P_{k+1} v_k.$$

The matrix P_k represents the coarse-to-fine interpolation of corrections from level $k-1$ to level k. On the other hand if the approximation v_k is deemed inadequate, either another relaxation is performed, via

(2.8) $$v_k \leftarrow v_k + \omega_k (F_k - L_k v_k),$$

or else control shifts to a problem on the next-coarser level $k-1$. The relaxation parameter ω_k on level k is chosen to damp preferentially those error components which are not represented on coarser grids. The right-hand-side of the coarser grid problem is obtained from

(2.9) $$F_{k-1} = R_k (F_k - L_k v_k).$$

The matrix R_k represents the fine-to-coarse residual transfer from level k to level $k-1$.

The natural interpolation operators in the present context represent trigonometric interpolation. They have the useful property that R_k is the adjoint of P_k. Numerous multi-grid investigations have determined that it is desirable for the coarse grid discretization operators to satisfy

(2.10) $$L_{k-1} = R_k L_k P_k.$$

This can be implemented approximately but efficiently by modifying the usual pseudo-spectral computation of $(av_x)_x$. On the finest level $k = K$ the pointwise values of $a(x)$ are retained but on the coarser levels $\tilde{a}(x)$ is used instead, where $\tilde{a}(x)$ is obtained via trigonometric interpolation on the finest grid values of $a(x)$. This filtering of the coefficients may be viewed as a de-aliasing procedure. Further details on the interpolation operators are given in [9].

An essential part of any multi-grid algorithm is a specific control structure that determines when attention is shifted from one grid to another. See [7] for some flow charts and [10] for additional variations and assessments.

To appreciate why the number of iterations is independent of $N = N_K$, consider the eigenvalues of the constant coefficient case. The objective of the relaxation scheme on level K is to minimize $|\nu(\lambda)|$ only for $\lambda \in [(1/16)N^2,(1/4)N^2]$. The upper bound on $|\nu(\lambda)|$ for this minimax problem is called the smoothing rate and is denoted by μ. A simple calculation reveals that $\mu = 3/5$. This is substantially less than 1 and, perhaps more importantly, it is independent of N. A similar result obtains for the two-dimensional version as indicated in Table 1 for several $n \times n$ grids. Of course, the work on the coarser levels $k < K$ should also be counted. The relaxations there are much cheaper and the same smoothing rate applies. A more subtle issue is whether the various interpolations magnify some error components. The numerical results in [9] suggest that this effect is relatively insignificant.

Table 1. Convergence Rates for Fourier Richardson Iteration in Two-dimensions

n	Single-grid Spectral Radius	Multi-grid Smoothing Rate
4	0.333	0.333
8	0.895	0.636
16	0.980	0.719
32	0.996	0.751
64	0.999	0.765
∞	1.000	0.778

When Dirichlet boundary conditions are applied to equation (2.1), Chebyshev polynomials are more appropriate than Fourier series. The additional complication here is that $\lambda_N = O(N^4)$ whereas $\lambda_{N/2} = O(N^2)$. The ratio of these two numbers determines the smoothing rate. (This ratio may be termed the multi-grid condition number.) Since this multi-grid condition number grows dramatically with N, the straightforward use of Richardson iteration leads to a smoothing rate which tends rapidly to 1.

The cure is to pre-condition the iteration by applying

$$v \leftarrow v + \omega H^{-1}(F-Lv), \tag{2.11}$$

where H is some readily invertible approximation to L. An obvious choice for H is a finite difference approximation H_{FD} to equation

(2.1) as proposed in [3] and [4] for single-grid iterations. However, in more than one-dimension these finite difference approximations are themselves costly to invert. A more desirable choice is an approximate LU-decomposition of H_{FD}, i.e., H is taken as the product of a lower triangular matrix L and an upper triangular matrix U. In one such type of preconditioning, denoted by H_{LU}, L is identical to the lower triangular portion of H_{FD} and U is chosen so that the two super diagonals of LU agree with those of H_{FD}. In [9] an alternative pre-conditioning, denoted here by H_{RS}, was proposed in which the diagonal elements of L were altered from those of H_{FD} to ensure that the row sums of H_{RS} and H_{FD} were identical.

Table 2. Extreme Eigenvalues for Pre-conditioned Chebyshev Operator in Two-dimensions

n	$H_{FD}^{-1}L$		$H_{LU}^{-1}L$		$H_{RS}^{-1}L$	
	λ_{min}	λ_{max}	λ_{min}	λ_{max}	λ_{min}	λ_{max}
4	1.000	1.757	0.929	1.717	1.037	1.781
8	1.000	2.131	0.582	2.273	1.061	2.877
16	1.000	2.305	0.224	2.603	1.043	4.241
24	1.000	2.361	0.111	2.737	1.031	5.379

The essential properties of all three types of pre-conditioning are shown in Table 2 for the constant coefficient problem. The total number of grid points $N = n^2$. The eigenvalues shown there were computed numerically. In order to assess the effectiveness of these pre-conditionings in multi-grid calculations, one also needs to know the smallest "high frequency" eigenvalue. The numerical results indicate that this is 1.22 for H_{FD} and H_{LU} and 1.45 for H_{RS}, essentially independent of n. The relevant condition numbers are given in Table 3. Both H_{LU} and H_{RS} require only $O(N)$ operations to invert. Thus, we reach the striking conclusion that although H_{RS} is more effective for single-grid iterations, H_{LU} is noticeably superior in the multi-grid context. Moreover, H_{FD} offers little improvement in smoothing rate over H_{LU}. Since H_{FD} is much costlier to invert, H_{LU} is the preferred multi-grid pre-conditioning. Table 4 indicates the convergence rates which can be obtained with the approximate LU-decomposition. Numerical evidence suggests that λ_{max} and hence the number of multi-grid iterations grows slowly as $N^{1/16}$ [11].

The tables in this section indicate the theoretical performance of SMG on constant coefficient problems. The numerical calculations reported in [9] confirm that these convergence rates are achieved in practice, even for variable coefficient problems. Recent calculations for the H_{LU} pre-conditioning confirm that it, too, behaves as predicted [11].

Table 3. Condition Number for Pre-conditioned Chebyshev Operator in Two-dimensions

n	Single-grid		Multi-grid	
	$H_{LU}^{-1}L$	$H_{RS}^{-1}L$	$H_{LU}^{-1}L$	$H_{RS}^{-1}L$
4	1.85	1.72		
8	3.91	2.71	1.79	2.07
16	11.62	4.07	2.12	2.92
24	24.66	5.22	2.26	3.79

Table 4. Convergence Rates for Chebyshev Richardson Iteration in Two-dimensions

n	Single-grid Spectral radius	Multi-grid Smoothing Rate
4	0.264	0.298
8	0.462	0.283
16	0.605	0.358
24	0.678	0.387

Spectral multi-grid methods can certainly be applied to a wider set of problems than covered by equation (2.1) with periodic or Dirichlet boundary conditions. Effective methods exist for other boundary conditions, such as Dirichlet in one direction and periodic in the other. Non-self-adjoint and nonlinear problems, including systems of equations, can also be handled. Results for single grid calculations will be presented elsewhere ([12]).

3. Semi-implicit Time-stepping Methods for Navier-Stokes Equations. An important source of implicit variable coefficient spectral equations is semi-implicit time-stepping algorithms for evolution equations with spectral spatial discretizations. Efficient iterative schemes are especially needed for Chebyshev spectral methods due to their severe explicit time-step limitations and the expense of direct solutions of the implicit equations in all but the simplest cases.

The incompressible Navier-Stokes equations are an important application. The rotation form equations for two-dimensional channel flow are

$$u_t - v(v_x - u_y) + P_x = (\mu u_x)_x + (\mu u_y)_y \tag{3.1}$$

$$v_t + u(v_x - u_y) + P_y = (\mu v_x)_x + (\mu v_y)_y \tag{3.2}$$

$$u_x + v_y = 0, \tag{3.3}$$

with periodic boundary conditions in x and no-slip boundary conditions at $y = \pm 1$. The variable P denotes the total pressure. The viscosity μ is presumed to depend upon y.

A useful discretization employs Fourier series in x and Chebyshev series in y. The pressure gradient term and the incompressibility constraint are best handled implicitly. So, too, are the vertical diffusion terms because of the fine mesh-spacing near the wall. The variable viscosity prevents the standard Poisson equation for the pressure from decoupling from the velocities in the diffusion term. The algorithm described in [13] appears to be a good starting point. A Crank-Nicolson approach is used for the implicit terms and Adams-Bashforth for the remainder. After a Fourier transform in x, the equations for each wavenumber k have the following implicit structure

$$\hat{u} - \tfrac{1}{2}\Delta t(\mu\hat{u}_y)_y + \tfrac{1}{2}\Delta t ik\hat{P} = \cdots \tag{3.4}$$

$$\hat{v} - \tfrac{1}{2}\Delta t(\mu\hat{v}_y)_y + \tfrac{1}{2}\Delta t\hat{P}_y = \cdots \tag{3.5}$$

$$ik\hat{u} + \hat{v}_y = 0. \tag{3.6}$$

Fourier transformed variables are denoted by hats, the subscript y denotes a Chebyshev pseudospectral derivative, and Δt is the time increment.

The algorithm in [13] was devised for constant viscosity, in which case the equations (3.4) - (3.6) can be reduced to essentially a block-tridiagonal form. This cannot be done in the present, more general situation. We advocate solving these equations iteratively after applying a finite difference pre-conditioning.

The interesting physical problems have high Reynolds number, i.e., low viscosity. Thus the first derivative terms in equations (3.4) - (3.6) predominate. The effective pre-conditioning of them is crucial. Four possibilities have been considered. The eigenvalues of pre-conditioned iterations for the model scalar problem $u_x = f$ with periodic boundary conditions on $[0,2\pi]$ are given for each possibility in Table 5. The term $\alpha\Delta x$ is the product of a wavenumber α and the grid spacing Δx. It falls in the range $0 \leq |\alpha\Delta x| \leq \pi$. For the staggered grid case the discrete equations (3.4) - (3.6) are modified so that the velocities and the momentum equations are defined at the cell faces $y_j = \cos(\pi j/N)$, $j=0,1,\cdots,N$, whereas the pressure and the continuity equation are defined at the cell centers $y_{j-1/2} = \cos(\pi(j-\tfrac{1}{2})/N)$, $j=1,\cdots,N$. Fast cosine transforms enable interpolation between cell faces and cell centers to be implemented efficiently. The staggered grid for the Navier-Stokes equations has the advantage that no artificial boundary condition is required for the pressure at the walls.

Table 5. Pre-conditioned Eigenvalues for One-dimensional First Derivative Model Problem

Pre-conditioning	Eigenvalues
Central Differences	$\frac{\alpha\Delta x}{\sin(\alpha\Delta x)}$
One-sided Differences	$e^{-i(\alpha\Delta x/2)} \frac{\alpha\Delta x/2}{\sin\left((\alpha\Delta x)/2)\right)}$
High Mode Cut-off	$\begin{cases} \frac{\alpha\Delta x}{\sin(\alpha\Delta x)} & 0 \leq \lvert\alpha\Delta x\rvert \leq (2\pi/3) \\ 0 & (2\pi/3) < \lvert\alpha\Delta x\rvert \leq \pi \end{cases}$
Staggered Grid	$\frac{(\alpha\Delta x)/2}{\sin\left((\alpha\Delta x)/2\right)}$

The actual eigenvalues for pre-conditioned iterations of equations (3.4) - (3.6) are displayed in Figures 1 and 2. The model problem estimates the eigenvalue trends surprisingly well considering that it is just a scalar equation, has only first derivative terms and uses Fourier series rather than Chebyshev polynomials.

The preceding results indicate that the staggered grid leads to the most effective treatment of the first derivative terms. The condition number of the pre-conditioned system is reasonably small and no resolution is lost by a high mode cut-off. (Although it is possible to devise a high-mode cut-off which avoids the small eigenvalues shown in the figures, some of the spectral resolution is thereby lost.) A simple and effective iterative scheme for this system with its complex eigenvalues is a minimum residual method. At a Reynolds number of 7500 each iteration reduces the residual by almost an order of magnitude.

This semi-implicit technique has several obvious extensions. It is easily applied to incompressible flow over a flat plate in the context of the parallel flow assumption. Pre-conditioned eigenvalues for this situation are shown in Figure 3. A substantial increase in the allowable time-steps can be achieved by treating the mean streamwise advection term in a semi-implicit fashion. This is easily implemented. Adding a third dimension with periodic boundary conditions is trivial, aside from storage and run-time considerations. Treating no-slip boundary conditions in two directions and/or including more of the advection terms in a semi-implicit manner is more

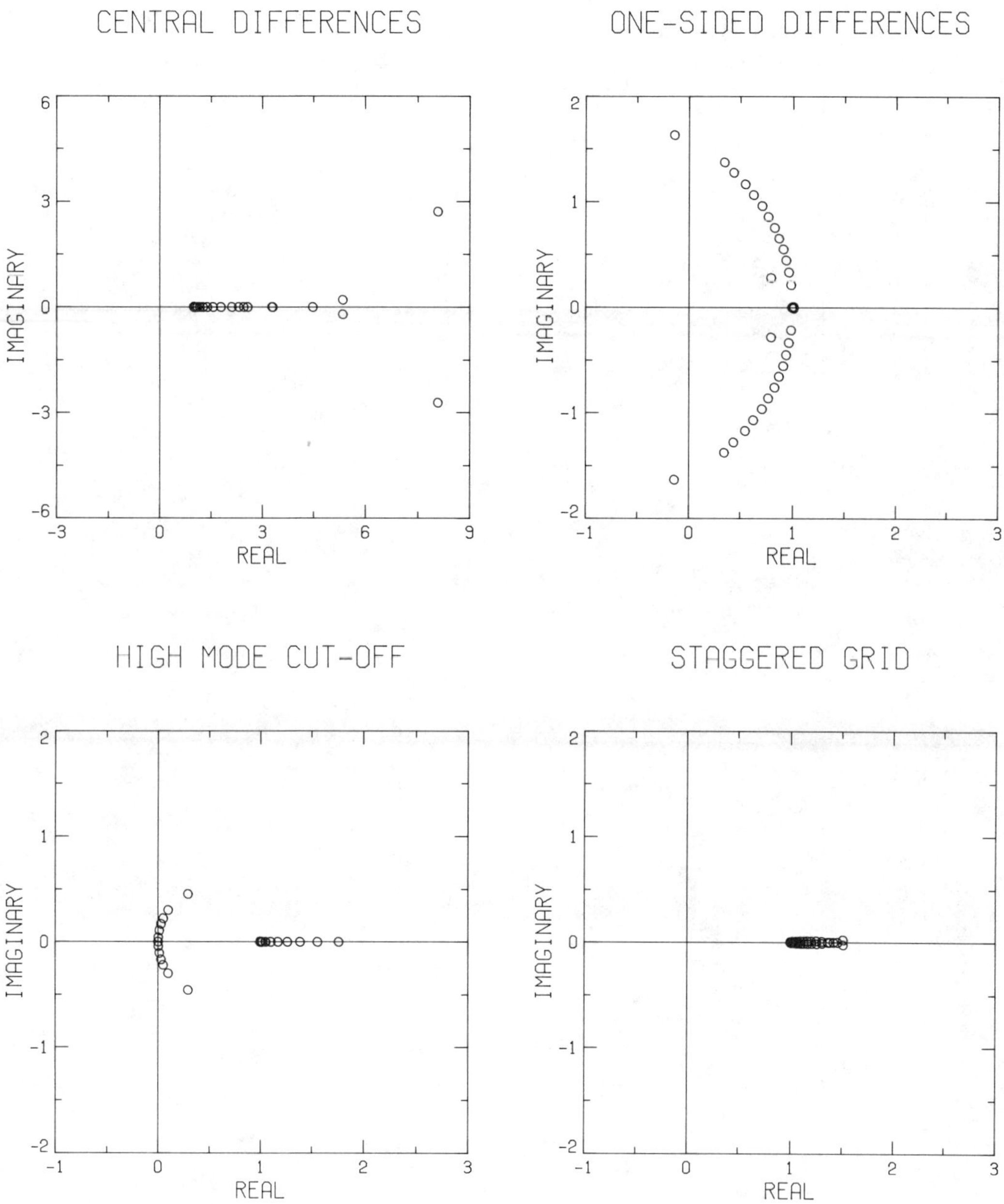

Figure 1. Eigenvalues of the pre-conditioned matrices for semi-implicit channel flow when the streamwise wave number k = 1. The grid is 32×17, the Reynolds number is 7500 and the CFL number is 0.10. Note the different scale used for the central differences pre-conditioning results.

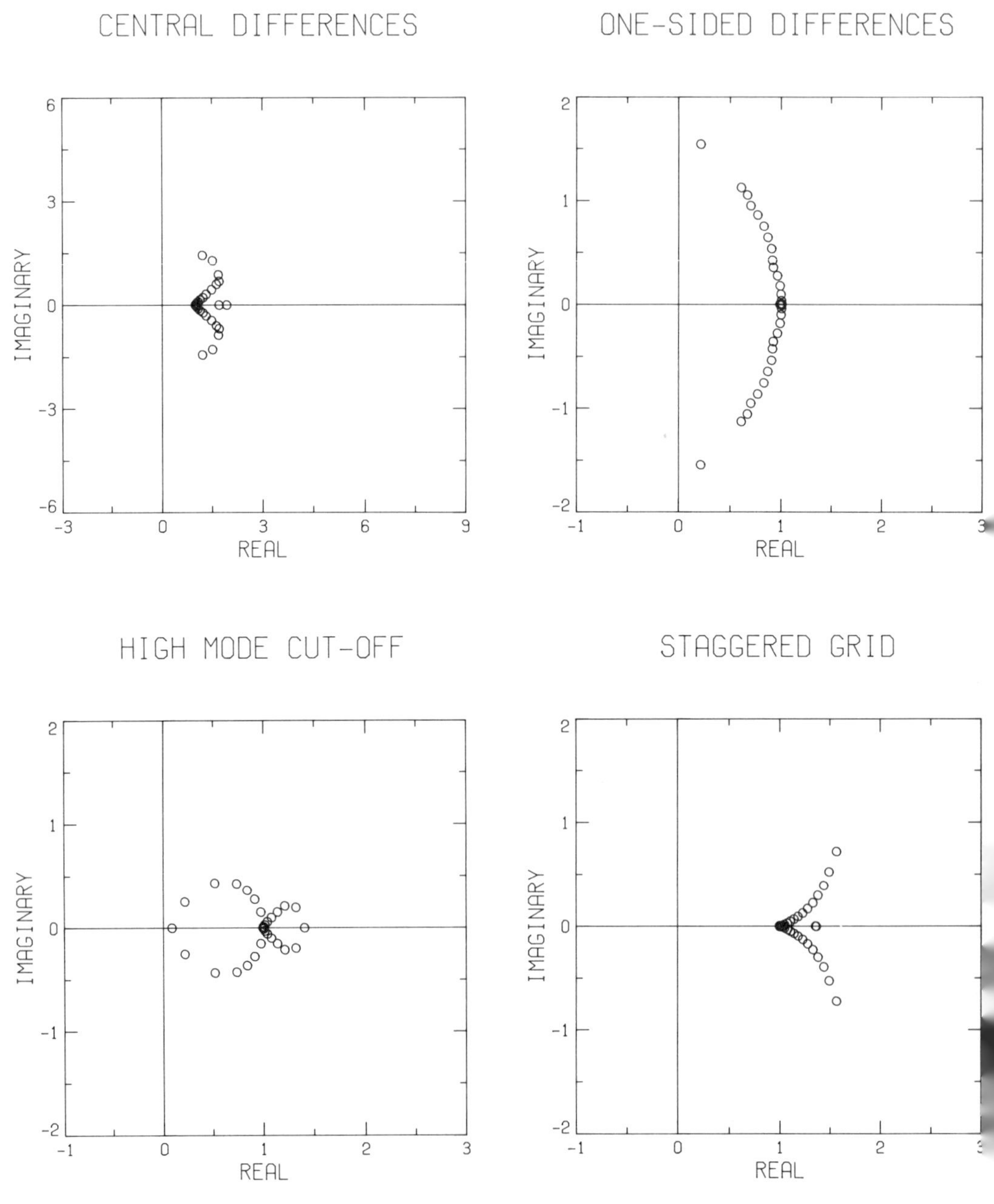

Figure 2. Eigenvalues of the pre-conditioned matrices for semi-implicit channel flow when the streamwise wave number k = 10. The grid is 32×17, the Reynolds number is 7500 and the CFL number is 0.10. Note the different scale used for the central differences pre-conditioning results.

difficult. Here, however, one can employ the approximate LU-decomposition described in section 2.

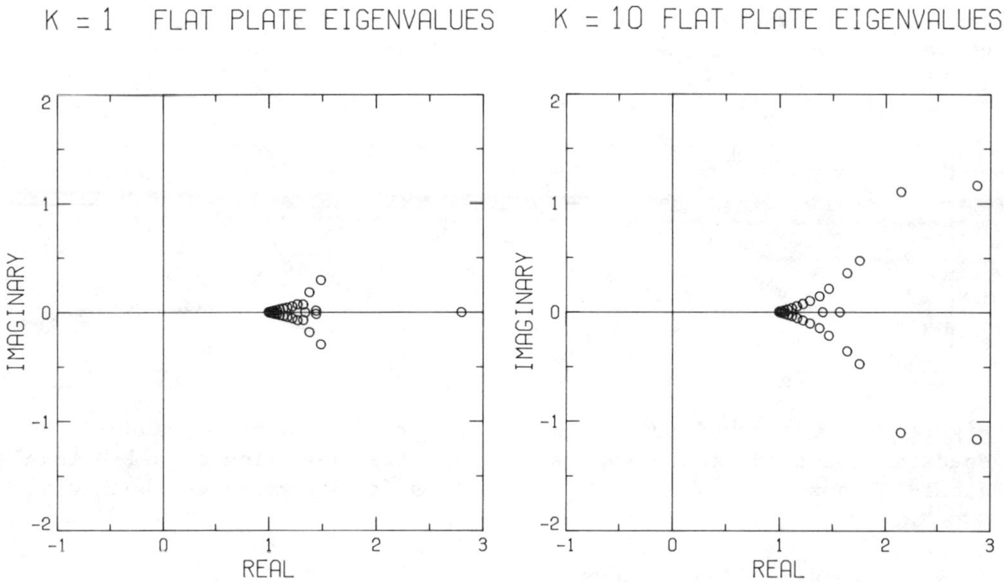

Figure 3. Eigenvalues of the staggered grid pre-conditioned matrices for semi-implicit flat plate flow. The grid is 32×17, the Reynolds number is 7500 and the CFL number is 0.10.

Further details are discussed in [14]. That report also contains numerical examples using production codes for the channel and flat plate problems in both constant viscosity and variable viscosity situations.

4. Pseudospectral Solutions to Compressible Flows.

4.1 Quasi-one-dimensional flows. Recent investigations ([15], [16],[17]) of one-dimensional problems indicate that spectral methods may provide a promising approach to compressible flows with shocks. In these calculations the shock wave is "captured" and a kind of filtering is applied to deal with the oscillations resulting from the sharp discontinuity. The goal of the filtering is to suppress the oscillations without degrading the accuracy in the smooth but structured regions of the flow. Simple flows such as those represented by piecewise linear profiles are not very demanding tests since the series representations of such functions forgive a number of filtering crimes.

Figures 4 and 5 reproduce the results presented in [15] for reasonably demanding flows. The first of these figures refers to the standard test problem of a quasi-one-dimensional nozzle flow. The second figure pertains to a rather unusual, but highly structured, astrophysical flow problem. The rapid decompression region behind the shock is especially challenging. Note that the computed shock is quite sharp and that the complex flow structure is preserved.

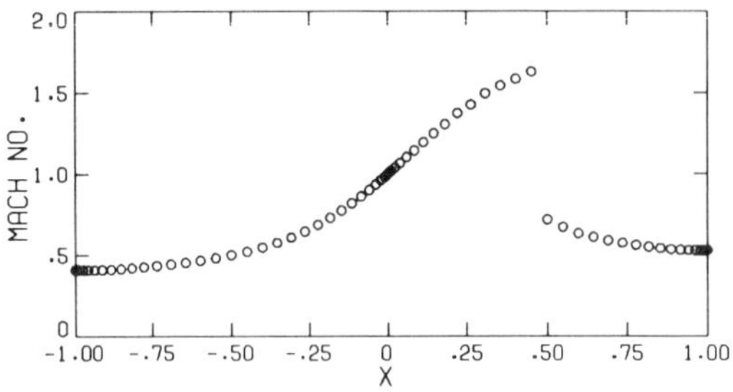

Figure 4. Chebyshev pseudo-spectral solution of transonic quasi-1-D nozzle flow.

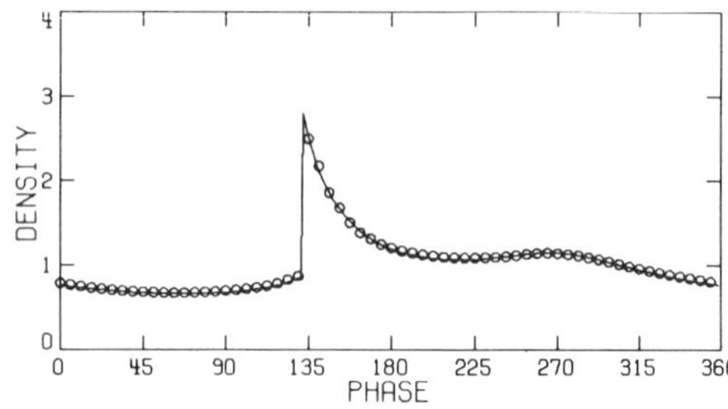

Figure 5. Fourier pseudo-spectral solution of a 1-D model of a forced galactic shock wave.

4.2 Two-dimensional flows.

Interaction of a shock wave with an entropy spot and a vortex. The two-dimensional compressible Euler equations present a more challenging test for the spectral methods. The shock "capturing" techniques of the one-dimensional flow problems are not as successful in the two-dimensional case, and the filtering methods presently used (to deal with the Gibbs phenomena) affect the accuracy of the solution. It stands to reason that while applying pseudospectral methods to complex shock interaction problems, the spectral accuracy can be maintained by "tracking" or fitting the shock. In such cases the relevant governing equations are not necessarily cast in conservation form, and they are solved in the transformed or computational domain where the shock becomes a coordinate line. The unsteady, two-dimensional, compressible, Euler equations in the computational plane (X, Y) are written in the form,

$$Q_T + \underline{A}Q_X + \underline{B}Q_Y = 0 \tag{4.1}$$

where $Q = [P,u,v,S]^T$ and

$$\underline{A} = \begin{bmatrix} U & \gamma X_x & \gamma X_y & 0 \\ a^2X_x/\gamma & U & 0 & 0 \\ a^2X_y/\gamma & 0 & U & 0 \\ 0 & 0 & 0 & U \end{bmatrix} \qquad \underline{B} = \begin{bmatrix} V & \gamma Y_x & \gamma Y_y & 0 \\ a^2Y_x/\gamma & V & 0 & 0 \\ a^2Y_y/\gamma & 0 & V & 0 \\ 0 & 0 & 0 & V \end{bmatrix} .$$

The natural logarithm of the pressure, the speed of sound, and the entropy are represented by P, a, and S, respectively, and γ is the ratio of specific heats. The velocity in the Cartesian x and y directions are u and v, respectively. All variables are normalized with respect to reference conditions at downstream infinity, as in [18]. The contravariant velocity components are defined by

$$U = X_t + uX_x + vX_y \quad \text{and} \quad V = Y_t + uY_x + vY_y .$$

Subscripts denote partial derivatives with respect to the independent variables.

The coordinate transformation is defined as follows:

$$X = \frac{x - h(t)}{x_s(y,t) - h(t)}$$

$$Y = \frac{\tanh(\alpha y) + 1}{2}$$

$$T = t,$$

where $x = h(t)$ is some left boundary of the interaction region and $x = x_s(y,t)$ is the shock wave front.

The computational domain is thus $(X,Y) \in [0,1] \times [0,1]$. Note the stretching (with parameter α) that has been used to handle the infinite extent of the lateral coordinate y. If the relative shock Mach number M_s is sufficiently high ($M_s > 2.08$), the flow upstream of the shock remains supersonic. In this case, the left boundary corresponds to a supersonic inflow, and all dependent variables can be prescribed on it. However, if the relative shock Mach number is low, then radiation-type boundary conditions are used at the left boundary. On the right, the computational region is bounded by the shock wave. Downstream of the shock the flow field is given analytically. The flow field immediately upstream of the shock, as well as the shape and velocity of the shock, are evaluated such that the Rankine-Hugoniot jump conditions and the compatibility condition reaching the shock wave from the upstream side are simultaneously satisfied.

Let k denote the time level and let Δt be the time step increment. The time discretization of eq. (4.1) is then as follows:

$$\tilde{Q} = [1 - \Delta t L^k] Q^k,$$

$$Q^{k+1} = \tfrac{1}{2} [Q^k + (1 - \Delta t \tilde{L}) \tilde{Q}],$$

where the spatial operator L represents an approximation to $\underline{A}\partial/\partial X + \underline{B}\partial/\partial Y$. In the pseudospectral method, the solution Q is first expanded as a double Chebyshev series,

$$Q(X,Y,T) = \sum_{p=0}^{M} \sum_{q=0}^{N} Q_{pq}(T)\tau_p(\xi)\tau_q(\eta),$$

where

$$\xi = 2X - 1 \quad \text{and} \quad \eta = 2Y - 1,$$

and τ_p and τ_q are the Chebyshev polynomials of degrees p and q. The derivatives appearing in the spatial operators are then evaluated as

$$Q_X = 2 \sum_{p=0}^{M} \sum_{q=0}^{N} Q_{pq}^{(1,0)} \tau_p \tau_q,$$

where

$$Q_{pq}^{(1,0)} = \frac{2}{c_p} \sum_{\substack{m=p+1 \\ m+p \text{ odd}}}^{M} mQ_{mq},$$

and

$$c_0 = 2, \quad c_p = 1, \; p > 0.$$

The evaluation of the shock wave shape and velocity followed the same procedure outlined above including spectral evaluation of the derivatives on the upstream side of the shock are expressed as Chebyshev expansions. At the left boundary, all variables were specified for supersonic inflow. For the case of subsonic inflow, the two velocity components and the entropy were specified, while the pressure was computed from a quasi-one-dimensional characteristic.

The pseudospectral method has a tendency to develop slowly growing oscillations. Because of the global nature of this method they are spread over the entire flow field rather than being confined to the vicinity of sharp gradients. The underlying smooth solution can be recovered by a variety of filtering techniques. The results presented here were obtained by applying a von Hann window filter (see [15] for details) every 160 time steps.

Figure 6 shows a plane shock wave about to interact with a hot spot (situated in a quiescent field) with the temperature distribution σ given by

$$\sigma = k \exp\{-[(x-x_0)^2 + (y-y_0)^2]/2r^2\},$$

where $k = 0.25$, $r = 1.25$, $x_0 = 0.5$ and $y_0 = 0$. The initial shock position is $x = 0$, and its initial Mach number is 3. Figure 7 displays vorticity contours at time $t = 0.2$ when the shock wave has passed over the hot spot. See [18] for more details on the physics and [19] for comparisons with finite difference calculations.

Figure 8 shows the velocity field for a single vortex about to interact with a shock wave traveling initially with speed $M_s = 3$. The downstream conditions here are obtained by assuming a constant

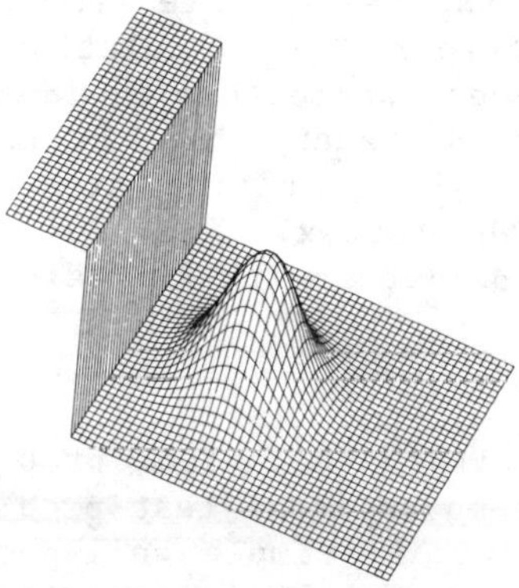

Figure 6. Surface plot of entropy for a hot spot and an initial advancing Mach 3 shock wave.

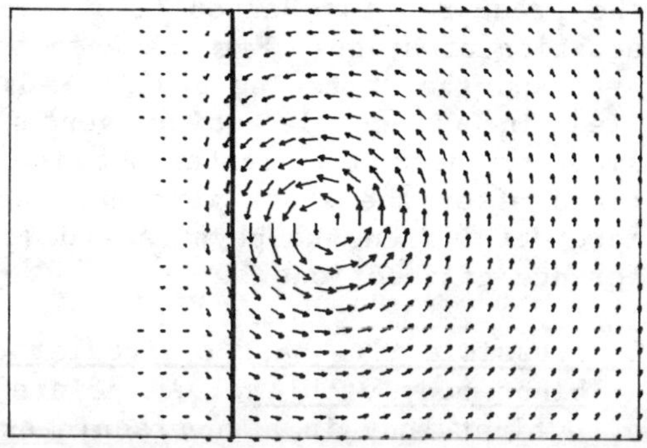

Figure 8. Velocity vectors for a vortex. Solid vertical line denotes an advancing Mach 3 shock wave.

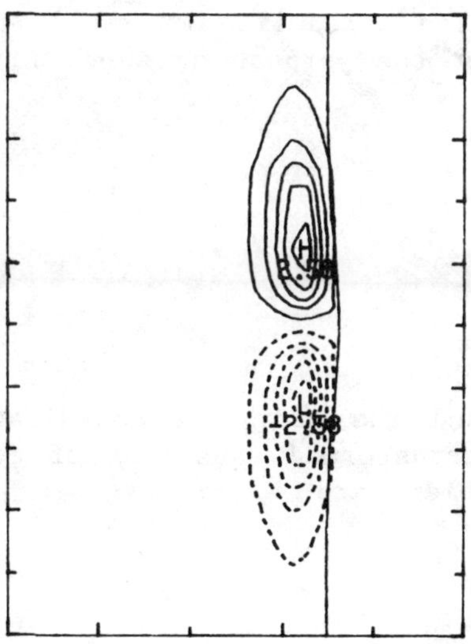

Figure 7. Vorticity contours from pseudospectral calculation for a hot spot after interaction with a Mach 3 shock wave (solid line).

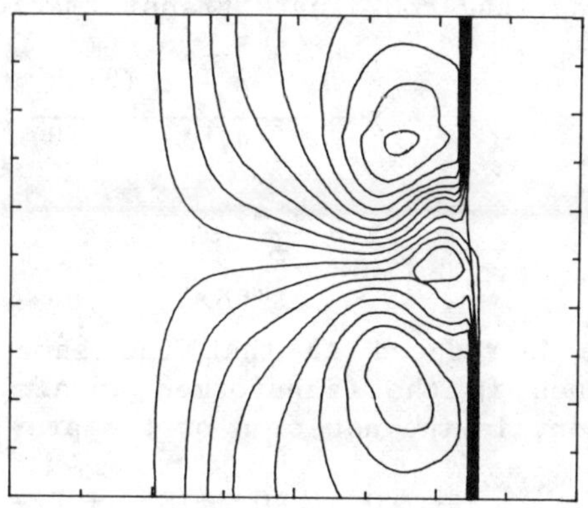

Figure 9. Pressure contours from pseudospectral calculation for a vortex after interaction with a Mach 3 shock wave.

density field, calculating the velocity from the stream function,

$$\phi = \frac{\kappa}{2\pi} \log \sqrt{r^2 + (x-x_0)^2 + (y-y_0)^2} ,$$

the pressure from Bernoulli's relation, and the temperature from the equation of state. For the case shown in Figure 8, the circulation $\kappa=2$ and the softening scale $r=0.1$. This model approaches an idealized incompressible point vortex at large distances but is much smoother near the center. Figure 9 shows the resulting pressure field after the shock wave has passed over the vortex. See [20] for more details on the physics and [19] for comparisons with finite difference calculations.

Compressible flow past a circular cylinder.

Blunt body problem. As pointed out in [21] the classical problem of a blunt body in a supersonic stream has been an ideal test problem for numerical methods as it provides a relatively simple well-posed transonic problem with nontrivial initial and boundary conditions. The present pseudospectral method like most common methods obtains the steady state solution as the time asymptotic solution of the unsteady Euler equations which are written in the cylindrical polar coordinate (r,θ) system. The physical domain of interest consists of the known body $r = r_b(\theta)$, the unknown shock location, $r = r_s(\theta,t)$, the axis of symmetry (the front stagnation streamline $\theta = \pi$) and the outflow boundary $\theta = \pi - \bar{\theta}_{max}$. For the purpose of shock fitting, the coordinate transformation

$$X = \frac{r - r_b(\theta)}{r_s(\theta,t) - r_b(\theta)}$$

$$Y = \frac{\pi - \theta}{\bar{\theta}_{max}}$$

is introduced so that the shock wave and the body are coordinate lines in the transformed domain. The transformed equations of motion, in the notation of the previous problem, are

$$Q_T + \underline{A}Q_X + \underline{B}Q_Y + R = 0,$$

where $Q = [P,u,v,S]^T$ and

$$\underline{A} = \begin{matrix} U & \gamma X_r & (\gamma/r)X_\theta & 0 \\ (a^2/\gamma)X_r & U & 0 & 0 \\ (a^2/\gamma)(1/r)X_\theta & 0 & U & 0 \\ 0 & 0 & 0 & U \end{matrix} ,$$

$$\underline{B} = \begin{bmatrix} V & \gamma Y_r & (\gamma/r)Y_\theta & 0 \\ (a^2/\gamma)Y_r & V & 0 & 0 \\ (a^2/\gamma)(1/r)Y_\theta & 0 & V & 0 \\ 0 & 0 & 0 & V \end{bmatrix},$$

and

$$R = \lfloor \gamma \frac{u}{r}, -\frac{v^2}{r}, \frac{uv}{r}, 0 \rfloor^T$$

with
$$U = X_r + uX_r + \frac{\gamma}{r} X_\theta$$

and
$$V = \frac{r}{r} Y_\theta.$$

The flow field variables P, u, v, and S are expanded in double Chebyshev series, and the solution technique is the same as for the previous problem.

The shock boundary $r = r_s(\theta,t)$ (i.e., $X = 1$) is computed using Rankine-Hugoniot jump conditions and the compatibility equation along the incoming characteristic from the high pressure side of the shock. At the symmetry line $\theta = \pi$ ($Y = 0$) the θ-component of velocity v is set equal to zero. On the body $r = r_b(\theta)$ (i.e., $X = 0$), the normal component of velocity, u, is zero. $\overline{\theta}_{max}$ is chosen so that the outflow boundary $Y = 1$ is supersonic, and hence no boundary conditions need be imposed.

Figure 10 shows the Mach number contours and the velocity vectors for a circular cylinder in a uniform stream at $M_\infty = 4$. The results are found to be in very good agreement with the tabulated values given in reference [22]. The coarse 8×8 mesh and the Chebyshev grid point distribution are evident in the velocity vector plot. Figure 11 displays the results for the linearly sheared free stream. This may be compared with Figure 12 where the finite difference results obtained on a 20×30 grid are shown.

Subsonic Flow Past a Circular Cylinder. In the case of a circular cylinder in the subsonic stream, it is expedient to map the infinite exterior domain onto the interior of a unit circle by the coordinate transformation

$$X = 1/r \qquad 0 \leq X \leq 1.$$

The dependent variables are then represented in terms of Chebyshev polynomials in X; Fourier representation in θ is appropriate as

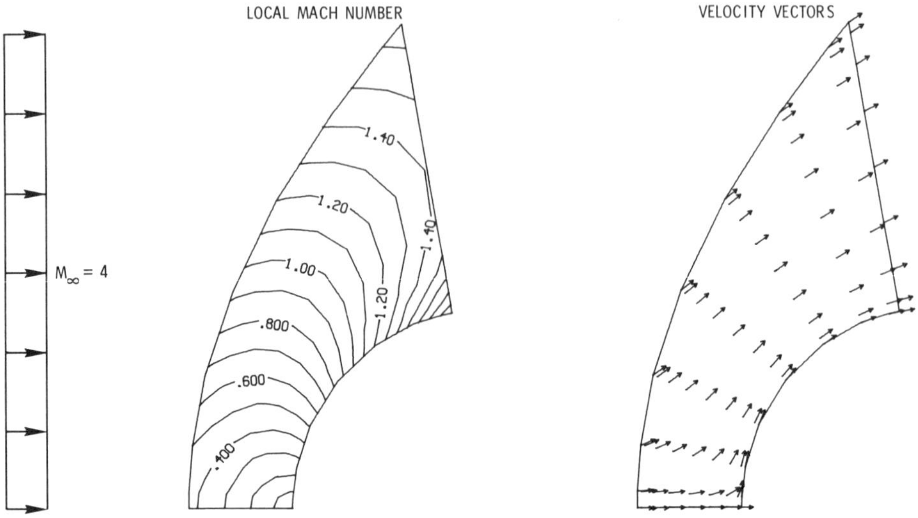

Figure 10. Pseudospectral solution on an 8×8 grid for a circular cylinder in a Mach 4 uniform stream.

the flow field is periodic with period 2π. However, one needs to consider only the interval $0 \leq \theta \leq \pi$ because of symmetry. For this semi-circle problem, the dependent variables can be expanded in terms of sine and cosine functions in θ; they may again be represented by Chebyshev polynomials in θ. These two different representations are found to yield practically identical results.

Figure 13 shows the Mach number contours for the flow past a circular cylinder at $M_\infty = 0.4$ computed by a finite difference technique (120×40 mesh) and the pseudospectral method (16×8). At this free stream Mach number the incipient critical flow is attained at the top of the cylinder. The results show very good agreement between the two numerical calculations. Further comparisons, other details and additional results for various free stream Mach numbers are reported in reference [23].

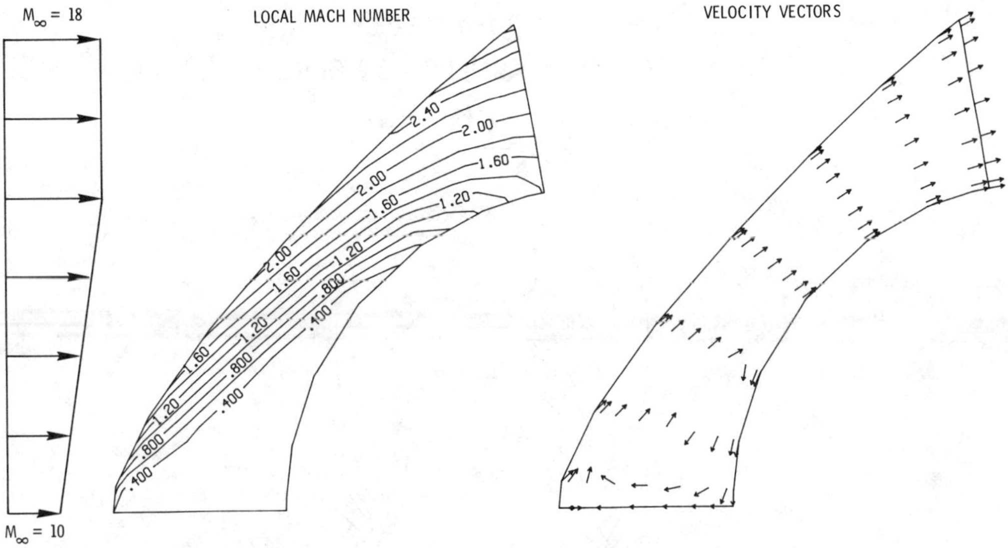

<u>Figure 11.</u> Pseudospectral solution on an 8×8 grid for a circular cylinder in a linearly sheared stream.

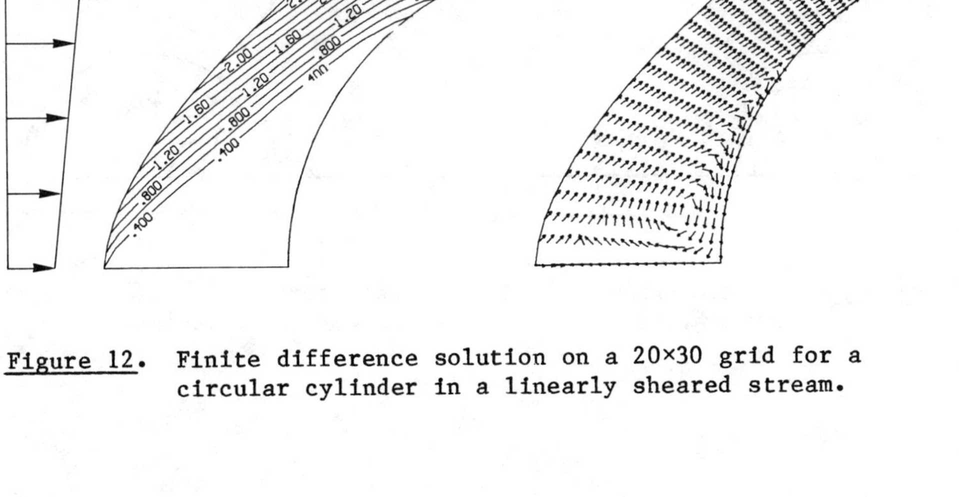

<u>Figure 12.</u> Finite difference solution on a 20×30 grid for a circular cylinder in a linearly sheared stream.

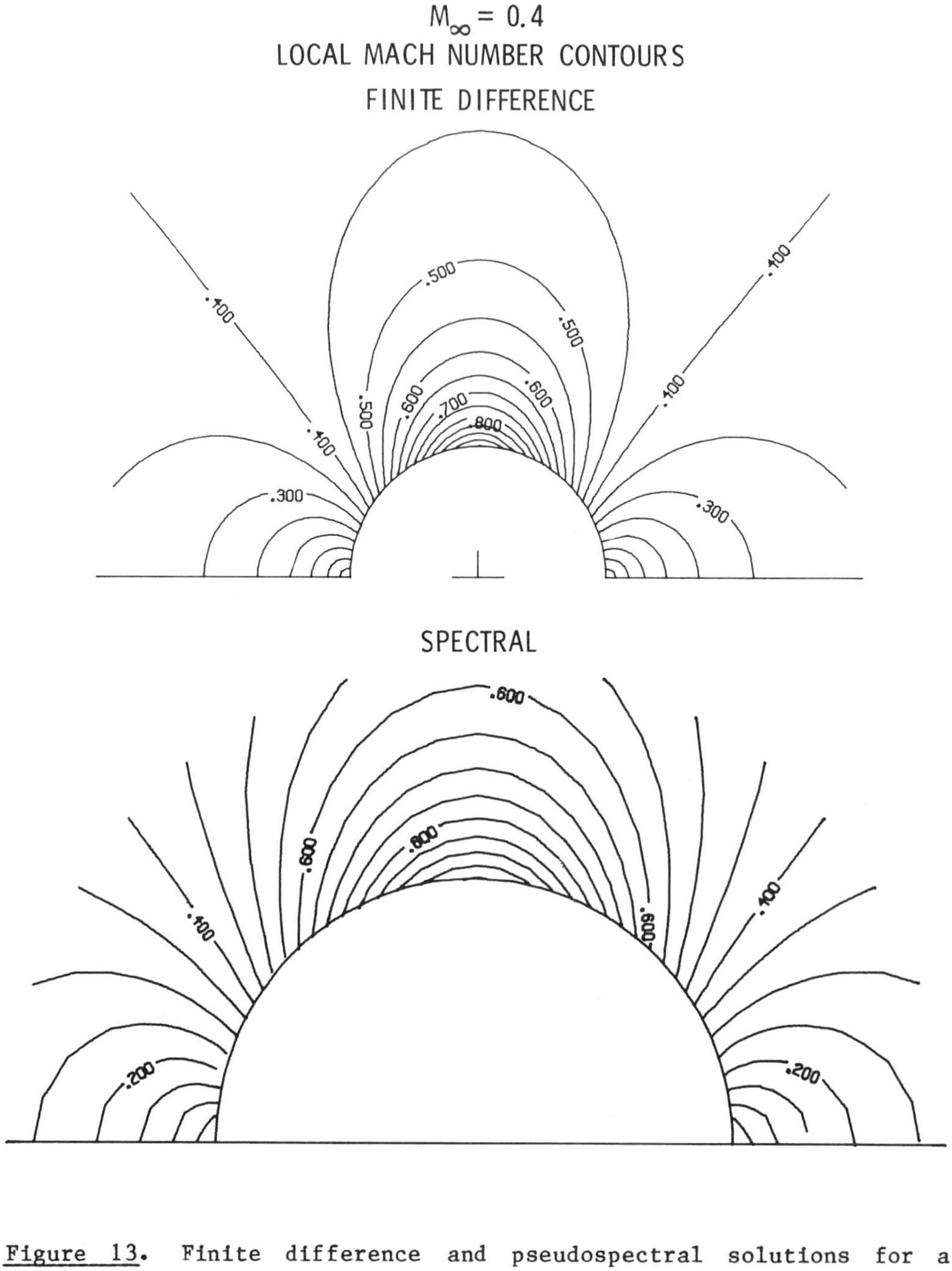

Figure 13. Finite difference and pseudospectral solutions for a circular cylinder in a Mach 0.4 uniform stream.

REFERENCES

[1] S. A. ORSZAG and M. ISRAELI, Numerical simulation of viscous incompressible flows, Ann. Rev. of Fluid Mechanics, 6, (1974), pp. 281-318.

[2] D. GOTTLIEB, and S. ORSZAG, Numerical Analysis of Spectral Methods: Theory and Applications, CBMS-NSF Regional Conference Series in Applied Mathematics, Society for Industrial and Applied Mathematics, 1977.

[3] S. A. ORSZAG, Spectral methods for problems in complex geometries, J. Comp. Phys., 37, (1980), pp. 70-92.

[4] Y. MORCHOISNE, Resolution of Navier Stokes equations by a space-time pseudo spectral method, La Recherche Aerospatial, 5, (1979), pp. 293-309.

[5] P. N. SWARTZTRAUBER, The methods of cyclic reduction, Fourier analysis and the FACR algorithm for the discrete solution of Poisson's equation on a rectangle, SIAM Rev., 19, (1977), pp. 490-501.

[6] R. P. FEDORENKO, A relaxation method for solving elliptic difference equations, Z. Vycisl. Mat. i Mat. Fix., 1, (1961), pp. 922-927.

[7] ACHI BRANDT, Multi-level adaptive solutions to boundary-value problems, Math. Comp., 31, (1977), pp. 333-390.

[8] R. A. NICOLAIDES, On multiple grid and related techniques for solving discrete elliptic systems, J. Comp. Phys., 19, (1975), pp. 418-431.

[9] T. A. ZANG, Y. S. WONG, and M. Y. HUSSAINI, Spectral multi-grid methods for elliptic equations, J. Comp. Phys., 48, (1982).

[10] H. LOMAX, editor, Multigrid Methods, NASA Conference Publication 2202, (1981).

[11] T. A. ZANG, Y. S. WONG, and M. Y. HUSSAINI, Spectral multi-grid methods for elliptic equations II,, to appear.

[12] Y. S. WONG, T. A. ZANG and M. Y. HUSSAINI, Efficient iterative solutions to spectral equations, to appear.

[13] P. MOIN and J. KIM, On the numerical solution of time-dependent viscous incompressible fluid flows involving solid boundaries, J. Comp. Physics, 35, (1980), pp. 381-392.

[14] M. R. MALIK, T. A. ZANG and M. Y. HUSSAINI, Efficient solution to semi-implicit spectral methods for Navier-Stokes equations, to appear.

[15] T. A. ZANG and M. Y. HUSSAINI, Mixed spectral-finite difference approximations for slightly viscous flows, Proc. of the 7th Intl. Conf. on Numerical Methods in Fluid Dynamics, (1981), pp. 461-466.

[16] D. GOTTLIEB, L. LUSTMAN and S. ORSZAG, Spectral calculations of one-dimensional inviscid compressible flows, SIAM J. Sci. Statis. Comput., 2, (1981), pp. 296-310.

[17] T. D. TAYLOR, R. B. MYERS and J. H. ALBERT, Pseudo-spectral calculations of shock waves, rarefaction waves and contact surfaces, Computers and Fluids, 9, (1981) pp. 469-473.

[18] T. A. ZANG, M. Y. HUSSAINI, and D. M. BUSHNELL, Numerical computations of turbulence amplification in shock wave interactions, AIAA paper 82-0293, Presented at the AIAA 20th Aerospace Sciences Meeting, January 11-13, (1982), Orlando, FL.

[19] M. D. SALAS, T. A. ZANG and M. Y. HUSSAINI, Shock-fitted Euler solutions to shock-vortex interactions, Proc. of the 8th Intl. Conf. on Numerical Methods in Fluid Dynamics, to appear.

[20] S. P. PAO and M. D. SALAS, A numerical study of two-dimensional shock vortex interaction, AIAA paper 81-1205, Presented at the AIAA 14th Fluids and Plasma Dynamics Conference, June 23-25, (1981), Palo Alto, CA.

[21] M. D. SALAS, Flow properties for a spherical body at low supersonic speeds, Presented at the Symp. on Computers in Aerodynamics, Polytechnique Institute of New York, June (1979).

[22] A. N. LYUBIMOV and V. V. RUSANOV, Gas flows past blunt bodies, (Translation of "Techeniya Gaza Okolo Tupykh Tel, Chast′ II: Tablitsy Termodinamicheskikh Funktsiy," "Nauka" Press, Moscow, 1970), NASA TT F-715, February (1973).

[23] M. Y. HUSSAINI, M. D. SALAS, and T. A. ZANG, Pseudospectral solution to compressible Euler equations, to appear.

SPECTRAL SIMULATIONS OF THE LAMINAR-TURBULENT TRANSITION PROCESS IN PLANE POISEUILLE FLOW +

L. KLEISER* AND U. SCHUMANN**

Abstract. Laminar-turbulent transition in plane Poiseuille flow is simulated using the Fourier-Chebyshev spectral method to integrate the three-dimensional time-dependent Navier-Stokes equations. For pressure computation, a new algorithm has been developed which enforces incompressibility and boundary conditions exactly in the discretised equations. This algorithm is not restricted to special geometries or numerical discretisation techniques. Detailed comparisons of numerical results have been made with vibrating-ribbon experiments of Nishioka et al., and very satisfactory agreement has been found up to the "spike" stage of the transition process. Visualisations of the three-dimensional flow structures developing in transition at subcritical and supercritical Reynolds numbers are presented.

+Work performed at Kernforschungszentrum Karlsruhe, Institut für Reaktorentwicklung

*Now with DFVLR, Institut für Theoretische Strömungsmechanik, 3400 Göttingen, Germany

**Now with DFVLR, Institut für Physik der Atmosphäre, 8031 Oberpfaffenhofen, Germany

1. Introduction.

1.1 Overview. Although spectral methods have a long history, their efficient use for complex nonlinear problems is a relatively new field. In the past decade, Orszag and coworkers have pioneered their application and introduced many fundamental ideas; for references see {1} and the other contributions of this volume. The basic principle of this numerical technique is to expand dependent variables in a finite series of known global, smooth functions. Usually the unknown coefficients are determined by some weighted residual method. The main advantage of spectral methods over finite difference or finite element methods is their high, often exponential convergence order. This makes their use particularly attractive for multi-dimensional problems with high accuracy requirements such as arise in computational fluid dynamics.

Spectral methods have been applied by the present authors to several problems including homogeneous turbulence {2}, shell-fluid interactions {3} and laminar-turbulent transition {4-6}. An overview over some earlier results has been given in {7}. In the present paper, we report on the investigation of the laminar-turbulent transition process in plane Poiseuille flow. For this the computer code CHANSON has been developed which is based on Fourier and Chebyshev polynomial expansions. A novel feature of our numerical method is the treatment of pressure and continuity equations which represents an intrinsic problem of incompressible flow computations. A general algorithm has been developed which enforces incompressibility and boundary conditions exactly in the discretised equations. The algorithm is not restricted to the channel flow geometry or the particular discretisation method of the present work. These matters will be discussed in sect. 2. Sect. 3 introduces the mathematical model used for transition simulations and the fundamentals of the numerical method. In sect. 4 some sample results taken from {5} are presented as well as results of new simulations. In the following, a brief introduction to the transition problem is given to motivate our approach to this investigation, and the main results obtained in {5} are summarized.

1.2 Laminar-turbulent transition. In his 1979 survey lecture on transition in shear flows, Stuart {8} characterized the (then) present situation as follows:

"(A) There is no scientific or rational theory of shear flow turbulence;
(B) There is no theory of transition to turbulence;
(C) There is no theory for the description of large-amplitude complex oscillations."

On the positive side, he notes that we have

"(i) Some excellent experiments, showing important features of some cases of transition to turbulence;

(ii) Several theoretical concepts...".

As to transition in plane Poiseuille flow, the experimental breakthrough has been achieved by Nishioka et al. {9-11}. Using the vibrating-ribbon technique, they have been able to make detailed measurements of the transition process. They have shown that in these experiments transition of the laminar flow into turbulence proceeds through a sequence of increasingly complex intermediate stages which are very similar to those observed earlier in flat plate boundary layer transition experiments {12}. Some characteristic features in this sequence are the development of a spanwise "peak-valley structure" in the initially two-dimensional ribbon-induced Tollmien-Schlichting wave, a "catastrophic" increase of the disturbance amplitude, instantaneous inflectional velocity profiles in connection with low-velocity "spikes", and the local breakdown of the regular wave motion into turbulent spots. As pointed out by Stuart, a theoretical description (to say nothing of an explanation) of the transition process has not been possible so far. This is mainly due to the highly nonlinear and three-dimensional nature of the problem.

In spite of these difficulties, important progress in understanding transition has been made in recent years; see the extensive review by Herbert {13}. Here numerical calculations using spectral methods have played an important role {14-17, 5}. We mention just two results. First, the behaviour of a single purely two-dimensional wave is now fairly well understood. Depending on Reynolds number, wave number and initial amplitude, the disturbance either decays or evolves into the periodic secondary flow calculated from the steady-state equations {14}. The development towards the secondary flow proceeds through a long quasi-steady (or quasi-equilibrium) state in a time (in dimensionless units) of the order of the Reynolds number {16,17,5}. These quasi-equilibria, which exist down to Reynolds number (based on channel half-width and centerline velocity) Re $\approx$ 2900 {14}, form the basis for the transition process. The second result to be mentioned is that the quasi-equilibria are strongly unstable against small three-dimensional disturbances {16,5}. For the strictly steady state this secondary instability has been demonstrated by a rigorous linear stability analysis {16,13}. With sufficiently large amplitudes of the initial disturbance "break-down" was observed in the simulations {15-17} down to Reynolds numbers of about 1000 which is also the lower limit for turbulence found in experiments.

However, the relation of the previous simulations {15-17} to the experimentally observed transition process has not been made evident. Most simulations were done at lower Reynolds numbers and larger disturbance amplitudes than present in the relevant experiments {9-11}. In particular, the use of the streamwise-periodic and timewise-amplified disturbance model, which is also applied in the present work, has been disputed by some researchers. For boundary layer transition, however, Wray and Hussaini {18} found excellent agreement, using this model, with measurements up to the 2-spike stage of transition.

On this background, we decided to simulate plane Poiseuille flow for a case with the physical parameters adjusted as closely as possible to those of the experiment {10,11} at Re = 5000 and oscillation frequency $\omega_r = 0.32$ where most comprehensive experimental data is available. An additional advantage is that here the disturbance amplitudes are small (of the order of a few percent) initially so that a relatively coarse resolution is sufficient. Detailed comparisons of numerical results with the measurements of mean velocity, rms fluctuations and instantaneous velocity distributions at subsequent stages of the transition process have been made, and a remarkable, even quantitative, agreement has been found up to the spike stage {5}. Among other things, the development of the peak-valley structure, longitudinal vortices, inflectional velocity profiles and spike-type signals have been reproduced. Having thus demonstrated that the calculations provide an accurate simulation of the experimentally observed phenomena, we have been able to document the development of three-dimensional flow field structures during the transition process. In particular, an array of "horseshoe vortices" similar to those observed in flow visualisation experiments of boundary layer transition {19,20} has been found (see 4.2). To a limited extent, parameter variations have been done to study the dependence of the secondary instability on Reynolds number, wavenumber (corresponding to ribbon frequency) and initial disturbance amplitudes {5}. In general, the instability is quite sensitive to changes in these parameters. An example is given in 4.3.

2. Treatment of Pressure and Continuity Equations.
In this section we describe a general algorithm for numerical computation of wall-bounded viscous incompressible flows using the primitive variables. This algorithm has been proposed in {4} for the channel flow problem and in {5} for the general case. The present formulation is given in terms of the influence (or capacitance) matrix technique {21}. Velocity $\underline{u}(\underline{x},t)$ and pressure $p(\underline{x},t)$ are solutions of the Navier-Stokes equations

$$\partial_t \underline{u} = \underline{M} - \nabla p/\rho + \nu\Delta\underline{u} \tag{2.1}$$

$$\text{div}\, \underline{u} = 0 \tag{2.2}$$

written in Cartesian coordinates for constant density ρ and viscosity ν. Here, $\underline{M}$ includes the convective terms and external accelerations. (2.1), (2.2) are to be solved in a flow domain R with, e.g., no-slip boundary conditions

$$\underline{u} = 0 \quad \text{on } \partial R, \tag{2.3}$$

the rigid wall boundary of R, together with initial conditions at t=0.

We assume that these equations are integrated in time treating $\underline{M}$ explicitly but the viscous terms $\nu\Delta\underline{u}$ implicitly. Further we assume that some suitable space discretisation is introduced to approximate the fields $\underline{u}$, p and the operators div, ∇ and Δ. For brevity we subsequently use the same symbols for the discrete quantities. Then, for each time step we have to solve the discretised Stokes equation

(2.4) $$\Delta\underline{u} - \lambda\underline{u} - \nabla q = -\underline{r}$$

together with (2.2), (2.3). Here, $\underline{r}$ is given, $\lambda > 0$ is a constant ($\lambda = 2/\nu\Delta t$ in our case) and the scalar q is proportional to the pressure. Eqs. (2.2)-(2.4) form a coupled system of four discretised scalar equations. It is this coupling which makes a direct solution of the system difficult. For decoupling, the common procedure is to take the (discretised) divergence of (2.4) in order to obtain the Poisson equation

(2.5) $$\Delta q = \operatorname{div} \underline{r} .$$

If boundary conditions for q were known, q could be determined from (2.5) and thereafter each component of $\underline{u}$. However, conditions for q on ∂R are not known a priori rather than determined implicitly by (2.2)-(2.4). In case of zero viscosity (inviscid flow) or explicit treatment of the viscous terms, it is possible to use the normal component of the momentum equations on ∂R to obtain a Neumann boundary condition for q. In the system (2.3)-(2.5), however, the normal component of (2.4)$|\partial R$ does not ensure (2.2) and would not eliminate the coupling between $\underline{u}$ and p. Instead, the necessary and sufficient condition which renders (2.3)-(2.5) equivalent to the original equations (2.2)-(2.4) reads

(2.6) $$\operatorname{div} \underline{u} = 0 \quad \text{on} \quad \partial R .$$

This can be seen {4,5} by noting that for a solution $\underline{u}$ of (2.3)-(2.5) the variable $d := \operatorname{div} \underline{u}$ obeys the equation $\Delta d - \lambda d = \overline{0}$ so that $d = 0$ everywhere if and only if $d = 0$ on ∂R.

Our algorithm to solve the so-called {21} "A-problem" (2.3)-(2.6) essentially consists of two solutions of a "B-problem" in which (2.6) is replaced by direct boundary conditions for q,

(2.7) $$q = g \quad \text{on} \quad \partial R .$$

In the first B-problem, we arbitrarily set $g = 0$ and compute the corresponding solution $\tilde{q}$, $\underline{\tilde{u}}$ from

(2.8a) $$\Delta\tilde{q} = \operatorname{div} \underline{r} , \quad \tilde{q} = 0 \quad \text{on } \partial R$$

and thereafter

(2.8b) $$\Delta\underline{\tilde{u}} - \lambda\underline{\tilde{u}} = \nabla\tilde{q} - \underline{r}, \ \underline{\tilde{u}} = 0 \quad \text{on } \partial R .$$

This solution satisfies (2.3)-(2.5) but has in general a nonzero residual $\tilde{w} = \operatorname{div} \underline{\tilde{u}}$ on ∂R. The correct boundary distribution g which ensures a zero residual, i.e., (2.6), is found with aid of an influence matrix. This matrix is calculated in advance as follows. Assume that in the discretised equations the boundary conditions are imposed at N points $\underline{x}_j$ (the algorithm is most easily explained for - but not restricted to - such a "collocation" situation; in general, N linear conditions for the unknown parameters in the discretised problem are given). We compute a basis of N solutions q_j, $\underline{u}_j$ of the "homogeneous" ($\underline{r} \equiv 0$) B-problem with boundary values $q_j(\underline{x}_i) = \delta_{ij}$, i.e. with pressure boundary values zero except for one single point (parameter). We define the matrix D by

$$(2.9) \qquad D_{ij} = (\operatorname{div} \underline{u}_j)(\underline{x}_i) ,$$

i.e. the j-th column of D consists of the vector of the residuals div $\underline{u}_j$ on ∂R. The inverse $C = D^{-1}$ is the required influence matrix. The amount of work for this precomputation is negligible compared to the overall effort if N is small compared to the total number of discrete points (or parameters) in R and the number of time steps.

With C known we now solve the second B-problem with boundary values g in (2.7) obtained from the residual $\tilde{w}$ of the solution of the first B-problem:

$$(2.10) \qquad g = - C \tilde{w} .$$

It is easy to verify {21} that the solution q, $\underline{u}$ of the second B-problem is the desired solution of the A-problem (2.3)-(2.6).

In the present application for the channel flow problem, after Fourier transform in the coordinates parallel to the walls a set of uncoupled one-dimensional two-point boundary value problems in the normal coordinate is obtained. Thus, for each Fourier mode, N = 2 and the algorithm is particularly simple. For arbitrary three-dimensional domains, N is large and storage of C may be prohibitive. In this case an iterative approach based on the implicit capacitance matrix concept {22} should be considered.

3. Numerical Method.

3.1 Mathematical Model. In this section we describe the main features of the numerical method which is implemented in the computer code CHANSON for solving the channel flow problem. We state the problem in non-dimensional quantities based on the channel half-width $\tilde{h}$ and on the maximum velocity $\tilde{U}_{max}$ of the undisturbed flow. $Re = \tilde{U}_{max} \tilde{h}/\tilde{\nu}$ denotes the Reynolds number, $U(x_3) = 1 - x_3^2$ the basic plane Poiseuille flow and $\underline{u}(\underline{x},t)$, $p(\underline{x},t)$ disturbances of this basic flow. Given an initial velocity disturbance $\underline{u}^o(\underline{x})$, we ask how this disturbance develops in time while the mean pressure gradient is kept fixed.

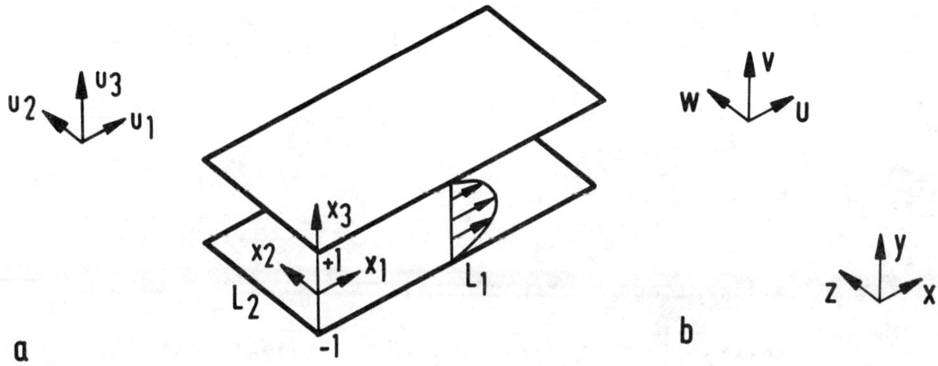

Fig. 1 Integration domain and coordinates in simulation (a) and experiment (b)

Integration domain and coordinate system are shown in fig. 1. In the horizontal streamwise (x_1) and spanwise (x_2) directions periodic boundary conditions are assumed, and no-slip boundary conditions are applied at the walls in $x_3 = \pm 1$. In our transition studies, the periodicity length L_1 is set equal to the Tollmien-Schlichting wavelength, and L_2 is usually taken from experiments. $\alpha_j = 2\pi/L_j$ denotes the fundamental wave numbers. The evolution of the disturbances is governed by the equations (2.1)-(2.3) with

(3.1) $$\underline{M} = \underline{u} \times \text{rot } \underline{u} - \underline{P}(\underline{u}),$$

(3.2) $$\underline{P}(\underline{u}) = U\partial_1 \underline{u} + u_3 \partial_3 U \underline{e}_1 ,$$

$\underline{e}_1$ = unit vector in direction x_1 and p now denoting the pressure head. We apply a Galilean transform to a reference frame $\xi_1 = x_1 - Ct$ moving with the phase velocity $C = c^{TS} = \omega_r^{TS}/\alpha_1$ of the Tollmien-Schlichting wave. For the initial stages of transition, where the Tollmien-Schlichting wave dominates, this gives the numerical advantages of reduced time-discretisation errors (as only the slow amplitude variation remains to be computed) and an increased stability limit (see below). In this simulation model a ξ_1-periodic disturbance develops in time, while in the vibrating-ribbon experiments a time-periodic flow develops in x_1. Therefore, downstream coordinate and time have to be interchanged for comparing numerical and experimental results.

3.2 Spatial discretisation and time-differencing. The numerical method is based on Fourier expansions in the horizontal directions,

$$(3.3)\quad \underline{u}(\underline{x},t) = \sum_{k_1}\sum_{k_2} \hat{\underline{u}}(k_1,k_2,x_3,t)\exp(ik_1\alpha_1x_1 + ik_2\alpha_2x_2)$$

and Chebyshev polynomial expansions of the Fourier coefficients $\hat{\underline{u}}$,

$$(3.4)\quad \hat{\underline{u}}(k_1,k_2,x_3,t) = \sum_{k_3=0}^{N_3} \hat{\hat{\underline{u}}}(k_1,k_2,k_3,t)\, T_{k_3}(x_3).$$

The sum in (3.3) extends over the domain

$$(3.5)\quad K = \{(k_1,k_2) \mid |k_j| < N_j/2,\ |k_1|/N_1 + |k_2|/N_2 < 2/3\}$$

which is illustrated in fig. 2. For details of the spectral method we refer to the monograph {1} and to {5,7}. Nonlinear terms are calculated by the pseudospectral approximation. Aliasing errors are reduced by applying the truncation (3.5) in the (k_1,k_2) plane. The dominating part $\underline{P}(\underline{u})$, eq. (3.2), is easily calculated directly in $\underline{k}$-space without aliasing errors in an order $N_1N_2N_3$ arithmetic operations because $U = 1 - x_3^2 = (T_o - T_2(x_3))/2$ is a polynomial of degree 2.

A number of time-differencing schemes are available in our code. These are given in tab. 1 where for ease of notation M stands for the convective terms (3.1) and L for the pressure and viscous terms. To avoid severe time step limitations due to the high resolution of the Chebyshev polynomials near the boundaries {1}, the viscous terms must be treated implicitly. We use the Crank-Nicolson method in all cases. The differencing schemes in Table 1 include the first order Euler method EU and the second order methods Adams-Bashforth (AD), Leapfrog (LE), modified Euler (EM), and modified Adams-Bashforth with corrector step (AC) after Gazdag {23}. EM requires two evaluations of the nonlinear terms M per time step (which is the most time-consuming part), all other methods need only one evaluation. An analysis of accuracy and stability of these schemes is given in {1,23}. In practice they are stable if a criterion of the type

$$(3.6)\quad \Delta t \leq 2 \,/\, (N_1\alpha_1 \max|U-C + u_1| + N_2\alpha_2 \max|u_2|)$$

is satisfied. As usually $|u_j| \ll \max U = 1$, we see that a non-zero C $(0 < C < 1)$ increases the admissible time step. Fig. 3 compares the maximum relative errors of the schemes obtained in a test calculation using an Orr-Sommerfeld eigensolution with the parameters Re = 1250, $\alpha_1 = 1$, $N_3 = 32$ and $C = 0$. The differences in the accuracy of the second order methods are not large. In some test cases the Leapfrog scheme showed the well-known 2 Δt-oscillations which lead not to instabilities but cannot be tolerated. EM needs twice the time of AD

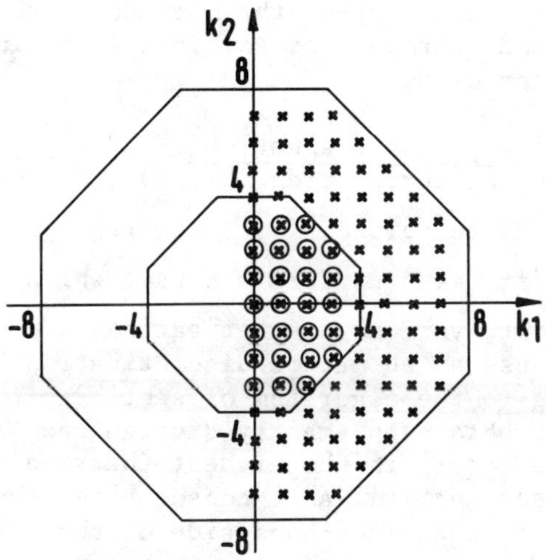

Fig. 2

Fourier modes retained in simulation with $N_1 = N_2 = 8$ (O) and with $N_1 = N_2 = 16$ (x)

$$\frac{u^{n+1} - u^n}{\Delta t} = M^n + \frac{1}{2}(L^{n+1} + L^n) \qquad \text{EU}$$

$$\frac{u^{n+1} - u^n}{\Delta t} = \frac{3}{2}M^n - \frac{1}{2}M^{n-1} + \frac{1}{2}(L^{n+1} + L^n) \qquad \text{AD}$$

$$\frac{u^{n+1} - u^{n-1}}{2\,\Delta t} = M^n + \frac{1}{2}(L^{n+1} + L^{n-1}) \qquad \text{LE}$$

$$\begin{aligned} \frac{\tilde{u}^{n+1} - u^n}{\Delta t} &= M^n + \frac{1}{2}(\tilde{L}^{n+1} + L^n) \\ \frac{u^{n+1} - u^n}{\Delta t} &= \frac{1}{2}(\tilde{M}^{n+1} + M^n) + \frac{1}{2}(L^{n+1} + L^n) \end{aligned} \qquad \text{EM}$$

$$\begin{aligned} \frac{\tilde{u}^{n+1} - u^n}{\Delta t} &= \frac{3}{2}\tilde{M}^n - \frac{1}{2}\tilde{M}^{n-1} + \frac{1}{2}(\tilde{L}^{n+1} + L^n) \\ \frac{u^{n+1} - u^n}{\Delta t} &= \frac{1}{2}(\tilde{M}^{n+1} + \tilde{M}^n) + \frac{1}{2}(L^{n+1} + L^n) \end{aligned} \qquad \text{AC}$$

Table 1 Time-differencing schemes

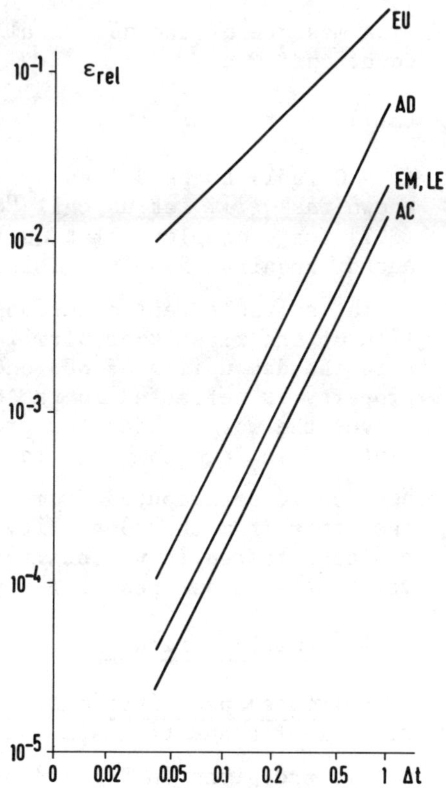

Fig. 3 Relative error in test problem

and is used only to start the calculation. AC gives the most accurate results but requires also more time and storage than AD. Therefore, in most transition simulations AD has been used.

3.3 Solution of discretised velocity-pressure equations. As mentioned at the end of section 2, for each Fourier mode (k_1,k_2) a boundary value problem on $-1 \leq x_3 \leq 1$ has to be solved. (2.6) now reads $\partial_3 u_3 (\pm 1) = 0$ so that the equations for q and u_3 form a subset which can be solved separately. At this point, various discretisations and solution procedures are possible. We use a tau method discretisation {1} and solve the equations according to the algorithm of sect. 2. It is important to note that in order to obtain an exactly divergence-free solution, the tau terms (i.e. the errors in the highest Chebyshev modal equations) must be taken into account {4}. As a consequence, the tau errors of the u_3-equation appear on the right-hand side of the q-equation for all Chebyshev modes, which leads to an additional coupling.

A measure of the numerical work involved is the number of Helmholtz equations

$$(3.7) \qquad f'' - \mu f = r , \quad f(\pm 1) = 0$$

($\mu > 0$ real) to be solved by the Chebyshev-tau method which is a well-known task {1}. Let us call "one unit" the work required to solve (3.7) for a complex solution f. Then the algorithm {4} to compute q and u_3 requires 5 units. This may be reduced to 4 units, e.g., by adding suitable real solutions with complex coefficients to the solution of the first B-problem instead of solving the second B-problem. Here the decoupling of odd and even Chebyshev modes is exploited (this property is reflected in the symmetry of the influence matrix which halves the work needed for precomputation). The remaining two components u_1, u_2 are computed in 2 units. Alternatively, only one component could be computed from the momentum equation (one work unit) and the other from the continuity equation. The present method {4} which has been tested in various verification calculations {5} always gave very satisfactory results.

4. Numerical results.

4.1 Case specifications. In this section we report some numerical results of transition simulations. The physical and numerical parameters are given in Table 2. ω^{TS} denotes the eigenvalue of the two-dimensional Tollmien-Schlichting eigensolution

$$(4.1) \qquad \underline{u}^{TS}(\underline{x},t) = \{\underline{\hat{u}}^{TS}(x_3) \exp [i(\alpha_1 x_1 - \omega^{TS} t)]\}_r$$

($\{z\}_r$ is the real part of a complex number z), and ω^{3D} the eigenvalue of a three-dimensional oblique wave

$$(4.2)\quad \underline{u}_{\pm}^{3D}(\underline{x},t) = \{\hat{\underline{u}}_{\pm}^{3D}(x_3)\exp[i(\alpha_1 x_1 \pm \alpha_2 x_2 - \omega^{3D}t)]\}_r$$

according to linear stability theory. The eigensolutions (4.1), (4.2) are calculated using a program by Herbert based on a Chebyshev spectral method {14}. In all of the runs reported here the initial distribution $\underline{u}^o(\underline{x})$ is composed of a two-dimensional Tollmien-Schlichting wave $\underline{u}^{TS}(\underline{x},0)$ and a pair $\underline{u}^{3D}(\underline{x},0) = (\underline{u}_+^{3D}(\underline{x},0) + \underline{u}_-^{3D}(\underline{x},0))/2$ of oblique waves, as in the Benney-Lin model:

$$(4.3)\quad \underline{u}^o(\underline{x}) = A^{TS}\underline{u}^{TS}(\underline{x},0) + A^{3D}\underline{u}^{3D}(\underline{x},0)\ .$$

The normalisation is $\max|u_1^{TS}| = \max|u_1^{3D}| = 1$ and zero phase difference of u_1^{TS} and u_1^{3D} at $t = 0$.

run	Re	α_1	α_2	ω^{TS}, ω^{3D}	$N_1 \cdot N_2 \cdot N_3$	Δt
1	5000	1.12	2.1	0.3156-0.0028 i 0.3637-0.0785 i	$\leq$ 32·32·40	$\geq$ 0.03
2	1000	1	2	0.3463-0.0421 i 0.5079-0.0962 i	8·8·40	0.2
3	3000	1	2	0.2923-0.0102 i 0.3866-0.0771 i	8·8·40	0.2
4	4000	1	2	0.2785-0.0049 i 0.3577-0.0722 i	8·8·40	0.2
5	5000	1	2	0.2681-0.0018 i 0.3363-0.0684 i	8·8·40	0.2
6	7000	1	2	0.2529+0.0017 i 0.3059-0.0631 i	8·8·40	0.2
7	10000	1	2	0.2375+0.0037 i 0.2760-0.0577 i	8·8·40	0.2
8	7000	1	2	same as run 6	16·16·64	.0625

Table 2 Specification of physical and numerical parameters

A^{TS} is set to 3 % and A^{3D} to 0.1 % throughout. The disturbance components u_1^{TS} and u_1^{3D} are antisymmetric in x_3. $\bar{u}_1$ denotes the streamwise mean disturbance (with the average taken over one wavelength $0 \le x_1 - Ct \le L_1$), $u_1' = \overline{(u_1 - \bar{u}_1)^2}^{1/2}$ the rms value of the fluctuation about this mean and u'_{1max} its maximum over x_3.

In 4.2 we report briefly on results of run 1 where the wavenumbers α_1 and α_2 have been adjusted to the experiments {10,11} at Re = 5000 and ribbon frequency $\tilde{f}$ = 72 Hz ($\omega_r^{TS} \approx 0.32$). For an extensive discussion of run 1 see {5}; a brief account has been given in {6}. In runs 2-8 the Reynolds number has been varied from 10^3 to 10^4. These results are presented in 4.3. Run 8, which is in the linearly unstable parameter range, will be discussed in more detail and compared with the subcritical run 1.

4.2 Some results of a simulation at Re = 5000. As in most parametric studies, we used the horizontal resolution $N_1 = N_2 = 8$ with time step $\Delta t = 0.2$ for the initial phase $t \le 80$ in run 1. This was shown to be sufficient in the early stages of the flow development at the present low initial amplitudes. The resolution was increased to $N_1 = N_2 = 16$ ($\Delta t = 0.0625$) in the range $80 \le t \le 116$ and to $N_1 = N_2 = 32$ ($\Delta t = 0.03$) for $116 \le t \le 128$. $N_3 + 1 = 41$ Chebyshev polynomials were retained throughout. The total CPU time for run 1 was 5 h on an IBM 3033 computer.

To obtain an overview over the flow development, we look at the evolution of u'_{1max}, fig. 4. At $t \approx 20$ the typical spanwise peak-valley splitting sets in. The fluctuation increases slowly first, but rapidly for $t \gtrsim 90$. As shown in {5,6} this is due to an exponential amplification of the three-dimensional ($k_2 > 0$) Fourier modes $\hat{u}_1(k_1,k_2)$, the largest of which exceed the magnitude of the two-dimensional ($k_2 = 0$) modes at $t \approx 90$. At this stage, inflection point profiles begin to form at the peak position $x_2 = 0$ and low velocity pulses appear in the instantaneous disturbance $u_1(x_1 - Ct)$. The 1-spike and 2-spike stages occur at $t \approx 122$ and $t \approx 125$ in run 1. In the experiment, the number of spikes increases rapidly afterwards and soon irregular fluctuations appear which indicate the breakdown of the well-ordered motion at these local positions. With the present spatial resolution, the simulation is not able to reproduce these multi-spike stages accurately, though it did not become unstable when continued up to $t = 134$. However, the detailed comparisons {5} with measurements demonstrate that the flow development agrees quite well at least up to the 1-spike

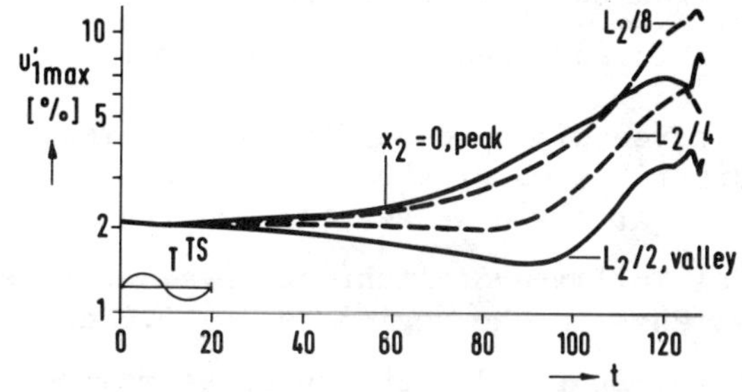

Fig. 4

Development of maximum rms fluctuation u'_{1max} at different spanwise positions in run 1

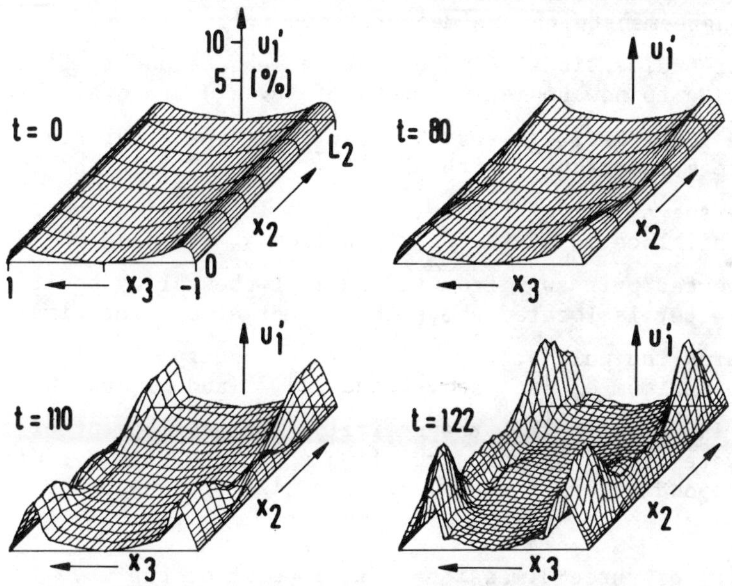

Fig. 5 Spatial structure of rms fluctuation u_1'

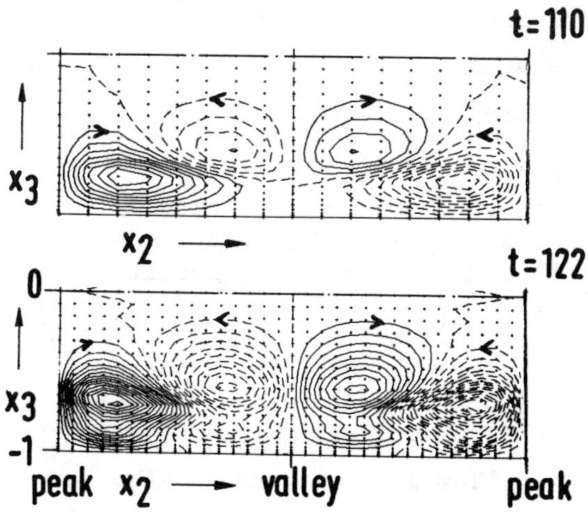

Fig. 6

Streamlines $\overline{\Psi}$ = const of mean secondary flow $(\overline{u}_2, \overline{u}_3)$. Increment of niveau lines is 0.2 %

stage (though there is some indication that N_3 should be increased beyond 40 in order to resolve the interior shear layers better).

As an example illustrating the three-dimensional flow development in run 1, fig. 5 shows the spatial distribution of the streamwise fluctuation $u_1'(x_2,x_3)$ at different stages. At t = 0 we see the amplitude distribution of the two-dimensional Tollmien-Schlichting wave. At later times, a strong dependence on x_2 appears in the region $|x_3| \gtrsim 0.6$ near the walls, the peak-valley structure. At t = 122 the spanwise maximum no longer occurs at $x_2 = 0$ (see fig. 4) but at $x_2 = \pm L_2/8$, in agreement with the measurements {11}.

During the three-dimensional development, a mean secondary flow $(\overline{u}_2,\overline{u}_3)$ evolves. Up to now no measurements of $(\overline{u}_2,\overline{u}_3)$ are available for plane Poiseuille flow. Fig. 6 gives the results of the simulation in terms of the stream function $\overline{\psi}$ with $\overline{u}_2 = \partial_3\overline{\psi}$, $\overline{u}_3 = -\partial_2\overline{\psi}$ in the lower half channel. We see a pair of counter-rotating longitudinal vortices near the lower wall located at the peak position $x_2 = 0$, and an additional, weaker vortex pair away from the wall at the valley position $x_2 = L_2/2$. The latter is located above the former at earlier times and extends towards the wall at t = 122. This secondary flow is relatively weak: the maximum of $|\overline{u}_2|$ is 1 % and 2.1 % (and of $|\overline{u}_3|$ 0.5 % and 1 %) at t = 110 and t = 122, respectively. The development of the double vortex system found by our simulation is in remarkable similarity to the corresponding experimental results {12} for the boundary layer.

An illustration of three-dimensional development of the wave is given in fig. 7. Fig. 7a shows a relief plot of the instantaneous disturbance u_1 in the plane $x_3 = -0.71$. At late times we observe a strong mean flow component with wavelength $L_2/2$. Niveau lines of $U + u_1$ at $x_3 = -0.73$ obtained in experiment and simulation are given in fig. 7b, c. These instantaneous flow structures are quite similar even at this late stage.

In flow visualization experiments of boundary layer transition discrete vortex structures have been observed {19,20}, so it is natural to look for those structures in the present simulation results. From projections of the velocity and vorticity fields at t = 122 the sketch of the horseshoe-vortex structure shown in fig. 8 has been constructed quantitatively. We find two discrete three-dimensionally warped vortices which have opposite sense of rotation and a distance of about $L_1/2$. The streamwise extent of the stronger vortex is about one wavelength L_1. The instantaneous (u_2,u_3) flow is considerably stronger

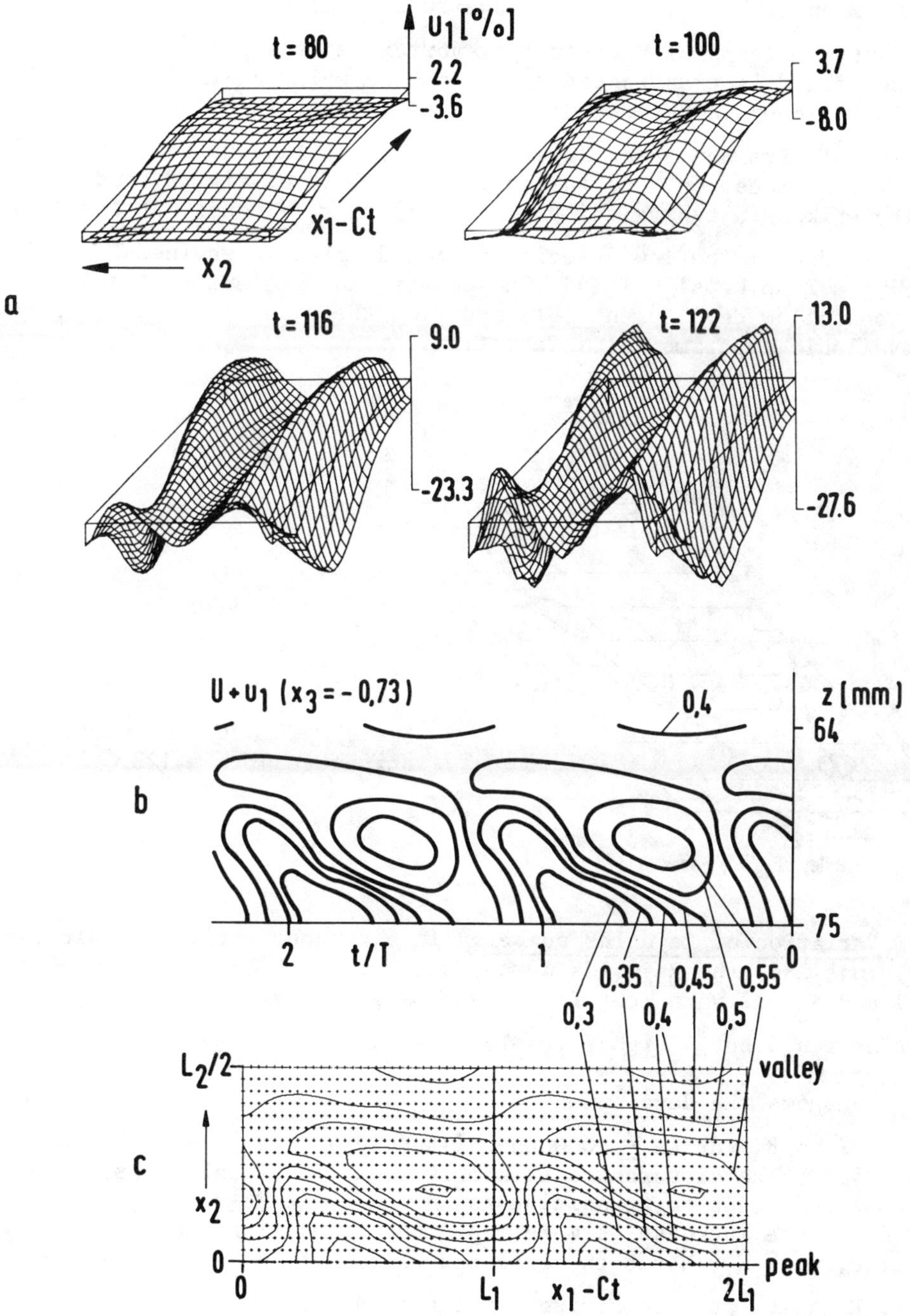

Fig. 7 Three-dimensional development of wave
(a) $u_1(x_3 = -0.71)$ (b) Instantaneous velocity $U+u_1(x_3 = -0.73)$ in experiment {10}, 1-spike stage, and (c) in simulation, t = 122

than the mean flow $(\bar{u}_2, \bar{u}_3)$; its magnitude is up to 7 % at t = 122. In the downstream loops of the stronger vortex, at the peak positions, low-speed fluid is transported away from the wall leading to a local low-velocity bump in the region $-0.8 \lesssim x_3 \lesssim -0.5$. This implies local velocity deficits and shear layers in the instantaneous velocity profiles at this location, as indicated in the left of fig. 8, and low-velocity spikes in the velocity $u_1(x_1 - Ct)$ at x_3 = const. At t = 125, the outer edge of the low-velocity region develops a waviness, thus producing a 2-spike signal {5}. In summary, the picture of the three-dimensional flow development obtained from the simulation is consistent with available experimental information.

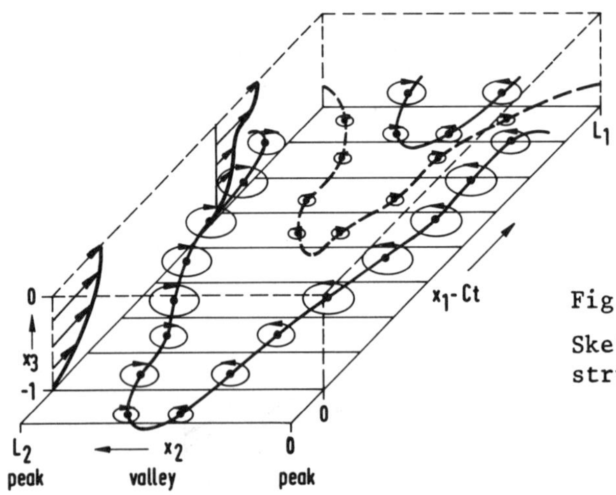

Fig. 8

Sketch of horseshoe vortex structure at t = 122

4.3 Variation of Reynolds number. In a further series of calculations, runs 2-8, the Reynolds number was varied while the wavenumbers $\alpha_1 = 1$ and $\alpha_2 = 2$ were kept fixed. Run 8 has the same physical parameters as run 6 but a higher resolution.

Fig. 9 shows the development of u'_{1max} at the peak position $x_2 = 0$ for runs 2 to 8. For Re $\leq$ 4000 the disturbance is exponentially damped. At Re = 5000, the disturbance is nearly neutral initially, but finally increases and leads to transition as in run 1. At Re = 7000 and Re = 10 000 the increase is more rapid and transition occurs earlier. Of course, with $N_1 = N_2 = 8$ the late stages cannot be resolved accurately. However, as a comparison of run 6 and run 8 at Re = 7000 shows, the low resolution results agree well up to t ≈ 100 shortly before breakdown. At Reynolds numbers of order 1000 the damping is very strong so that a much larger initial disturbance is necessary to obtain transition {15}.

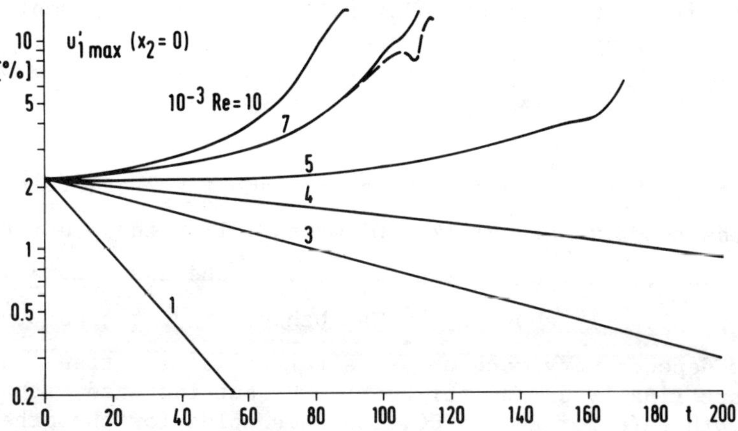

Fig. 9 Development of maximum rms fluctuation u'_{1max} at peak position $x_2 = 0$ in runs 2-7 (——) and in the higher resolution run 8 (- - -)

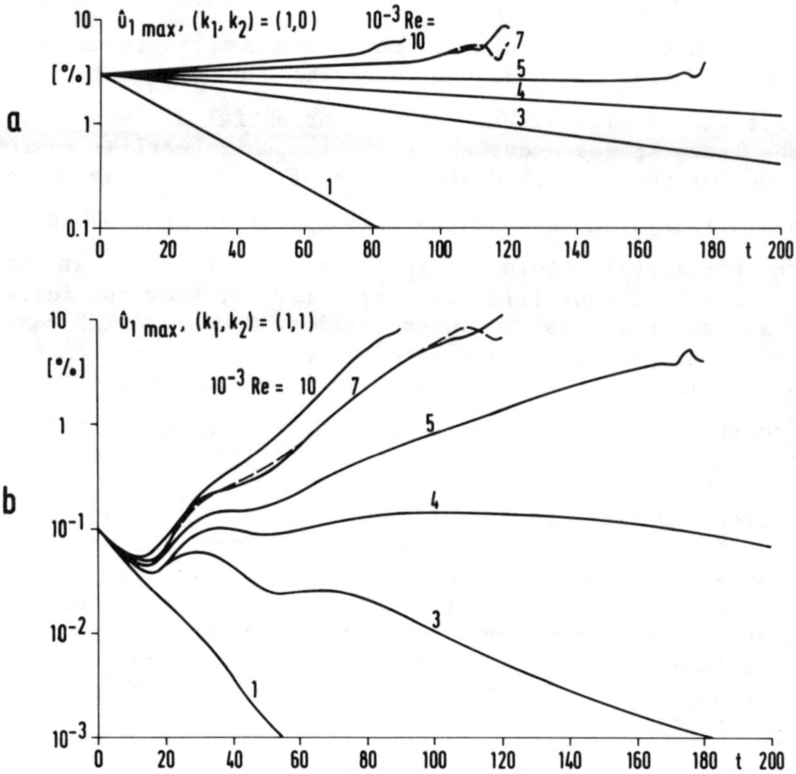

Fig. 10 Development of (a) 2-D and (b) 3-D Fourier amplitudes —— runs 2-7, - - - run 8

In fig. 10 we see the evolution of the maximum Fourier amplitudes defined as

$$(4.4) \qquad \hat{u}_{1max}(k_1,k_2) = f(k_1,k_2)\cdot \max_{x_3} |\hat{u}_1(k_1,k_2,x_3)|$$

with $f = 1$ if $k_1 = k_2 = 0$, $f = 4$ if $k_1 \cdot k_2 \neq 0$ and $f = 2$ otherwise. For $Re \leq 4000$, the two-dimensional ($k_2 = 0$) modes follow the prediction of the linear theory, decaying as $\exp(\omega_i^{TS} t)$. In the cases $Re \geq 5000$ the growth rate is slightly above ω_i^{TS}. The behaviour of the three-dimensional modes depends very much on Re: after an initial transient period, they are finally damped for $Re \leq 4000$, but increase with nearly constant growth rate for $Re \geq 5000$. The same holds for the other three-dimensional modes not shown here. Again, as in run 1, after the mode $(k_1,k_2) = (1,1)$ has grown to the magnitude of the mode (1,0), the latter increases also rapidly and shortly thereafter breakdown takes place. The nearly exponential growth of the three-dimensional modes reflects the secondary instability of the initially dominating two-dimensional Tollmien-Schlichting wave.

Let us now look at some results of run 8 and compare them with the subcritical run 1. Fig. 11 shows the development of u'_{1max} at the same spanwise positions as fig. 4. There are no essential differences, except that the later stages occur about 15 time units earlier in run 8. The same holds for the spatial distribution of $u_1'(x_2,x_3)$ shown in fig. 12. The oscillations with wave length $L_2/8$ appearing at $t = 110$ indicate that the horizontal resolution is no longer sufficient at this late stage. The velocity profiles at peak, fig. 13, show the inflection points and shear layers mentioned earlier. Compared with run 1, the shear layers are somewhat nearer to the walls. Disturbance signals u_1 are shown in fig. 14. The abscissa x_1-Ct increases to the left to correspond to the oscilloscope traces obtained in the experiment. Fig. 14 suggests that the spike stage sets in at $t \approx 110$.

In conclusion, no significant differences have been found between transition in the subcritical and supercritical case. The growth of the disturbances is somewhat accelerated in the supercritical case, but the essential features, in particular the three-dimensional flow structures, appear to be the same. This is another indication that the physical mechanisms dominating the transition process are quite different from those relevant for the small-amplitude disturbances of classical linear stability theory.

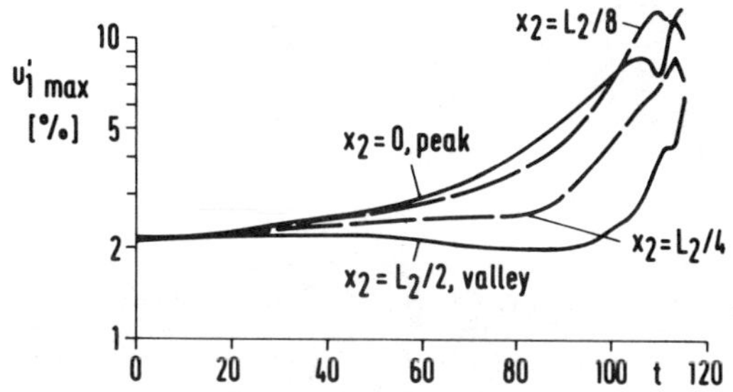

Fig. 11

Development of u'_{1max} at different spanwise positions x_2 in the supercritical case, run 8

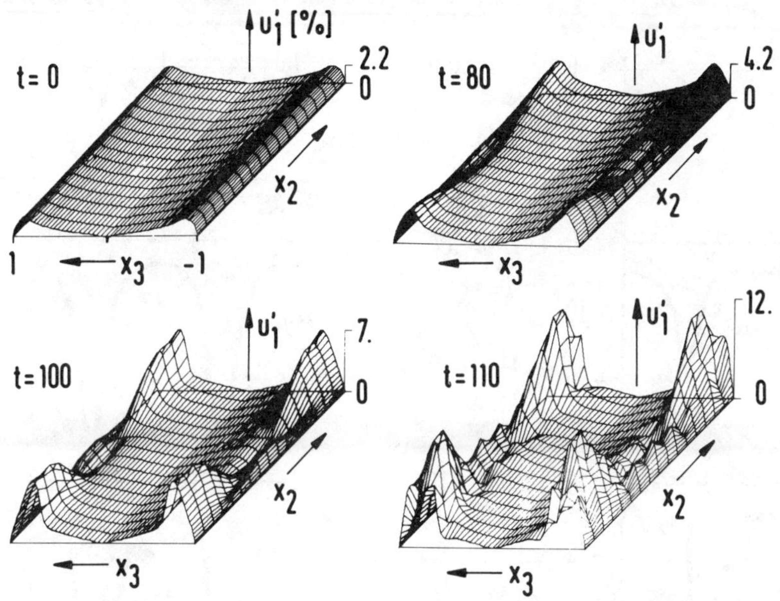

Fig. 12 Development of rms fluctuation u_1' in run 8

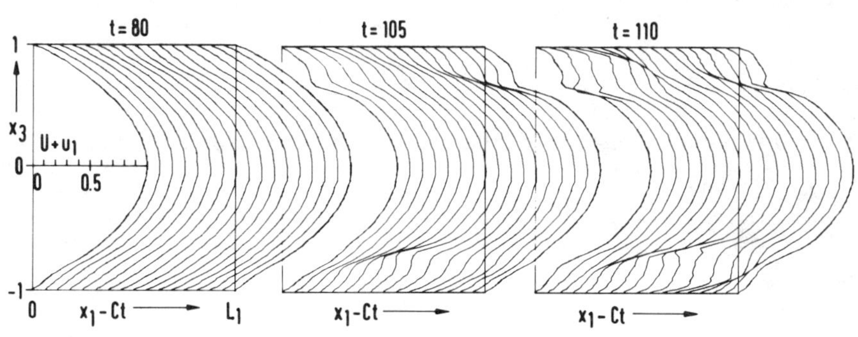

Fig. 13 Instantaneous velocity profiles at peak, run 8

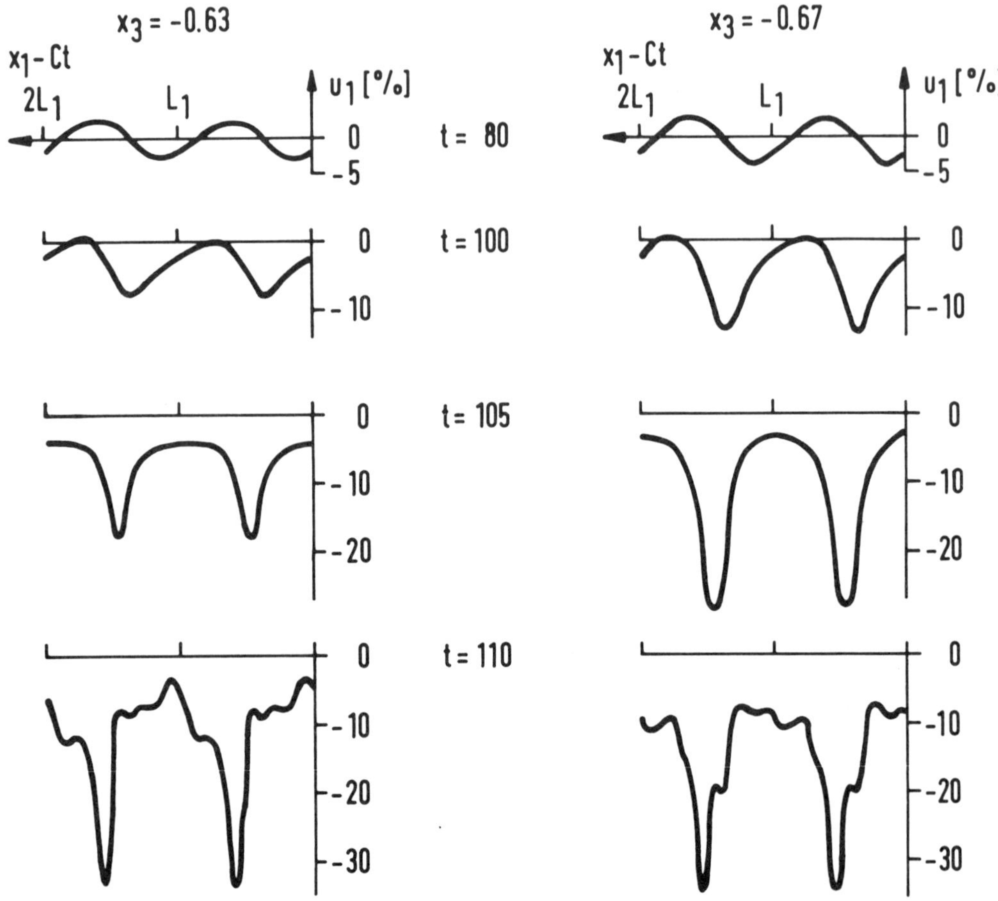

Fig. 14 Instantaneous disturbance u_1 $(x_1 - Ct)$ at peak, run 8

5. Summary and Conclusions.

A general algorithm has been developed for pressure computation in viscous incompressible flows in which no-slip wall boundary conditions and the continuity equation are satisfied exactly in the discretised equations. The algorithm has been implemented in the Fourier-Chebyshev spectral code CHANSON for plane channel flow and leads to an efficient numerical scheme. Basically, the continuity equation is replaced in the interior of the flow domain by the Poisson equation for the pressure, and from the continuity equation on the boundary the correct pressure boundary conditions are derived using the influence matrix technique.

The spectral code has been applied to investigate the laminar-turbulent transition process in plane Poiseuille flow {5,6} as observed in vibrating ribbon experiments {9-11}. Detailed comparisons with measurements have been made. To some extent, the influence of the parameters Reynolds number, amplitudes and wavenumbers of the initial disturbances on transition have been studied, and three-dimensional flow structures have been documented. We draw the following conclusions.

- Our numerical simulation results are in very satisfactory, quantitative agreement with experiments, at least up to the 1-spike stage of the transition process. This has been verified for practically all quantities for which measurements are available.

- The dominating three-dimensional flow structure is a system of horseshoe vortices quite similar to those inferred from flow visualization experiments in boundary layers {19,20}.

- In accordance with {13,16}, we find that a secondary instability of the two-dimensional ribbon-induced Tollmien-Schlichting wave against three-dimensional disturbances is responsible for transition.

- No essential differences have been found between the transition process in the subcritical and supercritical parameter range of linear stability theory.

- Our results conclusively demonstrate that the streamwise-periodic, timewise amplified mathematical model provides an appropriate description of the experimentally observed transition process in plane Poiseuille flow, as the results of Wray and Hussaini {18} did for the boundary layer case.

- For low initial disturbance amplitudes A, simulations with low horizontal resolution $N_1 = N_2 = 8$ are fully sufficient in the initial stages of transition, say, up to two oscillation periods before breakdown. Hence, parametric studies can be done relatively economically. A very crude estimate of the required number of points based

on the concept of Kolmogorov's inertial subrange gives
$N_i \sim A^{1/2} Re^{3/4}$.

The spatial resolution of up to $N_1 \cdot N_2 \cdot N_3 = 32 \cdot 32 \cdot 40$ we have used so far is not sufficient to simulate the multi-spike stages accurately. Future work should include additional parameter studies, further visualizations of the computed flow fields and inspection with respect to typical flow patterns (e.g. discrete vortex structures, inflection point profiles and streamline curvature), and high-resolution simulations to get beyond the 1-spike stage.

Comparing today's state with that commented upon by Stuart {8} only three years ago, we find that important progress has been made in understanding transition in shear flows in general, and in particular with respect to theoretical means "for the description of large-amplitude complex oscillations". This success has to be attributed, not least, to spectral simulations of transition.

REFERENCES

{1} D. Gottlieb, S.A. Orszag, Numerical analysis of spectral methods: theory and applications, NSF-CBMS Monograph No. 26, SIAM, Philadelphia, 1977.

{2} U. Schumann, G. S. Patterson, Numerical study of the return of axisymmetric turbulence to isotropy, J. Fluid Mech. 88 (1978), pp. 711-735.

{3} U. Schumann, Fast elliptic solvers and three-dimensional fluid-structure interactions in a pressurized water reactor, J. Comp. Phys. 36 (1980), pp. 93-127.

{4} L. Kleiser, U. Schumann, Treatment of incompressibility and boundary conditions in 3-D numerical spectral simulations of plane channel flows, in: E. H. Hirschel (ed.), Proc. Third GAMM-Conference on Numerical Methods in Fluid Mechanics, Vieweg, Braunschweig, 1980, pp. 165-173.

{5} L. Kleiser, Numerische Simulationen zum laminar-turbulenten Umschlagsprozeß der ebenen Poiseuille-Strömung, Dissertation, Karlsruhe, 1982, Kernforschungszentrum Karlsruhe, KfK 3271.

{6} L. Kleiser, Spectral simulations of laminar-turbulent transition in plane Poiseuille flow and comparison with experiments, 8 Int. Conf. Num. Meth. Fluid Dyn., Aachen, 1982.

{7} U. Schumann, G. Grötzbach, L. Kleiser, Direct numerical simulation of turbulence, in: W. Kollmann (ed.), Prediction methods for turbulent flows, Hemisphere, Washington, 1980, pp. 123-258.

{8} J. T. Stuart, Stability and transition: some comments on the problems, in: R. Eppler, H. Fasel (eds.), Laminar-turbulent transition, Springer, Berlin, 1980, pp. 1-13.

{9} M. Nishioka, S. Iida, Y. Ichikawa, An experimental investigation of the stability of plane Poiseuille flow, J. Fluid Mech. 72 (1975), pp. 731-751.

{10} M. Nishioka, S. Iida, S. Kanbayashi, An experimental investigation of the subcritical instability in plane Poiseuille flow (in Japanese), Proc. 10th Turbulence Symposium, Inst. Space Aeron. Sci., Tokyo Univ., 1978, pp. 55-62.

{11} M. Nishioka, M. Asai, S. Iida, An experimental investigation of the secondary instability, in: R. Eppler, H. Fasel (eds.), Laminar-turbulent transition, Springer, Berlin, 1980, pp. 37-46.

{12} P. S. Klebanoff, K. D. Tidstrom, L. M. Sargent, The three-dimensional nature of boundary-layer instability, J. Fluid Mech. 12 (1962), pp. 1-34.

{13} Th. Herbert, Stability of plane Poiseuille flow - theory and experiment, Virginia Polytechnic Institute and State University report VPI-E-81-35, to be published in Fluid Dynamics Transactions.

{14} Th. Herbert, Finite amplitude stability of plane parallel flows, AGARD-CP-224 (1977), pp. 3-1 to 3-10.

{15} S. A. Orszag, L. C. Kells, Transition to turbulence in plane Poiseuille and plane Couette flow, J. Fluid Mech. 96 (1980), pp. 159-205.

{16} S. A. Orszag, A. T. Patera, Subcritical transition to turbulence in plane channel flows, Phys. Rev. Lett. 45 (1980), pp. 989-993.

{17} S. A. Orszag, A. T. Patera, Subcritical transition to turbulence in planar shear flows, in: R. E. Meyer (ed.), Transition and Turbulence, Academic Press, New York, 1981, pp. 127-146.

{18} A. Wray, M. Y. Hussaini, Numerical experiments in boundary-layer stability, AIAA Paper 80-0275.

{19} F. X. Wortmann, Boundary-layer waves and transition, Springer Lect. Notes Physics 148 (1981), pp. 268-279.

{20} A. E. Perry, T. T. Lim, E. W. Teh, A visual study of turbulent spots. J. Fluid Mech. 104 (1981), pp. 387-405.

{21} U. Schumann, Fast elliptic solvers and their application in fluid dynamics, in: W. Kollmann (ed.), Computational fluid dynamics, Hemisphere, Washington, 1980, pp. 401-430.

{22} V. Faber, A. White, R. Sweet, An extension of the implicit capacitance matrix algorithm for solving Poisson equations on irregular regions, in: R. Vichnevetsky, R. S. Stepleman (eds.), Advances in computer methods for partial differential equations IV, IMACS, 1981, pp. 339-342.

{23} J. Gazdag, Time-differencing schemes and transform methods, J. Comp. Phys. 20 (1976), pp. 196-207.

APPLICATIONS OF SPECTRAL METHODS TO TURBULENT MAGNETOFLUIDS IN SPACE AND FUSION RESEARCH

DAVID MONTGOMERY*

Abstract. Recent and potential applications of spectral method computation to incompressible, dissipative magnetohydrodynamics are surveyed. Linear stability problems for one-dimensional, quasi-equilibria are approachable through a close analogue of the Orr-Sommerfeld equation. It is likely that for Reynolds-like numbers above certain as-yet-undetermined thresholds, all magnetofluids are turbulent. Four recent effects in MHD turbulence are remarked upon, as they have displayed themselves in spectral method computations: (1) inverse cascades; (2) small-scale intermittent dissipative structures; (3) selective decays of ideal global invariants relative to each other; and (4) anisotropy induced by a mean dc magnetic field. Two more conjectured applications are suggested. All the turbulent processes discussed are sometimes involved in current-carrying confined fusion magnetoplasmas and in space plasmas.

1. Introduction. Spectral method computation has had less impact on plasma physics and magnetohydrodynamics (hereafter: MHD) than it has on fluid mechanics. This is so despite the fact that there are an abundance of plasma problems which are technically more urgent, in an engineering sense, than the fluid mechanics problems to which the spectral method has typically been applied. For many of the plasma problems, there is little agreement on even the qualitative features of the solutions. It is not as much a matter of getting more accurate answers more efficiently as it is of uncovering new zeroth-order effects. This, too, is a stringent test of a numerical technique.

There has been an unfortunate gap in communication between investigators in plasma physics who must face these unsolved problems and the applied mathematicians who are in possession of the numerical expertise to deal with them. The following material is presented from the point of view of the plasma theory community and has a two-fold purpose: (1) to formalize, for spectral method numerical analysts, examples of central, unsolved problems which can be approached

---

*College of William and Mary, Williamsburg, VA 23185. Supported in part by NASA Grant NSG-7416 and U. S. Department of Energy Contract DE-AS05-76-ET53045

without the need to dig around in the untidy laboratory background of those problems; and (2) to remark upon some progress which has already been made by spectral techniques in the area of MHD turbulence.

Section 2 states the equations of incompressible MHD and their possible boundary conditions. They appear as a natural generalization of the Navier-Stokes equation of a viscous fluid, and define very similar classes of problems. Like the equations of a Navier-Stokes fluid, their properties will be seen to differ greatly between two- and three-dimensional flows. There is also a set of reduced MHD equations, the Strauss equations, which are intermediate between two and three dimensions and which are appropriate to the practically important case of a magnetofluid subjected to a very strong dc magnetic field.

Section 3 delineates a class of idealized MHD stability problems which parallel those related to the Orr-Sommerfeld equation for planar shear flows. No MHD problem has been studied as much as linear stability theory, but the scope and precision of the results achieved are no match for the corresponding ones for hydrodynamics.

Section 4 remarks upon some recent MHD turbulence results which have been obtained, and suggests some additional problems that seem ripe for investigation.

The need for such computations cannot be overemphasized. The plasmas of interest are often hot enough to burn up any probes which are inserted into them, and remote diagnostics are inadequate to pin down the internal MHD field variables. In many cases, all the reliable "experiments" we are going to have in the foreseeable future are likely to be those done on computers, solving what are believed to be the relevant dynamical equations. In this respect, the role computations have to play in deciding fundamental questions is potentially greater than in fluid mechanics, where the function of computers has often been in extending the accuracy of knowledge which is, in an experimental sense, nearly complete already.

2. Magnetofluid Description.

A. The MHD Equations. In this section, we write in dimensionless variables the simplest set of incompressible magnetofluid equations that could be called in some sense realistic. There are two basic solenoidal vector fields, the velocity field $\underset{\sim}{v} = \underset{\sim}{v}(\underset{\sim}{x},t)$ and the magnetic field $\underset{\sim}{B} = \underset{\sim}{B}(\underset{\sim}{x},t)$. They obey the following four partial differential equations:

$$(1) \qquad \frac{\partial \underset{\sim}{v}}{\partial t} + \underset{\sim}{v}\cdot\nabla\underset{\sim}{v} = \frac{1}{R}\nabla^2\underset{\sim}{v} - \nabla p + \underset{\sim}{B}\cdot\nabla\underset{\sim}{B}$$

(2) $$\nabla \cdot \underset{\sim}{v} = 0$$

(3) $$\frac{\partial \underset{\sim}{B}}{\partial t} + \underset{\sim}{v} \cdot \nabla \underset{\sim}{B} = \frac{1}{R_M} \nabla^2 \underset{\sim}{B} + \underset{\sim}{B} \cdot \nabla \underset{\sim}{v}$$

(4) $$\nabla \cdot \underset{\sim}{B} = 0.$$

Taking Eqs. (1) and (2) with the last term of Eq. (1) deleted just leaves us with the dynamics of a Navier-Stokes fluid, which appears in this light as a special case of magnetofluid dynamics. p is the total (dimensionless) pressure which results from taking the divergence of Eq. (1), using Eq. (2) to eliminate the time derivative, and solving the resulting Poisson equation.

R and R_M are two dimensionless numbers whose interpretation is not unique. Two cases are of particular importance. (1) If there is as much or more fluid kinetic energy than magnetic energy, velocities may be measured in units of a mean flow speed U, lengths in units of L, a macroscopic length scale over which the fields vary, and times in units of L/U, the "eddy turnover time." In this interpretation, $R \equiv UL/\nu$, where ν is the kinematic viscosity, and is of course the Reynolds number. $R_M \equiv UL/\eta$ is the magnetic Reynolds numbers, where η is the magnetic diffusivity, $c^2/4\pi\sigma$ in cgs units, with $c \equiv$ the speed of light, and σ is the electrical conductivity. Magnetic fields, in this interpretation, are measured in units of $U\sqrt{4\pi\rho}$, cgs units, where ρ is the (uniform) mass density. They may be $\lesssim U$, in magnitude. (2) Sometimes the magnetic field contains an externally-imposed dc part $\underset{\sim}{B}_0$ which is expected to remain larger than $\underset{\sim}{v}$ and the self-consistently-supported part of $\underset{\sim}{B}$. Then velocities may be measured in units of $C_A \equiv B_0/\sqrt{4\pi\rho}$, the Alfvén speed, and $\underset{\sim}{B}$ may be measured in units of B_0. In this interpretation, R_M becomes $S \equiv C_A L/\eta$, the Lundquist number (often, in the fusion literature, misleadingly called the magnetic Reynolds number), and R becomes $M \equiv C_A L/\nu$, which we might call the "viscous Lundquist number." This scheme has some advantages for nearly quiescent initial conditions, since the dimensionless numbers remain constant as the mean flow speed increases. The disadvantages are that $\underset{\sim}{v}$ and the variable part of $\underset{\sim}{B}$ may then remain $<< 1$, and though the time scale L/U, the eddy turnover time, is still important, it may be masked by the existence of the faster L/C_A, the "Alfvén transit time," which has no direct significance for the nonlinear behavior at all.

In summary, there are four possible Reynolds-like numbers which can be of importance: $R \equiv UL/\nu$, $R_M \equiv UL/\eta$, $S \equiv C_A L/\eta$, and

$M \equiv C_A L/\nu$; when the magnetic field is a sum of an externally-imposed part plus a part which is supported by internal currents, either part may be used to define C_A. Sorting out their roles in various problems has not been fully accomplished. They always occur in front of the only dissipative terms in the equations, and in the denominators, and which pair is appropriate depends upon the problem. The situation is more involved than the Navier-Stokes case, but not hopelessly so.

B. Two-Dimensional MHD. A simplified but still rich limit of Eqs. (1) - (4) is the two-dimensional one, with all variables independent of one spatial coordinate (z, say). In the most frequently considered geometry, $\underset{\sim}{B} = (B_x, B_y, 0)$, $\underset{\sim}{v} = (v_x, v_y, 0)$ and $\partial/\partial z \equiv 0$. For this case we can write, $\underset{\sim}{B} = \nabla\times A_z \hat{e}_z$, $\underset{\sim}{v} = \nabla\times\psi\hat{e}_z$, in terms of a vector potential $A_z\hat{e}_z$ and a stream function ψ. A and ψ determine all other variables in the problem. The vorticity $\underset{\sim}{\omega} = \omega\hat{e}_z = -\nabla^2\psi\hat{e}_z$ and the current density $\nabla\times\underset{\sim}{B} = \underset{\sim}{j} = j\hat{e}_z = -\nabla^2 A_z\hat{e}_z$ have only single components and are also solenoidal, so that $j = j(x,y,t)$ and $\omega = \omega(x,y,t)$ only. Eqs. (1) - (4) get replaced by the much simpler pair

$$\frac{\partial A_z}{\partial t} + \underset{\sim}{v}\cdot\nabla A_z = \frac{1}{R_M}\nabla^2 A_z \tag{5}$$

$$\frac{\partial \omega}{\partial t} + \underset{\sim}{v}\cdot\nabla\omega = \underset{\sim}{B}\cdot\nabla j + \frac{1}{R}\nabla^2\omega. \tag{6}$$

The two possible sets of meanings attached to the four various possible Reynolds-like numbers R, R_M, S, and M, still apply. A uniform constant dc magnetic field $\underset{\sim}{B}_0 = B_0\hat{e}_z$ can be included in $\underset{\sim}{B}$ without altering Eqs. (5) and (6), or affecting the dynamics in any way.

C. The Strauss Equations. A case intermediate between two and three dimensions is the Strauss limit, appropriate to the situation with a uniform mean dc magnetic field $\underset{\sim}{B}_0 = B_0\hat{e}_z$, but with some spatial variation with z, and a perpendicular variable magnetic field $\underset{\sim}{B}_\perp \equiv \nabla\times A_z\hat{e}_z$. This limit can be derived systematically in the case $B_\perp << B_0$. The result of a lengthy derivation [14] is that Eqs. (5) and (6) are modified to read

$$\frac{\partial A_z}{\partial t} + \underset{\sim}{v}_\perp\cdot\nabla A_z = \frac{1}{R_M}\nabla_\perp^2 A_z + B_0\frac{\partial\psi}{\partial z} \tag{7}$$

(8) $$\frac{\partial \omega}{\partial t} + \underset{\sim}{v}_{\perp} \cdot \nabla \omega = \frac{1}{R} \nabla_{\perp}^{2} \omega + \underset{\sim}{B}_{\perp} \cdot \nabla j + B_{0} \frac{\partial j}{\partial z},$$

where $\underset{\sim}{v}_{\perp} \equiv \nabla \times \psi \hat{e}_z$ also has only x and y components also, and, $\omega = -\nabla_{\perp}^{2}\psi$, $j = -\nabla_{\perp}^{2} A_z$, with $\nabla_{\perp} \equiv \hat{e}_x \frac{\partial}{\partial x} + \hat{e}_y \frac{\partial}{\partial y}$. The last two terms of Eqs. (7) and (8) represent the three-dimensionality. All variables in Eqs. (7) and (8) are functions of x, y, z, and t. The parallel fields, v_z and B_z, are essentially transported as passive scalars and their dynamics need not be discussed when discussing Eqs. (7) and (8). They evolve according to a pair of linear equations which have coefficients involving $\underset{\sim}{v}_{\perp}$ and $\underset{\sim}{B}_{\perp}$, but they do not react back on $\underset{\sim}{B}_{\perp}$ and $\underset{\sim}{v}_{\perp}$.

D. Boundary Conditions. There are three sets of severely idealized simple boundary conditions for MHD fluids. The conventional analogues of "free-slip" ideal fluid boundary conditions are $\underset{\sim}{v} \cdot \hat{n} = 0$, $\underset{\sim}{B} \cdot \hat{n} = 0$ at a rigid, perfectly conducting wall along which the magnetofluid flows freely. Here $\hat{n}$ is the unit normal to the wall. These are "ideal MHD" boundary conditions and are expected to avoid boundary-layer effects.

The analogue of "no-slip," non-ideal, fluid boundary conditions at a rigid, perfect conductor are that $\underset{\sim}{v} = 0$, $\hat{n} \cdot \underset{\sim}{B} = 0$, and $\hat{n} \times \underset{\sim}{j} = \hat{n} \times (\nabla \times \underset{\sim}{B}) = 0$. All three components of the velocity vanish at a rigid, perfectly-conducting wall for a non-ideal magnetofluid, as do the normal component of the magnetic field and the tangential components of the current. For a rigid, perfectly-conducting wall coated with a thin insulating layer, a situation encountered in experimental practice, $\underset{\sim}{B} \cdot \hat{n} = 0$ survives but $\hat{n} \times \underset{\sim}{j}$ is unconstrained. Problems in which magnetic field lines are permitted to penetrate the wall material become very intractable, although in many interesting experiments such wall penetration does occur.

For two dimensions, ideal boundary conditions make A and ψ constant around a closed conducting boundary, and non-ideal boundary conditions at a rigid perfect conductor add the conditions $\nabla_{\perp}^{2} A = 0$, $\hat{n} \cdot \nabla_{\perp} \psi = 0$. (For the Strauss equations the same conditions apply, but A and ψ may be functions of z.) For a non-ideal magnetofluid inside a rigid perfect conductor coated with a thin insulating layer, $\nabla_{\perp} \psi = 0$ on the boundary as does $\hat{n} \times \nabla_{\perp} A$, but $\nabla_{\perp}^{2} A$ is unconstrained. (Again, in the Strauss case, A and ψ may be functions of z.)

The foregoing three sets of MHD equations, then, accompanied by their three sets each of appropriate boundary conditions, define a far wider class of interesting problems than do the equations of a

Navier-Stokes fluid. They include Navier-Stokes flow as a special case. But they are not hopelessly more complicated than the Navier-Stokes equation. For many of these problems, even zeroth-order solutions have not been given, and it is fair to say that none of them have been studied with an intensity that approaches what has been brought to bear in fluid mechanics. We now turn to a consideration of two of these.

3. Stability of Quiescent Equilibria. A central concern in controlled fusion confinement research has been with the production of quiescent ($\underset{\sim}{v} \approx 0$) plasma states which will remain stably in place under the presence of (inevitable) small fluctuations. More theoretical plasma papers have been published about this topic than about anything else. However, much of the concern has been with device-specific situations with numerous complications, such as: toroidal geometry; compressibility; highly non-uniform dissipation coefficients (such as ν and η, which, strictly speaking, are tensors); chemical and radiative interactions with walls; inhomogeneous boundaries; impurities; etc. Drastic and uncontrolled approximations have been necessary to get any answers at all. There has been relatively little enthusiasm in the fusion community for the kind of accurate solutions of clean, simple situations analogous to those which the Orr-Sommerfeld equation requires. Yet rather spectacular success has been achieved for the Orr-Sommerfeld problem [1] by spectral methods, and a natural generalization of them to MHD stability seems a logical direction in which to proceed. In this section, we wish to define such a problem, though no results are as yet available.

The equilibrium we consider is a static one ($\underset{\sim}{v} = 0$) with a magnetic field $\underset{\sim}{B} = B_0(y)e_x$ which points in a direction perpendicular to its direction of variation. This is the magnetic analogue of plane shear flow, and has an electric current density $\underset{\sim}{j} = j_0(y)\hat{e}_z$, where $j_0(y) = -DB_0$, and $D \equiv d/dy$. For a constant resistivity, this is not strictly a static configuration, and $B_0(y)$ will resistively decay according to $B_0(y,t) = \exp(R_M^{-1} t\nabla^2)B_0(y,0)$. We will confine attention to dynamical processes which are fast on the time scale of this resistive decay, and will freeze $B_0(y)$ at its initial value. (If there are instability thresholds as η decreases, the solution will be inaccurate in the immediate vicinity of the threshold. It should be kept in mind that the very existence of such thresholds is still conjectural, however.)

If the MHD equations are linearized about this equilibrium, it is possible to generalize Squire's theorem, which says [2] that if a three-dimensional unstable mode exists, it is possible to find equivalent two-dimensional unstable modes with dissipation coefficients ν, η which are at least as high. Thresholds for instability may be sought by restricting attention to the two-dimensional case, with $\partial/\partial z \equiv 0$.

We put rigid parallel walls at the two planes $y = \pm 1$. The perturbed fields can be written as

$$(9) \qquad \underset{\sim}{v}^{(1)} = (u,v,w)$$

$$\underset{\sim}{B}^{(1)} = (b_x, b_y, b_z),$$

and without loss of generality, we may assme the x and t dependence of all perturbed field variables is $\sim \exp(i\alpha x - i\omega t)$, where α is an arbitrary wave number and ω is a complex eigenvalue to be determined. Discarding all products of perturbation quantities, we may manipulate the components of Eqs. (1) and (4) to yield:

$$(10) \qquad (D^2-\alpha^2)^2 v = -i\omega R(D^2-\alpha^2)v - i\alpha RB_0(D^2-\alpha^2)b_y + iR\alpha(D^2B_0)b_y$$

and

$$(11) \qquad (D^2-\alpha^2 + i\omega R_M)b_y = -i\alpha B_0 R_M v,$$

which are the analogues of Orr-Sommerfeld. The boundary conditions are that $v = 0$ at $y = \pm 1$, $Dv = 0$ at $y = \pm 1$, and $b_y = 0$ at $y = \pm 1$, if the infinite parallel planes are regarded as perfectly conducting. For the sake of comparison, Orr-Sommerfeld is, in [2] the same notation, $(D^2-\alpha^2)^2 v = i\alpha R\lfloor (U_0-\omega/\alpha)(D^2-\alpha^2)v - (D^2U_0)v\rfloor$, where $U_0 = U_0(y)$ is a shear flow velocity field in the x direction.

The non-viscous $(R = \infty)$ limit of Eqs. (10) and (11) has been studied to some extent [3]. As far as I am aware, they have never even been written down for finite R and R_M (which since zero streaming has been assumed for the equilibrium, are properly interpreted as M and S). Since most plasmas of current interest are more viscous than they are resistive, it would have made more sense to have studied the $R_M = \infty$ limit. We suspect that all eigenfunctions of the system (10), (11) are stable $(\text{Im}\ \omega_i \leqslant 0)$ for small enough R, R_M, and that they become unstable above certain thresholds as R, R_M are increased. Just what combination of R, R_M is important, we simply do not know. Russell Dahlburg [private communication] has shown that the ideal $(M \to \infty,\ S \to \infty)$ limit of Eqs. (10) and (11) is always stable. It is tempting to speculate that, because of the extent to which the equations of motion tie

$\underset{\sim}{v}^{(1)}$ and $\underset{\sim}{B}^{(1)}$ together, the stability threshold will be determined by the smaller of M and S. For plasmas which are more viscous than resistive, this will be M. But this is pure conjecture, as are the values the critical M or S will have (will they be of the order of the Reynolds numbers for shear flow instability, a few thousand?). It is a problem for which the spectral method seems made to order.

Once questions of linear stability are settled there is again the possibility that the true stability boundary will be determined by nonlinear and three-dimensional effects, as in the case of plane shear flows (cf. Orszag & Kells [4] and Orszag & Patera [5]) and those problems, too, are of a kind for which spectral method dynamical computation has been shown to be useful.

The magnetic field profiles $B_0(y)$ that might be investigated first are: (1) B_0 constant (this is analytically soluble and the results are simply damped Alfvén waves); (2) B_0 linear in y, or j_0 = const., the uniform current profile, which would be ruled out by imposing the boundary condition $\underset{\sim}{j}\times\hat{n} = 0$ at $y = \pm 1$; (3) B_0 parabolic, j_0 linear in y, which would be again ruled out by putting perfect conductors at $y = \pm 1$; and (4) B_0 cubic, j_0 parabolic and vanishing at $y = \pm 1$. This fourth case may be thought of as roughly analogous to plane Poiseuille flow, and mocks up most closely pinch effect magnetofluids used in a generation of confinement experiments.

Still a wider class of problems results if a uniform constant, external, magnetic field is imposed parallel to the planes but at an angle to $B_0(y)\hat{e}_x$. For these, it would seem that C_A ought to be computed in terms of the internally supported part of the magnetic field. We will not take up here the class of problems which results from the assumption that there is a non-zero flow in the equilibrium.

4. Magnetohydrodynamic Turbulence. Just as Eqs. (1) - (4) may be thought of as defining a stability problem (Eqs. (10), (11) plus boundary conditions), so they also define problems of transition and fully-developed turbulence, once those thresholds are significantly exceeded. The phenomena will be more various than Navier-Stokes phenomena because of the interplay of the magnetic field with the velocity field. They are also largely undiagnosed experimentally, partially for reasons already given: internal probes of the MHD variables cannot withstand the temperatures in many plasmas of current interest. A second, and unfortunate, reason is simple bureaucratic incomprehension of the fact that many of the effects which are troublesome at high temperatures could be explored on devices which are not hot enough to burn up probes. But in the thirty-year-plus

"crash program" to produce controlled fusion, there has been deemed to be insufficient time to explore these central, basic phenomena! Computers, are, and are likely to be for some time to come, the "experimental" arbiters of any theories of MHD turbulence we may construct.

In the following pages, we will remark upon four classes of turbulent MHD processes upon which some progress using spectral method computation has already been reported. These are:

(1) Inverse Cascade Processes
(2) Small Scale Dissipative Structures
(3) Selective Decay Processes
(4) Mean Magnetic Field Anisotropy.

Lengthy computational papers have been published on all of the above topics, and the reader is referred to them for details and graphs. In the following summary remarks, there will be no attempt to reproduce representative graphical documentation for the assertions, but the relevant references will be cited.

In addition to the above summaries, two probems will be remarked upon which appear at this moment to be ripe for numerical exploration by spectral methods. These are:

(5) Turbulence in Current-Carrying Magnetofluids and
(6) Current Boundary Layers.

It is to be stressed that the computations which have been reported have seldom, if ever, involved state-of-the-art numerical methods. The affordable spatial resolution has often been pathetically small. The spectral codes have been used as an off-the-shelf physics research tool, rather than with any deep concern for the place they occupy in applied mathematics. There is not one problem on the list that could not, or should not, be done better by knowledgeable applied mathematicians.

(1) Inverse Cascades [6] - [14]. It is well known that under some circumstances, small scale two-dimensional Navier-Stokes turbulence may feed the large scales, somewhat reversing the usual three-dimensional picture. Precise theory for this effect is in short supply, but extensive numerical computations have given repeated verification of its existence. Generally what are required are a small scale source of excitations and at least two global ideal invariants which emphasize differently the differing parts of wave number space. The latter requirement ensures that in the part of wave number space which is approximately conservative, spectral transfer in either direction in wave number space must be accompanied by a comparable amount in the opposite direction. What is basically a statistical-mechanical constraint on the allowed phase space guarantees that any band-limited stirring mechanism must send out something to both longer and shorter scales.

For three-dimensional MHD turbulence, the ideal invariants are

$$\varepsilon = \tfrac{1}{2} \int (\underset{\sim}{v}^2 + \underset{\sim}{B}^2) d^3x$$

$$H_M = \tfrac{1}{2} \int \underset{\sim}{A} \cdot \underset{\sim}{B} d^3x \tag{12}$$

$$H_C = \tfrac{1}{2} \int \underset{\sim}{v} \cdot \underset{\sim}{B} d^3x,$$

in the presence of either periodic or perfectly-conducting boundary conditions, where $\underset{\sim}{B} = \nabla \times \underset{\sim}{A}$, in terms of the vector potential $\underset{\sim}{A}$. For the same boundary conditions, the two-dimensional ideal invariants are

$$\varepsilon = \tfrac{1}{2} \int (\underset{\sim}{v}^2 + \underset{\sim}{B}^2) d^2x = \tfrac{1}{2} \int \lfloor (\nabla A_z)^2 + (\nabla \psi)^2 \rfloor d^2x$$

$$H_c = \tfrac{1}{2} \int \underset{\sim}{v} \cdot \underset{\sim}{B} d^2x = \tfrac{1}{2} \int \nabla \psi \cdot \nabla A_z d^2x \tag{13}$$

$$A = \tfrac{1}{2} \int A_z^2 d^2x,$$

where the notations are the same as in Section 2B. For the Strauss equations of Section 2C, the <u>energy</u> invariant ε and the <u>cross-helicity</u> H_c are still ideal invariants, and the magnetic helicity H_M degenerates into $H_M = B_0 \int A_z d^3x$.

For reasons that have been gone into in considerable detail elsewhere, in all three sets of equations, there appears to be a tendency for the magnetic quantities to accumulate, in forced situations, at the longest wavelengths. The effect has been clearly demonstrated in two- and three-dimensions, but at the time of this writing is still conjectural for the Strauss equations. It is a way of accounting for large-scale magnetic fields' spontaneous generation, as in the classic "dynamo problem." This is not the only situation where there is a need to account for the occurence of large-scale magnetic activity in a system in which small-scale turbulence is present; it may be, for example, that large-scale disruptions in tokamaks can be represented as similar to inverse cascades within the framework of the Strauss equations.

(2) <u>Small-scale Dissipative Structures</u> [15] - [17]. Navier-Stokes turbulence has been geometrically very elusive as far as forming a concrete, intuitive picture of it is concerned. Some success at statistical characterization of turbulent Navier-Stokes flows has

been achieved, but visualizing anything explicit about what is going on has proved frustrating. Despite its primitive state, MHD turbulence has pulled ahead in one department, at least, in providing a readily visualizable description for structures associated with dissipation. The picture is clearest in two dimensions, and may not be significantly different in three. A current filament concentrates itself near a magnetic X-point, and an associated flow pattern develops which sweeps magnetic flux symmetrically toward the X point where it is dissipated. There is outflow from the weak-field corners of the X point, sometimes at a significant fraction of the Alfvén speed, and the vorticity distribution is essentially that of a quadrupole. There is a strong asymmetry between the small-scale behavior of the magnetic and mechanical quantities. The general dissipative pattern when highly disordered initial conditions are present seems to be a random collection of such current filaments and associated X-points, which for decay situations gradually become fewer over time. One important unexplored question is that of the ratio of ohmic to viscous dissipation at high Reynolds numbers, and how this ratio varies with magnetic Prandtl number, ν/η.

(3) Selective Decay Processes [18], [19]. Not entirely unrelated to inverse cascades are the decay processes which are characterized by a more rapid decay of one global quantity than another. Geometrical constraints sometimes impose lower bounds to the ratio of the more rapidly to less rapidly decaying quantity, so that as time goes on, this ratio may approach its mimimum value. The complicated evolving turbulence heads, apparently, for a simple state which can be described by a variational principle. For instance, the minimum ε/A is achieved by Eqs. (5) and (6) whenever $\underset{\sim}{v} = 0$ and $\nabla^2 A_z + k^2 A_z = 0$, where k^2 is a positive constant. The minimum $|\varepsilon/H_M|$ is achieved by Eqs. (1) - (4) when $\nabla\times\underset{\sim}{B} = \lambda\underset{\sim}{B}$, and $\underset{\sim}{v} = 0$, the "force-free" state. There is evidence both experimental and theoretical that such "selective decays" may indeed occur. The typical attempt to test the hypothesis has involved following the evolution of a set of disordered initial conditions. The global ideal invariants are all observed to decay, but at unequal rates. In a typical two-dimensional MHD run [18], ε/A was observed to decay by more than a factor of three. In a three-dimensional situation, $|\varepsilon/H_M|$ approached to within about ten percent of its theoretical lower bound [19], though not by spectral method computation. In no case known to us has there been a turbulent computation in which ε/A or $|\varepsilon/H_M|$ did anything but decay.

The situation is complicated somewhat by the fact that there appears further to be an unexplained monotonic decay of $|\varepsilon/H_c|$ in two- and three-dimensions [20]. Two differing $t \to \infty$ asymptotic states for selectively decaying situations are compatible with what is known from high-Reynolds number computations so far, and they are not consistent with each other:

a. A state with no kinetic energy ($\underset{\sim}{v} = 0$), and ε/A(2D) or $|\varepsilon/H_M|$ (3D) minimized; and

b. A perfectly aligned state ($\underset{\sim}{v} = \pm\underset{\sim}{B}$) with perfect equipartition in which ε/A (2D), $|\varepsilon/H_c|$ (3D) are minimized, subject to the constraints that $\underset{\sim}{v} = \pm\underset{\sim}{B}$, pointwise. Further, rather demanding, computation will be required to sort out these possibilities.

(4) Mean Field Anisotropy [21]. Most laboratory magnetofluids of interest contain an externally-imposed dc magnetic field which selects a particular direction in space at each point. This destroys any possible isotropy of the turbulence, which is one of the most useful symmetries by which a turbulent tensor description can be simplified. The few laboratory measurements of MHD turbulence that have been made to indicate one ubiquitous feature: a large ratio ($\gtrsim 10$) of parallel to perpendicular correlation lengths, indicating a strong peaking of the $\underset{\sim}{k}$ vectors in the spectrum in a direction perpendicular to the mean field, [22], [23]. This inequality is also a key assumption in the derivation of the Strauss equations previously described. An important question is whether an initally isotropic turbulent MHD spectrum has some tendency, in the presence of a strong mean dc field, to develop an anisotropic $\underset{\sim}{k}$ spectrum which is sharply peaked perpendicularly to the direction of the mean field.

In an extensive set of 2D MHD computations (with a strong mean field in the plane of variation) Shebalin [21] has answered the question in the affirmative. What seems to be involved is an inhibited spectral transfer between Fourier modes with differing values of $\underset{\sim}{k}\cdot\underset{\sim}{B}_0$. In particular, in a resonant three-wave intéraction, $|\underset{\sim}{k}\cdot\underset{\sim}{B}_0|$ cannot increase, but $|\underset{\sim}{k}\times\underset{\sim}{B}_0|$ can. An initially isotropic distribution of $\underset{\sim}{k}$'s seems to elongate itself perpendicularly to the mean $\underset{\sim}{B}_0$ until something like the Kolmogoroff dissipation wave number is reached, so that the higher the Reynolds numbers, the more pronounced the anisotropy. Not paradoxically, if the dissipation is artificially (and unphysically) set equal to zero, the spectrum eventually re-isotropizes itself uniformly over the allowed Fourier modes due to higher order processes.

Now two processes are briefly remarked upon which seem ripe for investigation by spectral method computation. They are:

(5) Turbulence in Current-carrying Magnetofluids. Periodic boundary conditions preclude the existence of net fluxes of (say) electric current through any cross section of the region of computation. Since many of the most interesting laboratory plasmas carry

net currents, and as a consequence have non-vanishing circulations $\oint \underset{\sim}{B} \cdot d\underset{\sim}{\ell}$ around the perimeter, rectangular periodic boundary conditions stand in the way of numerical investigation of those turbulence effects (such as those for the Strauss equations involving inverse cascades) which require a net current. The expansion of the field variables in real Fourier sine series, as constrasted with complex Fourier series with terms $\sim \exp(i\underset{\sim}{k} \cdot \underset{\sim}{x})$, seems to pose no fundamental obstacles to spectral investigation of current-driven turbulence, but no results along these lines have as yet been reported.

(6) Current Boundary Layers. The boundary condition $\underset{\sim}{j} \times \hat{n} = 0$ at a conducting boundary demands that current densities which have no reason to be small in the interior of the magnetofluid must drop to zero at the boundary, reminiscent of the effect of a rough wall on a fluid velocity. The existence of a boundary layer of current gradients is to be expected near the boundary, just as a layer of vorticity is to be expected for a fluid flowing past a surface. Current gradients are widely (and probably correctly) believed to be a source of MHD activity ("tearing modes"), and it is not impossible that any confinement device with conducting interior boundaries will contain significant amounts of current-driven magnetic fluctuation associated with current-gradient boundary layers.

5. Summary. Fluid mechanics made a transition, in the wake of engineering achievements, into a sophisticated branch of applied mathematics. Magnetofluid mechanics may follow a somewhat different trajectory, in that the engineering achievements may well depend upon the prior development of a more sophisticated approach to the relevant questions of stability, transition, and turbulence than has characterized the first few decades of investigation. Spectral method computation could prove the most useful tool of all in this undertaking.

Acknowledgment

Helpful discussions of the material in this article with W. H. Matthaeus, G. Vahala, J. V. Shebalin, and R. Dahlburg are gratefully acknowledged. Equations (10) and (11) were derived jointly with R. Dahlburg.

REFERENCES

[1] S. A. ORSZAG, Accurate solution of the Orr-Sommerfeld stability equation, J. Fluid Mech. 50, (1971), pp. 689-703.

[2] C. C. LIN, The Theory of Hydrodynamic Stability, Cambridge University Press, Cambridge (U.K), 1955.

[3] See, e.g., J. WESSON, Finite resistivity instabilities of a sheet pinch, Nucl. Fusion 6, (1966), pp. 130-134.

[4] S. A. ORSZAG and L. C. KELLS, Transition to turbulence in plane Poiseuille and plane Couette flow, J. Fluid Mech. 96, (1980), pp. 159-205.

[5] S. A. ORSZAG and A. T. PATERA, Subcritical transition to turbulence in plane channel flows, Phys. Rev. Lett. 45, (1980), pp. 989-993.

[6] R. H. KRAICHNAN, Inertial ranges in two-dimensional turbulence, Phys. Fluids 10, (1967), pp. 1417-1423.

[7] R. H. KRAICHNAN and D. MONTGOMERY, Two-dimensional turbulence, Rept. Prog. Phys. 43, (1980), pp. 547-619.

[8] U. FRISCH, A. POUQUET, J. LEORAT, and A. MAZURE, Possibility of an inverse cascade of magnetic helicity in magnetohydrodynamic turbulence, J. Fluid Mech. 68, (1975), pp. 769-778.

[9] D. FYFE and D. MONTGOMERY, High beta turbulence in two-dimensional magnetohydrodynamics, J. Plasma Phys. 16, (1976), pp. 181-191.

[10] D. FYFE, G. JOYCE, and D. MONTGOMERY, Magnetic dynamo action in two-dimensional turbulent magnetohydrodynamics, J. Plasma Phys. 17, (1977), pp. 317-335.

[11] D. FYFE, D. MONTGOMERY, and G. JOYCE, Dissipative forced turbulence in two-dimensional magnetohydrodynamics, J. Plasma Phys. 17, (1977), pp. 369-398.

[12] A. POUQUET, On two-dimensional magnetohydrodynamic turbulence, J. Fluid Mech. 88, (1978), pp. 1-16.

[13] M. MENEGUZZI, U. FRISCH, and A. POUQUET, Helical and nonhelical turbulent dynamos, Phys. Rev. Lett. 47, (1981), pp. 1060-1064.

[14] D. MONTGOMERY, "Major disruptions, inverse cascades, and the Strauss equations," to be published in Physica Scripta,1982.

[15] W. M. MATTHAEUS and D. MONTGOMERY, Nonlinear evolution of the sheet pinch, J. Plasma Phys. 25, (1981), pp. 11-41.

[16] S. A. ORSZAG and C.-M. TANG, Small scale structure of two-dimensional magnetohydrodynamic turbulence, J. Fluid Mech. 90, (1979), pp. 129-143.

[17] W. H. MATTHAEUS, Reconnection in two-dimensions - localization of vorticity and current near magnetic X points, Geophys. Res. Lett. 9, (1982), pp. 660-663.

[18] W. H. MATTHAEUS and D. MONTGOMERY, Selective decay hypothesis at high mechanical and magnetic Reynolds number, Ann. N. Y. Acad. Sci. 357, (1980), pp. 203-219.

[19] S. RIYOPOULOS, A. BONDESON, and D. MONTGOMERY, Relaxation toward states of minimum energy in a compact torus, Phys. Fluids 25, (1982), pp. 107-115.

[20] R. GRAPPIN, U. FRISCH, A. POUQUET, and J. LEORAT, Alfvénic fluctuations as asymptotic states of MHD turbulence, Astronomy and Astrophysics 105, (1982), pp. 6-14.

[21] J. V. SHEBALIN, W. H. MATTHAEUS, and D. MONTGOMERY, "Anisotropy in MHD turbulence due to a mean magnetic field," submitted to J. Plasma Phys. 1982.

[22] D. C. ROBINSON and M. G. RUSBRIDGE, Structure of turbulence in the zeta plasma, Phys. Fluids 14, (1971), pp. 2499-2511.

[23] S. J. ZWEBEN, C. P. MENYUK, and R. J. TAYLOR, Small-scale magnetic fluctuations inside the macrotor tokamak, Phys. Rev. Lett. 42, (1979), pp. 1270-1274.

INHOMOGENEOUS FLOW CALCULATIONS BY SPECTRAL METHODS: MONO-DOMAIN AND MULTI-DOMAIN TECHNIQUES

Y. MORCHOISNE*

Abstract. The complexity of geometries and the accuracy of the time schemes are two important problems for spectral methods.

For a better time scheme accuracy, a spectral expansion in time is proposed.

For complex geometries, mono-domain or multi-domain mapping techniques are used.

Four examples are treated :

– the square domain with stream function formulation and spectral time derivatives,

– the cubic or square domain with velocity-pressure formulation and spectral time derivatives,

– the square duct with velocity-pressure formulation and finite difference time derivatives,

– the airfoil with velocity-pressure formulation, finite difference time derivatives and multi-domain technique.

I. SQUARE DOMAIN, STREAM FUNCTION FORMULATION [1]. The two dimensional (2-D) unsteady Navier Stokes (N.S.) equation is solved in a square domain by means of a space-time spectral method.

The stream function ψ is the unique solution of problem ($\mathcal{P}$) :

$$(1)\quad \begin{cases} g(\psi) = -\dfrac{\partial \Delta\psi}{\partial t} - U\dfrac{\partial \Delta\psi}{\partial x} - V\dfrac{\partial \Delta\psi}{\partial y} + \nu\,\Delta^2\psi = 0 \\ \text{within the space time domain } \Omega =]-1,+1[^2 \times]0,T] \\ \psi \text{ and } \dfrac{\partial\psi}{\partial n} \text{ given on the boundary } \Gamma \text{ of } [-1,+1]^2 \\ \psi(x,y,t=0) \text{ given for } (x,y) \in [-1,+1]^2 \end{cases}$$

where U and V are the velocity components :

$$U = \frac{\partial\psi}{\partial y} \qquad V = -\frac{\partial\psi}{\partial x}$$

* *Office National d'Etudes et de Recherches Aérospatiales (ONERA)*
92320 Châtillon (France).

The solution of ($\mathcal{P}$) will be obtained as the limit of a sequence of approximations

$$\psi^0, \psi^1 \dots \psi^n \dots$$

where each ψ^n satisfies the initial and boundary conditions of ($\mathcal{P}$) and is defined on the entire space-time domain Ω .

The variation $\delta\psi$ between two successive values of ψ satisfies homogeneous initial and boundary conditions. $\delta\psi$ is obtained by means of a Newton method :

$$\mathcal{L}^n(\delta\psi) + g^n = 0 \tag{2}$$

$$\psi^{n+1} = \psi^n + \sigma\,\delta\psi \tag{3}$$

where σ is an under relaxation coefficient calculated for stability of the iterative process,
where $\mathcal{L}^n$ is a pseudo-spectral space-time factorized approximation of the functional derivative of g,
and where g^n is a pseudo-spectral space-time approximation of $g(\psi^n)$.

We set :

$$\mathcal{L}^n = \mathcal{L}^n_t \cdot \mathcal{L}^n_x \cdot \mathcal{L}^n_y \tag{4}$$

We have :

$$g = \mathcal{H}(-\Delta) \tag{5}$$

with :

$$\mathcal{H} = \frac{\partial}{\partial t} + U^n\frac{\partial}{\partial x} + V^n\frac{\partial}{\partial y} - \nu\Delta \tag{6}$$

and :

$$-\Delta = -\frac{\partial^2}{\partial x^2} - \frac{\partial^2}{\partial y^2} \tag{7}$$

Factorized approximations of $\mathcal{H}$ and $-\Delta$ are (U^n and V^n assumed constant and given) :

$$\mathcal{H} \# \frac{1}{\varepsilon}\left(1+\varepsilon\frac{\partial}{\partial t}\right)\left(1+\varepsilon U^n\frac{\partial}{\partial x} - \varepsilon\nu\frac{\partial^2}{\partial x^2}\right)\left(1+\varepsilon V^n\frac{\partial}{\partial y} - \varepsilon\nu\frac{\partial^2}{\partial y^2}\right) \tag{8}$$

(9) $$-\Delta \# \frac{1}{\eta}\left(1-\eta\frac{\partial^2}{\partial x^2}\right)\left(1-\eta\frac{\partial^2}{\partial y^2}\right)$$

where ε and η are two factorization coefficients.

Then we have :

(10) $$\mathcal{L}_t^n = \frac{1}{\varepsilon\eta}\left(1+\varepsilon\frac{\partial}{\partial t}\right)$$

(11) $$\mathcal{L}_x^n = \left(1+\varepsilon U^n\frac{\partial}{\partial x} - \varepsilon\nu\frac{\partial^2}{\partial x^2}\right)\left(1-\eta\frac{\partial^2}{\partial x^2}\right)$$

(12) $$\mathcal{L}_y^n = \left(1+\varepsilon V^n\frac{\partial}{\partial y} - \varepsilon\nu\frac{\partial^2}{\partial y^2}\right)\left(1-\eta\frac{\partial^2}{\partial y^2}\right)$$

The variation $\delta\psi$ is calculated by solving :

(13) $$\mathcal{L}_t^n \cdot \mathcal{L}_x^n \cdot \mathcal{L}_y^n \,\delta\psi = -g^n .$$

Equation (13) gives us:

(14) $$\mathcal{L}_t^n a = -g^n$$

(15) $$\mathcal{L}_x^n b = a$$

(16) $$\mathcal{L}_y^n \delta\psi = b$$

Calculation of ψ^{n+1} as a function of ψ^n (on the entire space-time domain) consists in five steps :

1. Calculation of g^n
2. Calculation of a, solution of $\mathcal{L}_t^n a = -g(\psi^n)$ with initial condition at $t=0$.
3. Calculation of b solution of $\mathcal{L}_x^n b = a$ with boundary conditions for $x = \pm 1$.
4. Calculation of $\delta\psi$ solution of $\mathcal{L}_y^n \delta\psi = b$ with boundary conditions for $y = \pm 1$.
5. Calculation of ψ^{n+1} : $\psi^{n+1} = \psi^n + \sigma\delta\psi$.

<u>1. Calculation of g^n</u> . Along the t (or x or y) axis, ψ^n is expanded on a basis of first kind Chebyshev polynomials :

$$\Psi^n(x_\ell, y_m, \tau_p) = \sum_{k=0}^{K_t} \overline{\Psi}_k^\tau (x_\ell, y_m)\, T_k(\tau_p) \tag{17}$$

with $\tau_p = 2\dfrac{t_p}{T} - 1 \qquad (\tau_p \in [-1, +1])$

where $K_t + 1$ is the number of polynomials in the τ direction,
and where :

$x_\ell \in \{x_\ell\}$ collocation points along the x axis
$y_m \in \{y_m\}$ collocation points along the y axis
$\tau_p \in \{\tau_p\}$ collocation points along the τ axis.

Calculations of a first derivative with respect to τ (or x or y) is done as below :

– for given values of x and y the expansion in the τ -direction is obtained by means of a discrete Fourier transform as a function of the values at collocation points ;

– with a formal derivation we obtain the expansion of the first derivative ;

– by using an inverse Fourier transform the values of the first derivative are calculated at collocation points (for x and y given).

Example : Calculation of $\dfrac{\partial \Psi}{\partial t}$ at collocation points :

$$\frac{\partial \Psi}{\partial t} = \frac{2}{T}\frac{\partial \Psi}{\partial \tau}$$

$$\Psi(x_\ell, y_m, \tau_p) = \sum_k \overline{\Psi}_k^\tau (x_\ell, y_m)\, T_k(\tau_p)$$

$$\frac{\partial \Psi}{\partial \tau}(x_\ell, y_m, \tau_p) = \sum_k \overline{\Psi}_k^\tau (x_\ell, y_m)\, \frac{dT_k}{d\tau}(\tau_p)$$

$$= \sum_k \overline{\overline{\Psi}}_k^\tau (x_\ell, y_m)\, T_k(\tau_p)$$

with

$$\overline{\overline{\Psi}}_{k_1}^\tau = \alpha_{k_1 k_2}^{K_t}\, \overline{\Psi}_{k_2}^\tau \tag{18}$$

where α^{K_t} is a strictly triangular upper matrix (α^{K_t} is called the derivation matrix on the Chebyshev basis).

So, for given x_ℓ and y_m, we have three steps :

– calculation of $\overline{\Psi}_k^\tau (x_\ell, y_m)$

– calculation of $\overline{\overline{\Psi}}_k^\tau (x_\ell, y_m)$

– calculation of $\left\{ \dfrac{\partial \Psi}{\partial \tau}(x_\ell, y_m, \tau_p) \right\}$

The same technique is used for derivation along the x or y-axis.

2. Calculation of a. a is solution of :

$$\mathcal{L}_t^n a = -g^n$$

The initial condition for $\delta\psi$ is :

$$\delta\psi(x, y, t=0) = 0$$

and, by taking into account the relation

$$\mathcal{L}_x^n \cdot \mathcal{L}_y^n \, \delta\psi = a$$

we obtain :

(19) $$a(x, y, t=0) = 0 \quad \forall (x, y) \in [-1, +1]^2$$

Then we solve :

$$\mathcal{L}_t^n a = -g^n$$

with homogeneous initial conditions $(\tau = -1.)$

Along the time-axis we have :

$$\mathcal{L}_t^n = \frac{1}{\varepsilon\eta}\left(1 + \varepsilon \frac{\partial}{\partial t}\right)$$

and if α^K is the derivation matrix on the Chebyshev polynomial basis, we obtain :

(20) $$\left(I^{K_t} + \varepsilon' \alpha^{K_t}\right) \overline{a}^{\tau}(x_l, y_m) = -\varepsilon\eta \, \overline{g^n}^{\tau}(x_l, y_m), \quad \varepsilon' = \frac{2\varepsilon}{T}$$

where I^{K_t} is the unity matrix of order $K_t + 1$ and where $\overline{a}^{\tau}(x_l, y_m)$ and $\overline{g^n}^{\tau}(x_l, y_m)$ are the coefficients of the Chebyshev expansions of a and of g^n along the τ-axis for $x = x_l$ and $y = y_m$ $(x_l \in \{x_l\}$ and $y_m \in \{y_m\})$.

The initial condition $a(x_l, y_m, \tau = -1.) = 0$ is satisfied by replacing the last line of these relations (for $k = K_t$) by :

(21) $$\sum_{k=0}^{K_t} (-1)^k \, \overline{a}_k^{\tau}(x_l, y_m) = 0$$

3. Calculation of b. b is solution of :

$$\mathcal{L}_x^n b = a$$

The boundary conditions for $\delta\psi$ (in the x-direction) are homogeneous :

$$\begin{cases} \delta\psi(x=\pm 1, y, t) = 0 \\ \dfrac{\partial \delta\psi}{\partial x}(x=\pm 1, y, t) = 0 \end{cases}$$

Thanks to relation $\mathcal{L}_y^n \delta\psi = b$ we obtain :

$$\begin{cases} b(x=\pm 1, y, t) = \mathcal{L}_y^n \delta\psi(x=\pm 1, y, t) = 0 \\ \dfrac{\partial b}{\partial x}(x=\pm 1, y, t) = \dfrac{\partial}{\partial x}\mathcal{L}_y^n \delta\psi(x=\pm 1, y, t) \\ \qquad = \mathcal{L}_y^n \dfrac{\partial}{\partial x}\delta\psi(x=\pm 1, y, t) \\ \qquad = 0 \end{cases}$$

Then we have to calculate b solution of $\mathcal{L}_x^n b = a$ (a: given) with homogeneous boundary conditions :

$$\text{(22)} \begin{cases} b(x=\pm 1, y, t) = 0 \\ \dfrac{\partial b}{\partial x}(x=\pm 1, y, t) = 0 \end{cases}$$

Operator $\mathcal{L}_x^n$ is :

$$\mathcal{L}_x^n = \left(1 + \varepsilon U^n \frac{\partial}{\partial x} - \varepsilon \nu \frac{\partial^2}{\partial x^2}\right)\left(1 - \eta \frac{\partial^2}{\partial x^2}\right)$$

where $U^n = \dfrac{\partial \psi^n}{\partial y}$ is a function of x (and t and y also).

<u>Calculation of operator</u> $U^n \dfrac{\partial}{\partial x}$

$$\begin{cases} U^n(x_\ell, y_m, \tau_p) \# \displaystyle\sum_{k=0}^{K_x} \overline{U}_k^{n\,x}(y_m, \tau_p)\, T_k(x_\ell) \\ f(x_\ell, y_m, \tau_p) \# \displaystyle\sum_{k=0}^{K_x} \overline{f}_k^{\,x}(y_m, \tau_p)\, T_k(x_\ell) \end{cases}$$

$$U^n(x_l, y_m, \tau_p)\, f(x_l, y_m, \tau_p) = \left\{\sum_{k_1} \overline{U}^{n\,x}_{k_1}(y_m, \tau_p)\, T_{k_1}(x_l)\right\} \times \left\{\sum_{k_2} \overline{f}^{\,x}_{k_2}(y_m, \tau_p)\, T_{k_2}(x_l)\right\}$$

At collocation points $\{x_l\}$ we have :

$$T_{k_1}(x_l)\, T_{k_2}(x_l) = \frac{1}{2}\left(T_{\overline{k_1+k_2}} + T_{|k_1-k_2|}\right)(x_l)$$

with $$\overline{k_1+k_2} = \begin{cases} k_1+k_2 & \text{for } k_1+k_2 \leq K_x \\ 2K_x - k_1 - k_2 & \text{for } k_1+k_2 > K_x \end{cases}$$

by using the definition

$$\begin{cases} T_k(\cos\Theta_l) = \cos(k.\Theta_l) \\ \cos\Theta_l = x_l \quad , \quad \Theta_l = \dfrac{l.\pi}{K_x} \end{cases}$$

So :

$$(23) \quad U^n(x_l, y_m, \tau_p)\, f(x_l, y_m, \tau_p) \,\#\, \sum_{k_1, k_2} \beta^{K_x}_{k_1 k_2}(U^n) \times \overline{f}^{\,x}_{k_2}(y_m, \tau_p)\, T_{k_1}(x_l)$$

where $\beta^{K_x}(U^n)$ is a matrix function of $\{U^n(x_l, y_m, \tau_p)\}$ only.

The spectral form of operator $\mathcal{L}^n_x$ is then :

$$(24) \quad \left(I^{K_x} + \varepsilon\beta^{K_x}(U^n)\alpha^{K_x} - \varepsilon\nu(\alpha^{K_x})^2\right)\left(1 - \eta(\alpha^{K_x})^2\right)\overline{b}^{\,x}(y_m, \tau_p) = \overline{a}^{\,x}(y_m, \tau_p)$$

For boundary conditions we replace the four last lines of (24) by (TAU method) :

$$(25) \quad \begin{cases} (\pm 1)^k\, \overline{b}^{\,x}_k(y_m, \tau_p) = 0 & \longleftarrow \quad b(\pm 1, y_m, \tau_p) = 0 \\ (\pm 1)^k\, k^2\, \overline{b}^{\,x}_k(y_m, \tau_p) = 0 & \longleftarrow \quad \dfrac{\partial b}{\partial x}(\pm 1, y_m, \tau_p) = 0 \end{cases}$$

and for each value of (y_m, τ_p) we have to solve a linear system for $\overline{b}_k^{\,x}$. Then by using a discrete Fourier transform, we obtain $\{b(x_\ell, y_m, \tau_p)\}$

4. Calculation of $\delta\psi$. $\delta\psi$ is solution of :

$$\mathcal{L}_y^n \, \delta\psi = b$$

with, as boundary conditions :

$$\begin{cases} \delta\psi(x, y = \pm 1, t) = 0 \\ \dfrac{\partial \delta\psi}{\partial y}(x, y = \pm 1, t) = 0 \end{cases}$$

The same technique, described above, is used and we obtain :

(26) $$\left(I^{K_y} + \varepsilon \beta^{K_y}(V^n)\, \alpha^{K_y} - \varepsilon \nu (\alpha^{K_y})^2\right)\left(1 - \eta (\alpha^{K_y})^2\right) \overline{\delta\psi}^{\,y}(x_\ell, \tau_p) = \overline{b}^{\,y}(x_\ell, \tau_p)$$

For boundary conditions we use the TAU method and we replace the four last lines by :

(27) $$\begin{cases} (\pm 1)^k \, \overline{\delta\psi}_k^{\,y}(x_\ell, \tau_p) = 0 & \longleftarrow \delta\psi(x_\ell, \pm 1, \tau_p) = 0 \\ (\pm 1)^k \, k^2 \, \overline{\delta\psi}_k^{\,y}(x_\ell, \tau_p) = 0 & \longleftarrow \dfrac{\partial \delta\psi}{\partial y}(x_\ell, \pm 1, \tau_p) = 0 \end{cases}$$

For each value of (x_ℓ, τ_p) we have to solve a linear system for $\overline{\delta\psi}_k^{\,y}$. Then by means of a discrete Fourier transform we obtain : $\{\delta\psi(x_\ell, y_m, \tau_p)\}$

5. Calculation of $\psi^{n+1} = \psi^n + \sigma \delta\psi$

An approximate value of the optimal under relaxation coefficient σ is given by stability and convergence conditions[1].

Remark :

For steady flow calculation the solution can be obtained directly through a simplified formulation of the method.

We suppress the term $-\dfrac{\partial \Delta\psi}{\partial t}$ in equation (1) and the factor $\mathcal{L}_t^n$ in relation (4), factors $\mathcal{L}_x^n$ and $\mathcal{L}_y^n$ remaining identical.

Each iteration (calculation of ψ^{n+1} as a function of ψ^{n} on the entire space domain) consists now in four steps :

1. Calculation of g^n

$$g^n = -U^n \frac{\partial \Delta \psi^n}{\partial t} - V^n \frac{\partial \Delta \psi^n}{\partial y} + \nu \Delta^2 \psi^n \tag{28}$$

2. Calculation of a solution of $\mathcal{L}_x^n a = -g^n$ with boundary conditions for $x = \pm 1$

3. Calculation of $\delta\psi$ solution of $\mathcal{L}_y^n \delta\psi = b$ with boundary conditions for $y = \pm 1$

4. Calculation of ψ^{n+1}

$$\psi^{n+1} = \psi^n + \sigma \delta\psi$$

II. VELOCITY-PRESSURE FORMULATION FOR CUBIC DOMAIN [2]. The 3-D unsteady Navier-Stokes equations are solved in a cube by means of a space-time spectral method.

The velocity-pressure (U-P) is the unique solution of problem ($\mathcal{P}$) :

(29)
$$\begin{cases} \begin{cases} g_i(U,P) \equiv \dfrac{\partial U_i}{\partial t} + \dfrac{\partial U_i U_j}{\partial x_j} - \nu \Delta U_i + \dfrac{\partial P}{\partial x_i} = 0 \\ \qquad\qquad i, j, k = 1, 2 \text{ or } 3 \\ g_4(U,P) \equiv \dfrac{\partial U_k}{\partial x_k} = 0 \end{cases} \\ \text{within the space-time domain} \\ \Omega =]-1,+1[^3 \times]0,T] \\ U \text{ given on } \Gamma \text{ boundary of } [-1,+1]^3 \text{ and,} \\ U \text{ given for } t=0 \text{ on } [-1,+1]^3. \end{cases}$$

where U is the velocity vector and x the coordinate vector (x_1, x_2, x_3)

Solution of ($\mathcal{P}$) will be obtained as the limit of a sequence of approximations in the entire space-time domain :

$$(U,P)^0, \; --- \; (U,P)^n --- \tag{30}$$

where each $(U,P)^n$ satisfies the initial and boundary conditions of $(\mathcal{P})$ and is defined on Ω.

The variation (u,p) between two successive values of (U,P) satisfies homogeneous initial and boundary conditions.

(u,p) is calculated thanks to a finite difference preconditionning technique and we obtain four variation equations :

$$h_i^n(u,p) + g_i^n = 0 \qquad i = 1, \ldots 4 \tag{31}$$

where h_i^n is a finite difference (in space and time) approximation of the functional derivative of g_i and where g_i^n is a pseudo-spectral space-time approximation of $g_i((U,P)^n)$.

For approximations h_i^n we use classical finite difference schemes :

$$(32)\quad \begin{cases} h_i^n(u,p) = D_t u_i + D_j(u_i U_j^n + u_j U_i^n) \\ \qquad\qquad\qquad - \nu D^2 u_i + D_i p \\ i,j = 1,2 \text{ or } 3 \\ h_4^n(u,p) = D_k u_k \qquad k = 1,\ 2 \text{ or } 3 \end{cases}$$

where D_t, D_i, D^2 are finite difference approximations of the continuous operators $\frac{\partial}{\partial t}$, $\frac{\partial}{\partial x_i}$, Δ and are computed on the space-time spectral collocation grid.

For pressure variation p, the divergence of the variation momentum equations give us :

$$D_i(D_t u_i + D_j(u_i U_j^n + u_j U_i^n)) - \nu D_i D^2 u_i$$
$$+ D_i D_i p + D_i g_i^n = 0$$
$$i,j = 1,2 \text{ and } 3$$

Taking into account the fourth equation :

$$D_k u_k + g_4^n = 0$$

we have :

$$D_t g_4^n + D_i D_j(u_i U_j^n + u_j U_i^n) + \nu D^2 g_4^n$$
$$+ D^2 p + D_i g_i^n = 0 \tag{33}$$

This Poisson equation is solved with a Neumann boundary condition obtained by multiplying the variation momentum equations by the external normal vector (n_1, n_2, n_3)

(34) $$n_i\{D_j(u_i U_j^n + u_j U_i^n) - \nu D^2 u_i + D_i p + g_i^n\} = 0$$

As we have :

(35) $$\Delta u = -\text{curl curl } u + \text{grad div } u$$
$$\Delta u = -\text{curl curl } u - \text{grad } g_4^n$$

we replace $D^2 u_i$ by $-R_i^2 u - D_i g_4^n$
where R_i^2 is the i-th component of the finite difference approximation of the operator curl curl

So we obtain :

(36) $$n_i\{D_j(u_i U_j^n + u_j U_i^n) + \nu R_i^2 u + \nu D_i g_4^n + D_i p + g_i^n\} = 0$$

Calculation of $(U, P)^{n+1}$ as a function of $(U, P)^n$ (on the entire space-time domain) consists in three steps :

1. calculation of g_i^n
2. calculation of (u, p)
3. $\begin{cases} U^{n+1} = U^n + \sigma u \\ P^{n+1} = P^n + \sigma p \end{cases}$

1. Calculation of g_i^n. In each space or time direction U and P are expanded on a basis of first kind Chebyshev polynomials.

For example :

(37) $$P^n(x_m, y_p, z_q, \tau_r) = \sum_{l=0}^{L} \overline{P_l^n}^{\tau}(x_m, y_p, z_q)\, T_l(\tau_r)$$

with $\tau = \frac{2t}{T} - 1 \quad (\tau \in [-1, +1])$

and where $x_m \in \{x_m\}$ collocation points on the x-axis.

As in the case for stream function formulation, the time or space derivatives are calculated by formal derivation of the expansions. The value of g_i^n are then calculated at collocation points with a great accuracy.

2. Calculation of (u, p). Equations to solve are :

— three modified momentum equations

$$(38)\quad \begin{cases} D_t u_i + D_j(u_i U_j^n + u_j U_i^n) - \nu D^2 u_i + D_i p + g_i^n = 0 \\ \text{with homogeneous boundary conditions} \\ \text{for } u_i : \quad u_i = 0 \text{ on } \Gamma \end{cases}$$

– one Poisson equation

$$(39)\quad \begin{cases} D_t g_4^n + D_i D_j (u_i U_j^n + u_j U_i^n) \\ \qquad + \nu D^2 g_4^n + D^2 p + D_i g_i^n = 0 \\ \text{with Neumann boundary condition for } p \\ n_i \{ D_j (u_i U_j^n + u_j U_i^n) + \nu R_i^2 u + \nu D_i g_4^n \\ \qquad + D_i p + g_i^n \} = 0 \end{cases}$$

The finite difference scheme (based on the space-time collocation points) used for approximating the partial differential equations is :

– first order in space and time,
– implicit for diffusion terms,
– implicit for pressure gradient,
– explicit for convection terms.

For each collocation time, due to boundary conditions for p, an iterative process is necessary :

i) calculation of velocity : equation (38),
ii) calculation of boundary condition for p and calculation of p: equation (39).

III. SQUARE DUCT FLOW [3]

The 3-D unsteady N.S. equations are solved in a domain $\Omega =]-1, +1[^2 \times]0, L[$ by means of a "finite difference in time" and "spectral in space" method.

This velocity-pressure (U, P) is the unique solution of problem $(\mathcal{P})$

$$(40)\quad \begin{cases} \dfrac{\partial U_i}{\partial t} + \dfrac{\partial U_i U_j}{\partial x_j} - \nu \Delta U_i + \dfrac{\partial P}{\partial x_i} = 0 \qquad i, j = 1, 2 \text{ or } 3 \\ \dfrac{\partial U_k}{\partial x_k} = 0 \\ \text{boundary conditions :} \\ U = 0 \quad \text{for } x = \pm 1 \text{ or } y = \pm 1 \end{cases}$$

$$\left\{\begin{array}{l} U \text{ given for } z=0 \\ \dfrac{\partial U}{\partial z}=0 \quad \text{for } z=L \\ \text{initial conditions :} \\ U \text{ given for } t=0 \end{array}\right.$$

A leap-frog (for convection) and Crank-Nicolson (for diffusion) scheme is used :

$$(41)\quad \left\{\begin{array}{l} \dfrac{U_i^{n+1}-U_i^{n-1}}{2\delta t}+\dfrac{\partial (U_i U_j)^n}{\partial x_j}-\dfrac{\nu}{2}\left(\Delta U_i^{n+1}+\Delta U_i^{n-1}\right) + \dfrac{\partial P^{n+1}}{\partial x_i}=0 \\ \dfrac{\partial U_k^{n+1}}{\partial x_k}=0 \end{array}\right.$$

where $U^n = U(n.\delta t)$

Equations (41) become

$$(42)\quad \frac{U_i^{n+1}}{2\delta t}-\frac{\nu}{2}\Delta U_i^{n+1}+\frac{\partial P^{n+1}}{\partial x_i}+A_i^n=0$$

with $A_i^n = -\dfrac{U_i^{n-1}}{2\delta t}+\dfrac{\partial (U_i U_j)^n}{\partial x_j}-\dfrac{\nu}{2}\Delta U_i^{n-1}$

The divergence of (42) gives us a Poisson equation for P:

$$(43)\quad \Delta P^{n+1}+\frac{\partial A_k^n}{\partial x_k}=0$$

to be solved with a Neumann boundary conditions :

$$n_i\frac{\partial P^{n+1}}{\partial x_i}+n_i\left(\frac{U_i^{n+1}}{2\delta t}-\frac{\nu}{2}\Delta U_i^{n+1}+A_i^n\right)=0$$

To take into account the continuity equation along the boundary we replace

$$\begin{aligned}\Delta U^{n+1} &= \operatorname{grad}\operatorname{div} U^{n+1}-\operatorname{curl}\operatorname{curl} U^{n+1} \\ &= -\operatorname{curl}\operatorname{curl} U^{n+1}\end{aligned}$$

and we obtain (on Γ):

$$(44)\quad n_i\left(\frac{\partial P^{n+1}}{\partial x_i}+\frac{U_i^{n+1}}{2\delta t}+\frac{\nu}{2}\left(\operatorname{curl}\operatorname{curl} U^{n+1}\right)_i+A_i^n\right)=0$$

For each time level $t=t_{n+1}=(n+1)\,\delta t$, an iterative process is used and $(U,P)^{n+1}$ is the limit of a sequence of approximations :

$$(U,P)_0^{n+1},\ (U,P)_1^{n+1} \ldots (U,P)_\ell^{n+1} \ldots$$

where ℓ is an iteration subscript and n a time superscript.

So we have the following algorithm :

time loop : superscript n

$$(U,P)_0^{n+1} = \ldots$$

iteration loop : subscript ℓ

$$(U,P)_{\ell+1}^{n+1} = (U,P)_\ell^{n+1} + \sigma.(u,p)$$

For $n+1$ given, calculation of $(U,P)_{\ell+1}^{n+1}$ as a function of $(U,P)_\ell^{n+1}$ consists of three steps described below.

1. Calculation of residues

$$(45)\quad \begin{cases} g_{i,\ell}^{n+1} = \dfrac{U_{i,\ell}^{n+1}}{2\delta t} - \dfrac{\nu}{2}\Delta U_{i,\ell}^{n+1} + \dfrac{\partial P_\ell^{n+1}}{\partial x_i} + A_i^n \\ g_{4,\ell}^{n+1} = \Delta P_\ell^{n+1} + \dfrac{\partial A_k^n}{\partial x_k} \qquad i,k = 1,2 \text{ and } 3 \end{cases}$$

These residues are calculated by means of Chebyshev polynomial expansions along the x, y and z axis.

2. Calculation of variation $(\underline{u},\underline{p})$. (u,p) is solution of the finite difference equations (46) and (47) :

$$(46)\quad \begin{cases} \dfrac{u_i}{2\delta t} - \dfrac{\nu}{2}D^2 u_i + D_i p + g_{i,\ell}^{n+1} = 0 \\ \text{with boundary conditions} \\ \begin{cases} u_i = 0 \quad \text{for } x = \pm 1 \text{ or } y = \pm 1 \text{ or } z = 0 \\ \dfrac{\partial u_i}{\partial x_3} + \dfrac{\partial U_{i,\ell}^{n+1}}{\partial x_3} = 0 \quad \text{for } z = L \end{cases} \qquad i = 1, 2 \text{ and } 3 \end{cases}$$

$$(47)\quad \begin{cases} D^2 p + g_{4,\ell}^{n+1} = 0 \\ \text{with boundary conditions} \\ n_i\left[D_i p + \dfrac{u_i}{2\delta t} + \dfrac{\nu}{2}R_i^2 u + \dfrac{\partial P_\ell^{n+1}}{\partial x_i} + \dfrac{U_{i,\ell}^{n+1}}{2\delta t} + \dfrac{\nu}{2}(\text{curl curl } U_\ell^{n+1})_i + A_i^n\right] = 0 \end{cases}$$

where D^2, D_i and R^2 are finite difference approximations of $\Delta, \dfrac{\partial}{\partial x_i}$ and curl curl.

As the boundary conditions for p are functions of u, equations (46) and (47) are to be solved simultaneously.

3. Calculation of $(U, P)^{m+1}_{\ell+1}$

$$(48)\quad \begin{cases} U^{m+1}_{\ell+1} = U^{m+1}_{\ell} + \sigma u \\ P^{m+1}_{\ell+1} = P^{m+1}_{\ell} + \sigma p \end{cases}$$

where σ is an under relaxation coefficient calculated to obtain convergence for the method.

IV. AIRFOIL FLOW, MULTI-DOMAIN TECHNIQUE [4]

Let ($\mathcal{P}$) be a second order problem defined on an open set $\Omega \subseteq \mathbb{R}^2$ with boundary Γ.

$$(49)\quad \begin{cases} \left.\begin{array}{l} \dfrac{\partial U}{\partial t} + C - \nu \Delta U + \operatorname{grad} P = 0 \\ \operatorname{div} U = 0 \end{array}\right\} \text{ in } \Omega \\ \left.\begin{array}{ll} U = U_0 & \text{on } \Gamma_D \\ \dfrac{\partial U}{\partial n} = T & \text{on } \Gamma_N \end{array}\right\} \text{ with } \Gamma = \Gamma_D \cup \Gamma_N \text{ and } \displaystyle\int_\Gamma U_0 \, d\sigma = 0 \text{ if } \Gamma_N = \emptyset. \end{cases}$$

where C is the convective term with components

$$c_i = \frac{\partial U_i U_j}{\partial x_j}$$

The unsteady problem ($\mathcal{P}$) is transformed in a sequence of steady problems ($\mathcal{P}^m$) by means of a first order time scheme implicit for diffusion, explicit for convection.

$$\frac{\partial U}{\partial t} \# \frac{U^m - U^{m-1}}{\delta t}$$

with $\quad U^m = U(m.\delta t) \qquad \delta t$ time step

$$(50)\quad \begin{cases} \left.\begin{array}{l} \dfrac{U^m - U^{m-1}}{\delta t} + \tilde{C}^m - \nu \Delta U^m + \operatorname{grad} P^m = 0 \\ \operatorname{div} U^m = 0 \end{array}\right\} \text{ in } \Omega \qquad \tilde{C}^m = 1.5\, C^{m-1} - 0.5\, C^{m-2} \\ U^m = U_0^m \quad \text{on } \Gamma_D \\ \dfrac{\partial U^m}{\partial n} = T^m \quad \text{on } \Gamma_N \end{cases}$$

The divergence of the momentum equation gives us a Poisson equation for the pressure P^m

(51) $$-\Delta P^m = \operatorname{div}\left(\tilde{C}^m - \frac{U^{m-1}}{\delta t}\right)$$

By multiplying by the unit vector, n, normal to Γ, we obtain :

(52) $$\frac{\partial P^m}{\partial n} = -n \cdot \left(\frac{U^m - U^{m-1}}{\delta t} + \tilde{C}^m - \nu \Delta U^m\right)$$

where $\frac{\partial P}{\partial n}$ is the normal derivative of P.

We take into account the relation $\operatorname{div} U^m = 0$
so we have

(53) $$\begin{aligned}\Delta U^m &= \operatorname{grad} \operatorname{div} U^m - \operatorname{curl} \operatorname{curl} U^m \\ &= -\operatorname{curl} \operatorname{curl} U^m\end{aligned}$$

and the boundary condition for pressure becomes:

(54) $$\begin{aligned}\frac{\partial P^m}{\partial n} &= -n \cdot \left(\frac{U^m - U^{m-1}}{\delta t} + \tilde{C}^m + \nu \operatorname{curl} \operatorname{curl} U^m\right) \\ &= S^m(U^m)\end{aligned}$$

Then problem ($\mathcal{P}^m$) is equivalent to problem ($\mathcal{Q}^m$) :

(55) $$\left\{\begin{array}{l} \left\{\begin{array}{ll} -\Delta U^m + \dfrac{U^m}{\nu \delta t} + \dfrac{\operatorname{grad} P^m}{\nu} = -\dfrac{\tilde{C}^m}{\nu} + \dfrac{U^{m-1}}{\nu \delta t} & \text{in } \Omega \\ \left.\begin{array}{l} U^m = U_0^m \text{ on } \Gamma_D \\ \dfrac{\partial U^m}{\partial n} = T^m \text{ on } \Gamma_N \end{array}\right\} & \text{with } \Gamma = \Gamma_D \cup \Gamma_N \text{ and } \displaystyle\int_\Gamma U_0 \, d\sigma = 0 \text{ if } \Gamma_N = \emptyset \end{array}\right. \\ \left\{\begin{array}{ll} -\Delta P^m = \operatorname{div}\left(\tilde{C}^m - \dfrac{U^{m-1}}{\delta t}\right) & \text{in } \Omega \\ \dfrac{\partial P^m}{\partial n} = -n \cdot \left(\dfrac{U^m - U^{m-1}}{\delta t} + \tilde{C}^m + \nu \operatorname{curl} \operatorname{curl} U^m\right) & \\ \qquad = S^m & \text{on } \Gamma \end{array}\right. \end{array}\right.$$

Due to boundary conditions for P^m, problems for P^m and U^m are connected and are to be solved simultaneously.

The closure $\overline{\Omega}$ of Ω is split into a set of subdomains $(\mathcal{C}_k)_{k\in\mathbb{K}}$. Let Ω_k be the interior of $\mathcal{C}_k$ and Γ_k the boundary of Ω_k for each $k \in \mathbb{K}$.

Solving (Q^m) amounts to solving the N (number of elements of $\mathbb{K}$) problems (Q^m_k) obtained by taking the restriction to Ω_k of the equations set on Ω.

Within Ω_k, (U^m_k, P^m_k) restriction to Ω_k of (U^m, P^m) is calculated according to a mapping technique with Dirichlet boundary conditions deduced from the boundary conditions on Γ or from continuity conditions.

The following Dirichlet boundary conditions are imposed explicitly :
– value of U^m on Γ_D : $U^m = U^m_0$
– continuity of U^m and P^m on $\Omega \cap \Gamma_k$

The following Neumann boundary conditions are imposed implicitly by means of an iterative scheme :

– value of $\dfrac{\partial U^m}{\partial n}$ on Γ_N : $\dfrac{\partial U^m}{\partial n} = T^m$,
– value of $\dfrac{\partial P^m}{\partial n}$ on Γ : $\dfrac{\partial P^m}{\partial n} = S^m$,
– continuity of $\dfrac{\partial U^m}{\partial n}$ and $\dfrac{\partial P^m}{\partial n}$ on $\Omega \cap \Gamma_k$.

For a given value of t ($t = t_m$), the iterative scheme is then :

1. In each open set Ω_k, calculation of (U^m_k, P^m_k) for given Dirichlet boundary conditions
2. On each Γ_k, calculation of (U^m_k, P^m_k) deduced from the boundary conditions on Γ or from continuity conditions.

1. <u>Calculation of (U^m_k, P^m_k) within Ω_k</u>

First we solve

$$(56)\quad \begin{cases} -\Delta P^m_k = \operatorname{div}\left(\tilde{C}^m_k - \dfrac{U^{m-1}_k}{\delta t} \right) & \text{in } \Omega_k \\ P^m_k \text{ given on } \Gamma_k , \end{cases}$$

then

$$(57)\quad \begin{cases} -\Delta U^m_k + \dfrac{U^m_k}{\nu\,\delta t} = -\dfrac{\tilde{C}^m_k}{\nu} + \dfrac{U^{m-1}_k}{\nu\,\delta t} - \dfrac{\operatorname{grad} P^m_k}{\nu} \\ U^m_k \text{ given on } \Gamma_k . \end{cases}$$

So we have to solve three Helmholtz problems (within each Ω_k) :

$$(58)\quad \begin{cases} -\Delta \phi_k^m + \lambda \phi_k^m = f_k^m & \text{in } \Omega_k \\ \phi_k^m \text{ given on } \Gamma_k . \end{cases}$$

We define a bijection G_k from $\overline{\Omega_k}$ onto $[-1,+1]^2$

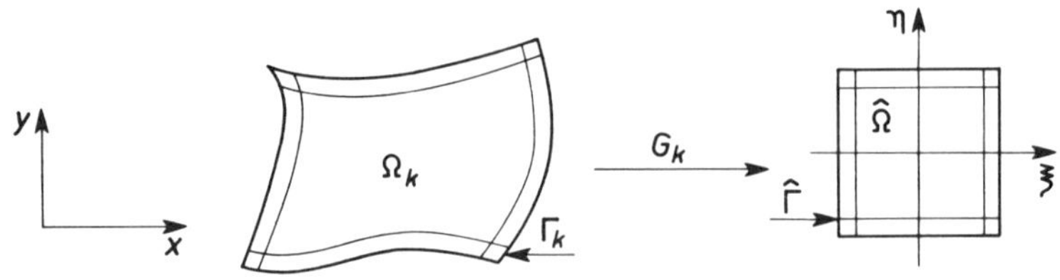

The equation (58) is transformed :

$$(59)\quad \begin{cases} -\hat{\Delta}^k \widehat{\phi_k^m} + \lambda \widehat{\phi_k^m} = \widehat{f_k^m} & \text{in } \hat{\Omega} \\ \widehat{\phi_k^m} \text{ given on } \hat{\Gamma} \\ \qquad \text{with } \hat{\phi}(\xi(x,y), \eta(x,y)) = \phi(x,y) \end{cases}$$

with

$$(60)\quad \begin{aligned} \hat{\Delta}^k \widehat{\phi_k^m} = {} & \partial_{\xi\xi} \widehat{\phi_k^m} \left[(\partial_x G_{k,1})^2 + (\partial_y G_{k,1})^2 \right] \\ & + \partial_{\eta\eta} \widehat{\phi_k^m} \left[(\partial_x G_{k,2})^2 + (\partial_y G_{k,2})^2 \right] \\ & + 2\partial_{\xi\eta} \widehat{\phi_k^m} \left[\partial_x G_{k,1} \partial_x G_{k,2} + \partial_y G_{k,1} \partial_y G_{k,2} \right] \\ & + \partial_\xi \widehat{\phi_k^m} \left[\partial_{xx} G_{k,1} + \partial_{yy} G_{k,1} \right] \\ & + \partial_\eta \widehat{\phi_k^m} \left[\partial_{xx} G_{k,2} + \partial_{yy} G_{k,2} \right] \end{aligned}$$

where $\partial_\xi = \dfrac{\partial}{\partial \xi}, \; --- \; .$

Equation (59) is solved by means of a pseudo-spectral Chebyshev method with a finite difference preconditioning technique.

Let $L_{S,k}$ be a spectral approximation and $L_{DF,k}$ be a finite difference approximation of the continuous operator $-\hat{\Delta}^{k}+\lambda$.

Solution of (59) is obtained as the limit of a sequence of approximations

$$\widehat{\phi}^{m}_{k,0} \quad \widehat{\phi}^{m}_{k,1} \cdots \widehat{\phi}^{m}_{k,p} \quad \cdots$$

each $\widehat{\phi}^{m}_{k,p}$ satisfying the boundary conditions on $\widehat{\Gamma}$. Calculation of $\phi_{k,p+1}$ as a function of $\widehat{\phi}^{m}_{k,p}$ is done as below :

i) resolution of

$$L_{DF,k}(\varphi) = -L_{S,k}(\widehat{\phi}^{m}_{k,p}) + \widehat{f}^{m}_{k} \tag{61}$$

with homogeneous boundary conditions

ii) under relaxation of the variation :

$$\widehat{\phi}^{m}_{k,p+1} = \widehat{\phi}^{m}_{k,p} + \sigma\varphi \tag{62}$$

2. Calculation of (U^{m}_{k}, P^{m}_{k}) on Γ_k.

Let us assume that approximation $(U^{m}_{k}, P^{m}_{k})^{*}$ of (U^{m}_{k}, P^{m}_{k}) has already been calculated.

Let $k \in IK$, $X \in \Gamma_k$ and $\{\Omega_{k_1}, \cdots \Omega_{k_X}\}$ be the set of subdomains that have X as a boundary point.

We set $IK(X) = \{k_1, \cdots k_X\}$ and $\mathrm{Ind}(X)$, the number of elements of $IK(X)$

a) X is an internal boundary point $(X \in \Omega)$.

A new approximation $(U^{m}_{k}, P^{m}_{k})^{**}$ of (U^{m}_{k}, P^{m}_{k}) on Γ_k is then :

$$\left\{\begin{aligned} U^{m**}_{k}(X) &= \frac{1}{\mathrm{Ind}(X)} \sum_{j \in IK(X)} \left(U^{m*}_{j} - c\,\frac{\partial U^{m*}_{j}}{\partial n_j} \right)(X) \\ P^{m**}_{k}(X) &= \frac{1}{\mathrm{Ind}(X)} \sum_{j \in IK(X)} \left(P^{m*}_{j} - c\,\frac{\partial P^{m*}_{j}}{\partial n_j} \right)(X) \end{aligned}\right. \tag{63}$$

where C is a constant and $\frac{\partial \phi}{\partial n_j}$ is the normal derivative of ϕ on Γ_j.

On Γ_k clearly formula (63) leads to continuity of (U^m, P^m).

At convergence, as the quantities $|U^{m**} - U^{m*}|_{\Gamma_k}$ and

$|P^{m**} - P^{m*}|_{\Gamma_k}$

vanish,

$$\left|\sum_{j\in K(X)} \frac{\partial U_j^m}{\partial n_j}\right| \quad \text{and} \quad \left|\sum_{j\in K(X)} \frac{\partial P_j^m}{\partial n_j}\right|$$

tend to zero and we obtain the continuity of normal derivatives of (U^m, P^m).

b) X is on external boundary point $(X \in \Gamma)$

For velocity :

$$\cdot\, x \in \Gamma_D \;:\; U_{k(X)}^{m**} = U_0^m(X)$$

$$\cdot\, x \in \Gamma_N \;:\; \text{relation } \frac{\partial U^m}{\partial n} = T^m$$

(64)

$$U_{k(X)}^{m**} = \frac{1}{\mathrm{Ind}(X)} \sum_{j\in K(X)} \left(U_j^{m*}(X) - c\,\frac{\partial U_{j(X)}^{m*}}{\partial n_j} \right) + c\,T_{(X)}^m$$

For pressure : relation (54)

(65)
$$P_{k(X)}^{m**} = \frac{1}{\mathrm{Ind}(X)} \sum_{j\in K(X)} \left(P_j^{m*}(X) - c\,\frac{\partial P_{j(X)}^{m*}}{\partial n_j} \right) + c\,S^{m*}(X)$$

with
$$S^{m*} = -n\cdot\left(\frac{U^{m*} - U^{m-1}}{\delta t} + \tilde{C}^m + \nu \,\mathrm{curl}\,\mathrm{curl}\, U^{m*} \right)$$

where n is the external unity normal vector at point X of boundary Γ.

V. RESULTS

<u>1. Square domain, stream function formulation.</u> Two problems have been treated :

– the Taylor-Green unsteady flow,

– the steady flow in a square cavity with "regularized" boundary conditions.

<u>a) Taylor-Green unsteady flow.</u> An exact solution of Navier-Stokes equation, Ψ_{th} is given by :

(66)
$$\Psi_{th} = \frac{1}{\lambda} \cos(\lambda x) \cos(\lambda y)\, e^{-2\nu\lambda^2 t}.$$

We calculate an approximation of Ψ_{th} as the limit of a sequence : $\Psi^0, \Psi^1, \ldots \Psi^m \ldots$

starting from Ψ^0 :

$$\psi^0(x,y,t) = \psi_{th}(x,y,t) + \varepsilon t (1-x^2)^2 (1-y^2)^2$$

such as :

$$\psi^0 = \psi_{th} \quad \text{and} \quad \frac{\partial \psi^0}{\partial n} = \frac{\partial \psi_{th}}{\partial n}$$

for $x = \pm 1$ or $y = \pm 1$

and $\psi^0 = \psi_{th}$ for $t = 0$.

In the calculation space-time domain $[-1, +1]^2 \times [0, 0.25]$ we use 9 Chebyshev polynomials in each space direction and 5 in the time direction.

With $\nu = 2.$ $(Re = 1.)$, $\lambda = \pi/2$ and $\varepsilon = .3$, an accuracy of 1% for the vorticity at the center of the domain is obtained by means of ten iterations (10 s on a CDC 7600 computer, see fig. 1). On figure 2, isobaric lines and iso-vorticity lines are shown.

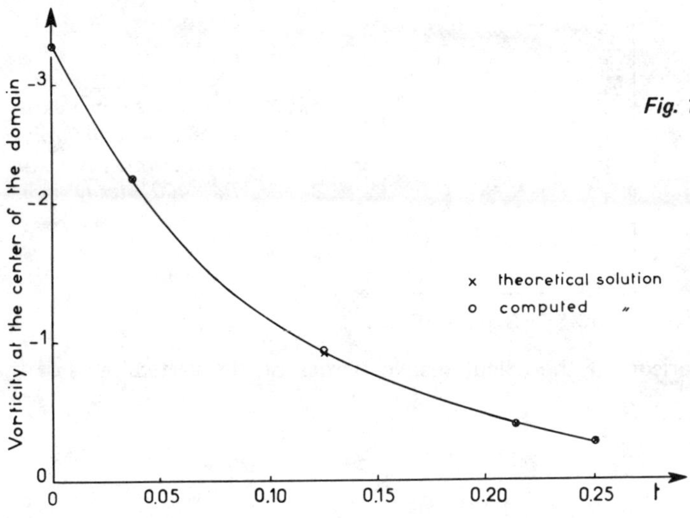

Fig. 1 – Taylor-Green flow, stream function formulation 9.9.5 harmonics in x.y.t, $R_e = 1.$

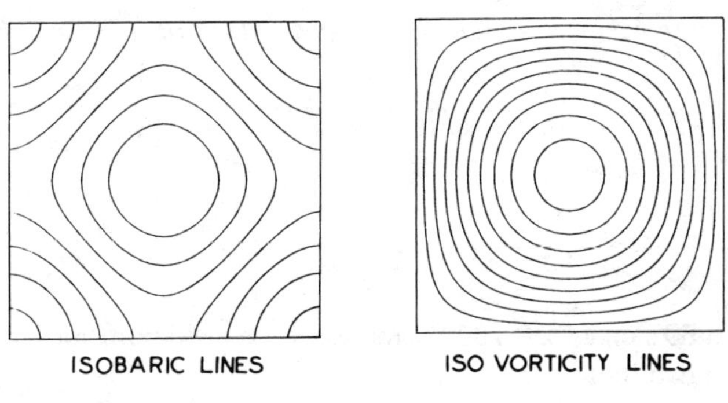

Fig. 2 – Taylor-Green flow, stream function formulation 9.9.5 harmonics in x.y.t, $R_e = 1.$, $t = 1.$

b) Regularized square cavity. Instead of the classical discontinuous boundary conditions, we use :

$$
(67) \quad \begin{cases} u = 0 \\ v = 0 \end{cases} \quad \text{for } x = \pm 1 \text{ or } y = +1
$$

$$
\begin{cases} u = -(1-x^2)^2 \quad (\text{instead of } u = -1) \\ v = 0 \end{cases} \quad \text{for } y = -1
$$

For steady state we compare our method and a finite element one.

For the finite element method, we have 64 elements (Q2 type, stream function-vorticity formulation, [5]).

For the pseudo-spectral method we use 33 x 33 harmonics with a steady formulation.

On figure 3, stream lines and iso-vorticity lines are shown.

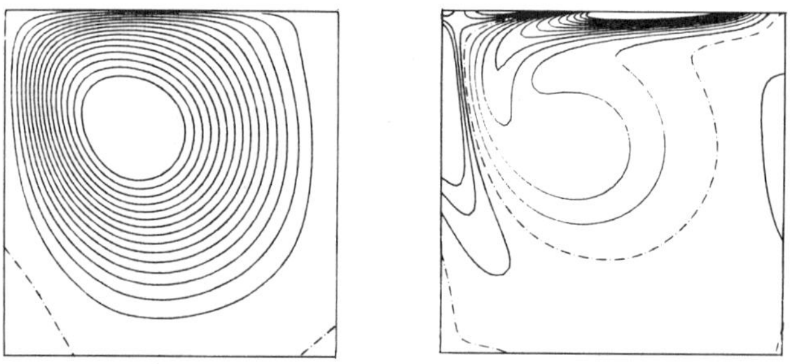

STREAM LINES ISO VORTICITY LINES

Fig. 3 — Regularized square cavity, stream function formulation 33.33 harmonics in x.y, $R_e = 400$, steady state.

On figure 4, a comparison of the calculated vorticities on the vertical axis of the cavity is presented.

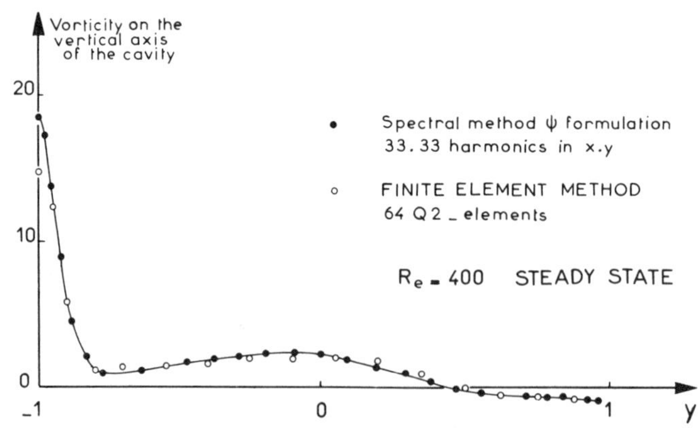

Fig. 4 — Regularized square cavity.

Computing time is less than 50 s on a CDC 7600 computer for a maximum vorticity relative variation of 10^{-4} between two iterations.

2. Cubic or square domain, velocity pressure formulation. Two problems have been considered :

- Taylor-Green analytical solution,
- Two dimensional flow in a regularized square cavity.

a) Taylor-Green analytical solution. We set :

$$U = \text{curl}\ \psi \tag{68}$$

where the components of vector ψ are :

$$\left\{\begin{array}{l} \psi_1 = 0 \\ \psi_2 = \alpha(t)\cos(\lambda x)\cos\left[\lambda(y+z)/\sqrt{2}\right] \\ \psi_3 = -\psi_2 \end{array}\right. \qquad \begin{array}{l} \nu = .1\ (R_e = 20.) \\ \text{with } \lambda = \pi/2 \\ \alpha(t) = \dfrac{\exp(-2\nu t\lambda^2)}{2\lambda} \end{array} \tag{69}$$

Starting from a perturbated solution, we calculate the unsteady flow with 9 x 9 x 9 x 5 harmonics in $x \times y \times z \times t$.

On figure 5, isobaric lines are drawn for $t = 0$ and $t = 1$.

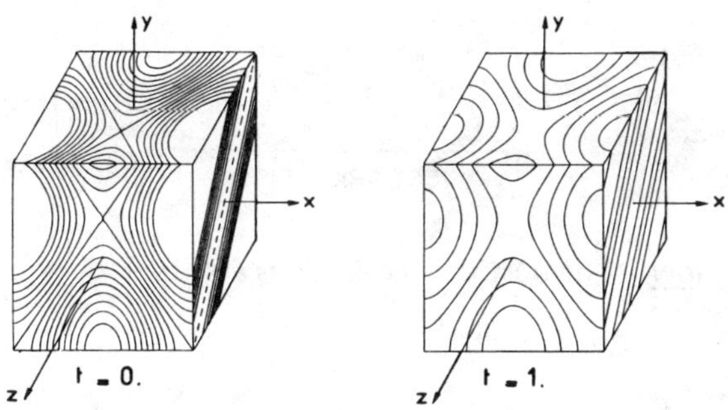

Fig. 5 – Cubic domain 9.9.9.5 harmonics in x.y.z.t,
$R_e = 20.$

b) Regularized square cavity. With 17 x 17 x 5 harmonics in $x \times y \times t$, we compute the flow in a square domain with boundary conditions (67).

Comparison with the results of the stream function method (33 x 33 harmonics in $x \times y$) is good (see Fig. 6).

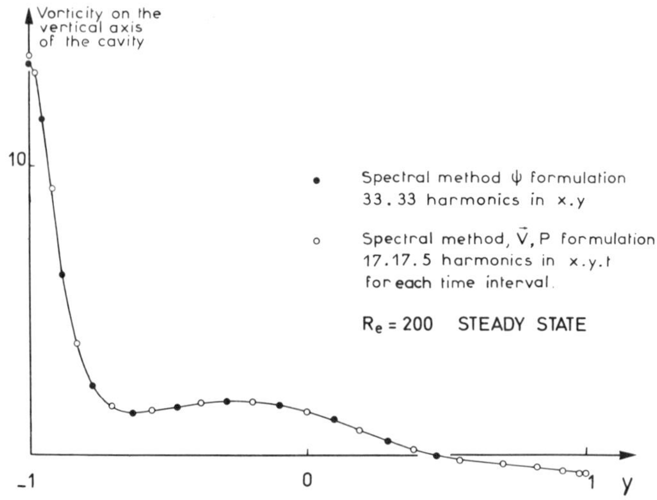

Fig. 6 — Regularized square cavity.

Isobaric lines and isovorticity lines are shown on figure 7.

Fig. 7 — Regularized square cavity velocity-pressure formulation 17.17.5 harmonics in x.y.t, $R_e = 200$, steady state.

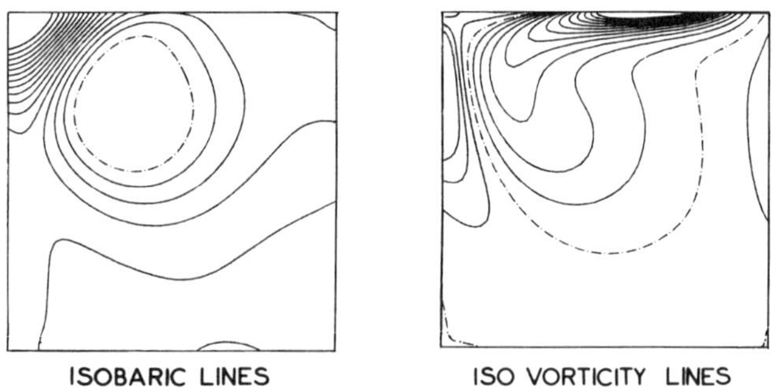

3. Square duct flow. The upstream conditions are (for $z = 0$):

(70) $$\begin{cases} U_1 = 0 \\ U_2 = 0 \\ U_3 = (1 - x^{2q})(1 - y^{2q})(1 + f(t)\,U_3^a(x,y)) \end{cases}$$

where

(71) $$f(t) = \lambda\,(1 - \cos(\omega_t t))$$

and where U_3^a is a random function such as :

(72) $$\max_{x,y} |U_3^a(x,y)| < 1$$

Calculations are done with :

$$(73)\quad \begin{cases} \nu = 1./300 & \\ q = 1 & \text{see relation (70)} \\ \lambda = 0.1 & \\ \omega_t = 20\pi & \text{see relation (71)} \\ \delta t = 0.4 & \\ L = 4 & \end{cases}$$

Figures 8, 9 and 10 show the variation of isobaric lines (in plan $y=0$) for $t=$ .25, 1. and 2.

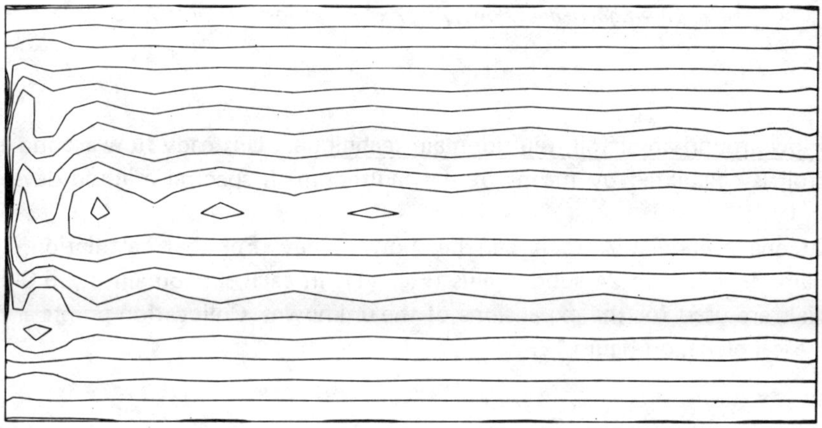

Fig. 8 – Square duct flow, 17 x 17 x 17 - Iso-vorticity lines T = .25

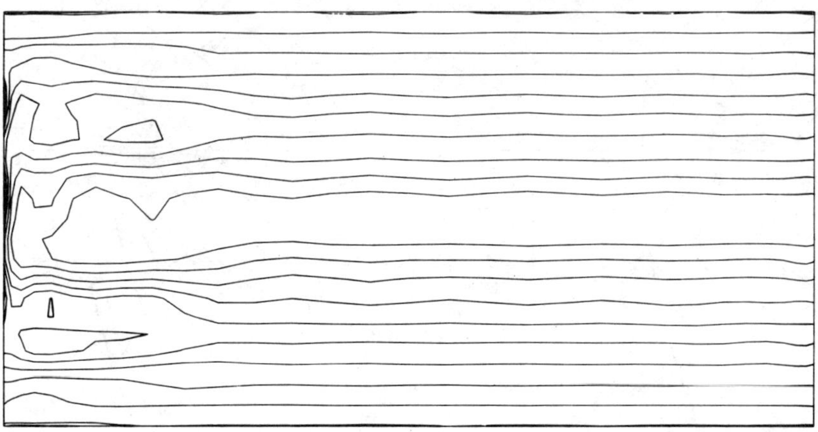

Fig. 9 – Square duct flow, 17 x 17 x 17 - Iso-vorticity lines T = 1.

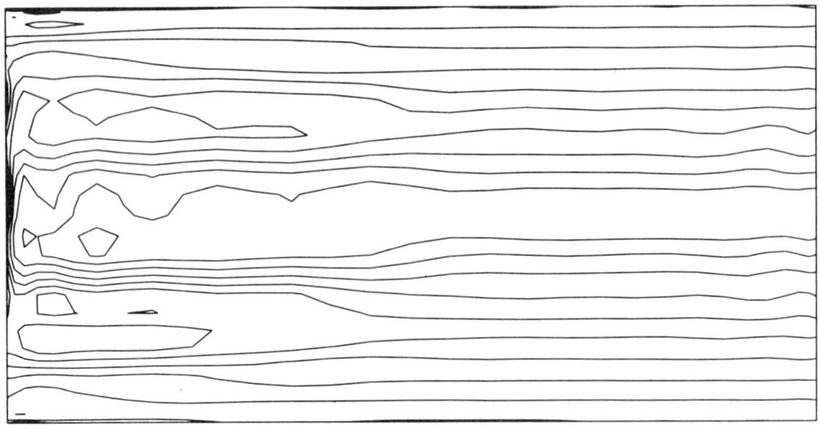

Fig. 10 – Square duct flow, 17 x 17 x 17 - Iso-vorticity lines T = 2.

4. Flow around an airfoil, multidomain technique. Unsteady flow around a NACA 0012 airfoil is calculated by means of the multi-domain spectral method developed in paragraph IV.

Only one preliminary result can be shown today. For that calculation, the outside domain is split in 24 subdomains (Fig. 11). In each subdomain 5 x 5 Chebyshev polynomials are used for the expansions of the unknowns. Collocation points are drawn, in the physical plane, on figure 12.

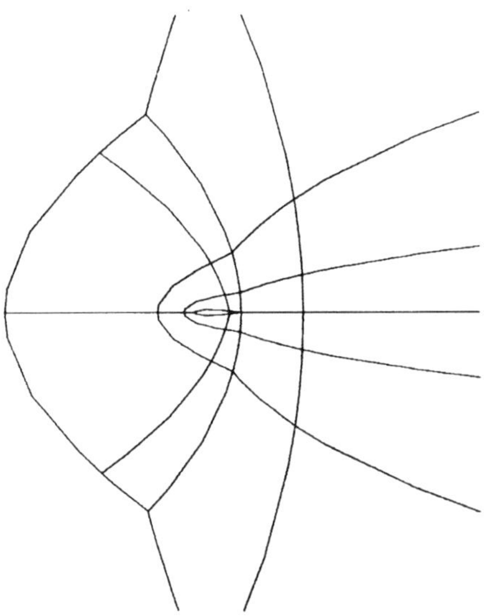

Fig. 11 – Subdomains.

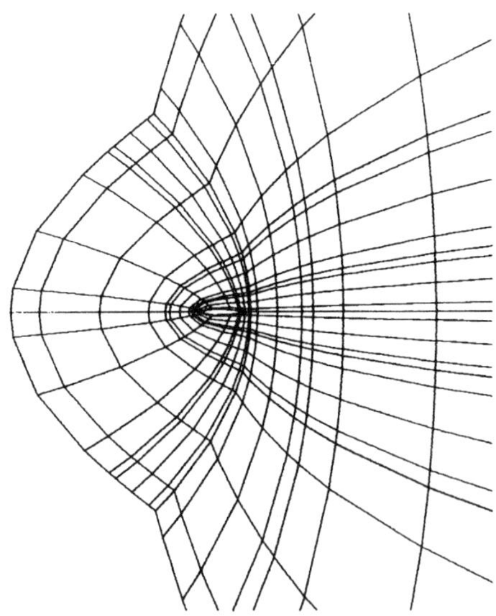

Fig. 12 – Collocation points.

For a Reynolds number of 100, a zero angle of attack and for steady state isobaric lines are shown in figure 13.

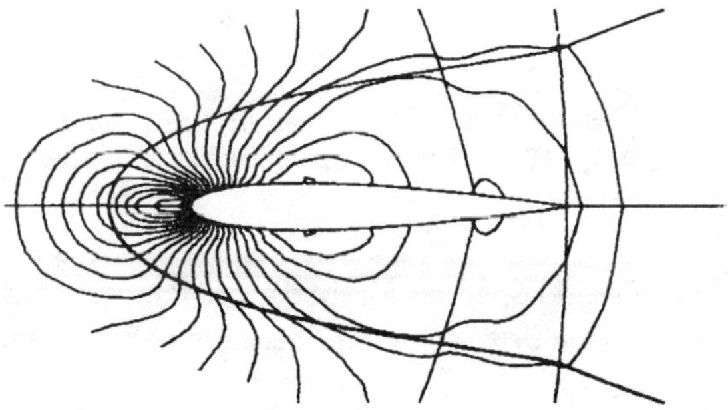

Fig. 13 — Isobaric lines.

VI. CONCLUSIONS. Two main problems are to be solved for the development of spectral methods :

– accuracy of the time scheme,
– flexibility for complex geometries.

For time accuracy, a square-time pseudo-spectral approach is implemented for 2-D and for 3-D flows. Other techniques can be used :

– Taylor expansions in time [6],
– high order finite difference scheme,

but spectral time expansions seem to be competitive even if such a method has to be improved.

For complex geometry, a multi domain spectral method is developed. That technique certainly could be more efficient with a better connexion algorithm between subdomains. Anyway the proposed method seems very accurate and has flexibility of the finite elements method. Due to subdomain technique there is no memory size problem and the extension to 3-D unsteady flow will be easy.

REFERENCES

[1] MORCHOISNE, Y. "Resolution of Navier-Stokes Equations by a Space-Time Pseudo Spectral Method". La Recherche Aérospatiale, No 1979-5, p. 293-306 (English translation).

[2] MORCHOISNE, Y. "Pseudo-Spectral Space-Time Calculations of Incompressible Viscous Flows". AIAA, 19th Aerospace Sciences Meeting, Saint-Louis, AIAA paper No 81-0109 (1981).

[3] MORCHOISNE, Y. "Pseudo-Spectral Methods for Homogeneous or Inhomogeneous Flows". Proceedings of the 3rd Trubulent Shear Flow, Symposium Davis-California, chap. 19, p. 1-5, 1981.

[4] METIVET, B., MORCHOISNE, Y. "Multi-Domain Spectral Technique for Viscous Flow Calculation". Proceedings of the 4th GAMM Conference on Numerical Methods in Fluid Mechanics (H. Viviand, Ed.). Vieweg 1982, p. 207-219.

[5] BREDIF, M. "On a Resolution of the Navier-Stokes Equations by Finite Elements". Ph. D. Thesis and ONERA Publication 1980 (English translation by ESA, to be published).

[6] ROY, Ph. "Resolution of Navier-Stokes Equations by a Method of High Accuracy in Space and Time". La Recherche Aérospatiale No 1980-6, p. 373-385 (English translation).

SMOOTHING FOR SPECTRAL METHODS

STANLEY OSHER*

Abstract. The purpose of this article is mainly to demonstrate, by means of simple examples, that some kind of smoothing must be an essential part of any spectral method.

Spectral methods have, in principle, infinite order of accuracy, if the true solution is smooth. I shall review some old work, done jointly with Majda and McDonough [1], which shows how drastically the situation changes when discontinuities are present. The error pollutes the solution globally if no smoothing is used, and pollutes it in a very large region, even if smoothing based on the finite spectral transform is used. A more drastic type of smoothing will remove this error.

Smoothing is, in general, necessary for two reasons:

a. Accuracy
b. Stability

1. Error Analysis for Smooth Solutions.

I shall begin with a periodic one-dimensional very simple problem:

$$u_t = au_x, \tag{1.1}$$

to be solved for $-\pi < x \leq \pi$, $0 < t$, with a, some fixed constant, and initial data

$$u(x,0) = \phi(x).$$

The Fourier method proceeds by breaking up the x interval via

$$x_\nu = \nu h, \qquad \nu = 0, \pm 1, \ldots, \pm N,$$

*Mathematics Department, UCLA, Los Angeles, CA 90024. Support was provided by NASA Contract No. NAS1-15810 for the research performed in residence at the Institute for Computer Applications in Science and Engineering, NASA Langley Research Center, Hampton, VA. 23665.

with $(2N+1)h = 2\pi$. For any $u(x)$, one finds its trigonometric interpolant

$$u^h(x) = (I^h u)(x) ,$$

as follows. First one computes the finite Fourier transform:

$$(1.2)\qquad (Fu)(j) = \tilde{u}(j) = 1/(2N+1) \sum_{\nu=-N}^{N} e^{-ijx_\nu} u(x_\nu)$$
$$j = 0, \pm 1, \ldots, \pm N.$$

Then one obtains the interpolant:

$$(1.3)\qquad u^h(x) = F^{-1}(\tilde{u}) = \sum_{j=-N}^{N} e^{ijx}\tilde{u}(j).$$

It follows that

$$u^h(x_\nu) = u(x_\nu) \quad \text{for} \quad \nu=0,\pm 1,\ldots,\pm N.$$

The approximate solution to the Cauchy problem (1.1) could be obtained by solving exactly the finite dimensional problem:

$$(1.4)\qquad \frac{\partial u^h}{\partial t} = a\,\frac{\partial u^h}{\partial x}$$
$$u^h(x,0) = \phi^h(x).$$

The solution is

$$(1.5)\qquad u^h(x,t) = \sum_{j=-N}^{N} \tilde{\Phi}(j)e^{aijt+ijx}.$$

To find the <u>true</u> solution one needs the <u>true</u> Fourier coefficients:

$$(1.6)\qquad \hat{u}(j) = 1/2\pi \int_{-\pi}^{\pi} e^{-ijx}u(x)dx.$$

One then has:

$$(1.7)\qquad e(x) = \sum_{j=-\infty}^{\infty} \hat{u}(j)e^{ijx}.$$

For (1.1), the true solution is thus the infinite sum:

$$(1.8)\qquad u(x,t) = \sum_{j=-\infty}^{\infty} \hat{\phi}(j)e^{aijt+ijx}.$$

Thus we have a simple expression for the error:

$$u - u^h = \sum_{|j|>N} \hat{\phi}(j)e^{aijt+ijx} + \sum_{|j|\leq N} [\hat{\phi}(j)-\tilde{\phi}(j)]e^{aijt+ijx} = E_I + E_{II}. \tag{1.9}$$

The relationship between $\hat{\phi}(j)$, the true Fourier coefficients, and $\tilde{\phi}(j)$, the finite Fourier coefficient is easily seen to be

$$\tilde{\phi}(j) = \sum_{\mu=-\infty}^{\infty} \hat{\phi}(j+\mu(2N+1)). \tag{1.10}$$

See Kreiss-Oliger [3].

Thus, we may rewrite (1.9):

$$u - u^h = \sum_{|j|>N} \hat{\phi}(j)e^{aijt + ijx} + \sum_{|j|\leq N} e^{aijt + ijx} \Big(\sum_{\substack{\mu=-\infty \\ \mu\neq 1}}^{\infty} \hat{\phi}(j+\mu(2N+1))\Big). \tag{1.11}$$

Now if $\phi(x)$ is a smooth function, i.e., $\phi(x) \in C^\infty$, then for any $K>0$, there exists universal constants C_k, so that

$$|\hat{\phi}(j)| \leq C_K(1+|j|)^{-K}.$$

This, together with (1.11), implies

$$\|u-u^h\| \leq C_p h^p, \tag{1.12}$$

in any reasonable norm, as $p \downarrow 0$. (We use the convention that universal constants are always denoted $C_{\text{subscript}}$). This explains the "infinite" order accuracy.

2. Deterioration Due to Discontinuities. Suppose, on the other hand, ϕ is not smooth at $x = 0$. As an example, we take

$$\phi(x) = |x|^\gamma \sigma(x), \ \gamma > 0, \tag{2.1}$$

with $0 \leq \sigma \leq 1$, a smooth function, $\sigma \equiv 1$ if $|x| < \pi - 2\varepsilon$, $\sigma \equiv 0$ if $|x| \geq \pi - \varepsilon$, for small $\varepsilon > 0$. Then $\hat{\phi}(j)$ satisfies

$$\Big|\Big(\frac{\partial}{\partial j}\Big)^\alpha \hat{\phi}(j)\Big| = O\big((1+|j|)^{-1-\gamma-\alpha}\big) .$$

The approximate solution has a global error.
Let R_δ be a region of smoothness for the exact solution

$$R_\delta = \{(x,t)/|x+at| > \delta > 0\}.$$

In [1], we show the following global error estimate.

Lemma 1

$$(1) \qquad \max_{(x,t)\,\varepsilon\,R_\delta} |u(x,t) - u^h(x,t)| \leq C_\delta\, h^{1+\gamma} ,$$

$$(2) \qquad \lim_{h\downarrow 0}\ \max_{(x,t)\,\varepsilon\,R_\delta} \left(\frac{|u(x,t) - u^h(x,t)|}{h^{1+\delta}}\right) \geq C > 0 .$$

For a Heaviside function type of initial data:

$$\phi(x) = (\mathrm{sgn}\ x)\ \rho(x),$$

with $\phi(0) = 0$, the global error is $0(h^2)$. This anamoly occurs because of a cancellation in the finite Fourier series for this function. If we took any nonzero value for $\phi(0)$, the error would be $0(h)$, as expected.

3. A Simple and Inadequate Smoothing Technique. For smooth $\phi(x)$ defined as in the previous section, we filter the initial data and use

$$(3.1) \qquad u^h(x,0) = \sum_{j=-N}^{N} \phi(jh)\ e^{ijx}\tilde{\phi}(j).$$

Note, we are still using the finite Fourier transform of the initial data. Then

$$(3.2) \qquad u - u^h = \sum_{j=-\infty}^{\infty} e^{ij(x+at)}\big(1-\sigma(jh)\big)\hat{\phi}(j)$$

$$- \sum_{|j|\leq N} e^{ij(x+at)}\phi(jh) \sum_{|\nu|\geq 1} \hat{\phi}\big(j+\nu(2N+1)\big) = E_I + E_{II}.$$

It is shown in [1] that

$$\max_{(x,t)\,\varepsilon\,R_\delta} |E_I| \leq C_\lambda h^\lambda.$$

for any $\lambda > 0$.

If $\phi(x)$ is the Heaviside function, then we also show that, in R_δ, E_{II} is within $0(h^\lambda)$, for any λ, of the solution to the true Cauchy problem (1.1), with initial data:

$$(3.3) \qquad u(x,0) = h^2[C_1\delta'(0) + hC_2\delta''(0) + \ldots] .$$

Here $\delta(x)$ is the Dirac delta function, and δ' is its distribution derivative, etc.

Since there is no coupling for the problem (1.1), the support of this distribution does not spread into R_δ when we solve the problem. Thus for this decoupled case we luckily have infinite order accuracy in R_δ. But this result is false in more complicated cases.

Suppose, for example, we consider the coupled hyperbolic system

$$(3.4) \qquad \begin{pmatrix} u_1 \\ u_2 \end{pmatrix}_t = \begin{pmatrix} 1 & 0 \\ 0 & -1 \end{pmatrix} \begin{pmatrix} u_1 \\ u_2 \end{pmatrix}_x + \begin{pmatrix} 0 & 1 \\ -1 & 0 \end{pmatrix} \begin{pmatrix} u_1 \\ u_2 \end{pmatrix},$$

with initial data

$$\begin{pmatrix} u_1 \\ u_2 \end{pmatrix} = \begin{pmatrix} \phi \\ 0 \end{pmatrix},$$

for ϕ the Heaviside function.
Then, there is a global deterioration of accuracy within the range of influence of the orgin.

Let

$$R_\delta^{(1)} = \{(x,t) \mid -1 + \delta < \frac{x}{t} < 1-\delta\} .$$

Then the analogous expressions for E_I and E_{II} yield the results

$$|E_I| = O(h^\lambda) \quad \text{for all} \quad \lambda > 0 \quad \text{in} \quad R_\delta^{(1)} .$$

However, E_{II} is, within $O(h^3)$, the solution of the Cauchy problem (3.4) with initial data

$$(3.5) \qquad u(x,0) = h^2 c_1 \begin{bmatrix} \delta'(0) \\ 0 \end{bmatrix}, \; c_1 \neq 0.$$

Because of the coupling through the lower order terms, the entire shadow region, $R_\delta^{(1)}$, feels the influence of this initial data and

$$\overline{\lim_{(x,t) \in R_\delta^{(1)}}} \frac{|E_{II}|}{h^2} > C .$$

Thus, we have a global deterioration of accuracy.

In earlier joint work with Majda [2], we analyzed this large error phenomenon for dissipative finite-difference schemes. Since such (usually simpler) methods do not strive for infinite order accuracy, their comparable deterioration might not trouble the finite-difference user, as much as it might disturb the proponent of spectral methods.

4. A Drastic Smoothing Technique For constant coefficient linear hyperbolic problems, we can remove the large region of low accuracy by using and smoothing the true Fourier coefficients of the initial data. Let

$$u^h(x,0) = \sum_{j=-N}^{N} \sigma(jh)\, e^{ijx}\, \hat{\phi}(j). \tag{4.1}$$

Actually, if $\phi(x) = \phi^{(1)} + \phi^{(2)}$, with $\phi^{(1)}$ discontinuous and $\phi^{(2)}$ smooth, we can take if convenient:

$$u^h(x,0) = \sum_{j=-N}^{N} \sigma(jh) e^{ijx} \hat{\phi}^{(1)}(j) + \sum_{j=-N}^{N} \sigma(jh) e^{ijx}\, \tilde{\phi}^{(2)}(j)]. \tag{4.2}$$

We then get infinite order of accuracy in regions where the true solution is smooth.

This procedure is somewhat impractical for nonlinear shock problems, when spontaneous shocks develop, because their true Fourier series is unknown. The research at ICASE of Gottlieb and Lustman, involving fitting in spectral variable space, is designed to overcome this.

5. Stability In multidimensions, periodic linear hyperbolic system with variable (say t independent, for convenience only) coefficients can be written:

$$\begin{aligned} u_t &= \Big(\sum_{\nu=1}^{d} A_\nu(x) \frac{\partial}{\partial x_\nu} + B(x) \Big) u \\ &= Lu. \end{aligned} \tag{5.1}$$

The naive (unsmoothed) Fourier-method approach to this problem, would be to solve:

$$\begin{aligned} u^h_t &= \Big(\sum_{\nu=1}^{d} A_\nu(x)\, F^{-1} \cdot i \cdot j_\nu + B(x)\, F^{-1} \Big) \tilde{u}(j) \\ &= L^h u^h. \end{aligned} \tag{5.2}$$

There is some controversy as to the stability of this method for general variable coefficient problems, [3], [4].

This naive method, however, cannot work for nonlinear problems. As an example, we consider

$$u_t = -\Big(\frac{u^2}{2}\Big)_x \, . \tag{5.3}$$

The naive approach would be: Solve

$$\frac{\partial u^h}{\partial t} = \frac{-F^{-1} \cdot i \cdot j F(u^2)}{2} . \tag{5.4}$$

Take, say, 3 grid points (N=1), with initial data

$$u^h(x_{-1},0) = a, \quad u^h(0, 0) = 0, \quad u^h(x_1,0) = -a. \tag{5.5}$$

Then

$$u^h(x,0) = \frac{-2a}{\sqrt{3}} \sin x$$

$$\left(u^2(x,0)\right)^h = \frac{2a^2}{3}(1-\cos x).$$

The solution to (5.4) is of the form:

$$u^h(x_{-1},t) = a(t), \quad u^h(0,t) = 0, \; u^h(x_1,t) = -a(t) \tag{5.6}$$

with

$$\frac{da}{dt} = \frac{a^2}{2\sqrt{3}} .$$

Thus $a(t)$ becomes infinite in finite time, and this method is impractical.

Another approach might be to stabilize, using smooth cutoff filters, e.g.

$$\frac{\partial u^h}{\partial t} = -F^{-1} \cdot i \cdot j \rho(jh) F\left(\frac{u^2}{2}\right). \tag{5.7}$$

This method would give the same solution if the initial data is either an expansion shock or a true shock, both moving at zero speed. This means $u^h(x,0) = \operatorname{sgn} x$, gives the same solution as $u^h(x,0) = -\operatorname{sgn} x$.

Thus, a method which approximates u^2, using $(F^{-1}\rho(jh)Fu)^2$ i.e., one which "sees" u, not just u^2, must be used.

For linear hyperbolic problems, to assure stability, we use

$$u_t^h = \left(\sum_{\nu-1}^{d} A_\nu(x)\, F^{-1} \cdot i \cdot j_\nu + B(x)\, F^{-1} \right) \rho(jh)\, \tilde{u}(j), \tag{5.8}$$

where ρ is the usual type of cutoff function. If we solve (5.8) with initial data $u^h(x,0)$ we have in [1], a very general stability theorem (See [1] for precise details).

Theorem 1.

$$\|u^h(\cdot,t)\|_s \leqslant C_s\, e^{Kt} \|u^h(\cdot,t)\|_s .$$

and

Corollary. If $u(x,0)$ is smooth, then we have infinite order accuracy, globally, for (5.4).

6. Convergence. We recommend solving our stabilized Fourier method (5.4), with the drastically smoothed initial data

$$(6.1) \qquad u^h(x,0) = \sum_{j_d=-N}^{N} \cdots \sum_{j_1=-N}^{N} \sigma(jh)\hat{\phi}(j)e^{ij\cdot x},$$

where $\rho(x)\ \sigma(x) \equiv \sigma(x)$.

A very general convergence theorem was proven in [1]. In a region R_δ, which is essentially any region of smoothness of the true solution to (5.1), we have:

Theorem 2. For any $\lambda > 0$ and any s, $|s| \geqslant 0$ there is a constant $C_{\delta,s,\lambda}$ such that

$$\sup_{(x,t)\,\varepsilon\, R_\delta} |\partial_x^s(u-u^h)| \leqslant C_{\delta,s,\lambda} h^\lambda .$$

Thus, this method is truly infinitely order accurate. The precise details are given in [1].

REFERENCES

[1] A. MAJDA, J. MCDONOUGH, and S. OSHER, The Fourier method for nonsmooth initial data, Math. Comp., 32, (1978), pp. 1041-1081.

[2] A. MAJDA and S. OSHER, Propagation of error into regions of smoothness for accurate difference approximations to hyperbolic equations, Comm. Pure Appl. Math, 30, (1977), pp. 671-705.

[3] H. O. KREISS and J. OLIGER, Comparison of accurate results for the integration of hyperbolic equations, Tellus, 24, (1972), pp. 199-215.

[4] D. GOTTLIEB, S. A. ORSZAG, and E. TURKEL, Stability of pseudospectral & finite difference method for variable coefficient problems, Math. of Comp., 27, No. 156, October (1981).

SPECTRAL METHODS FOR ATMOSPHERIC LASER CALCULATIONS

JOSEPH E. PASCIAK*

Abstract. Spectral methods for a nonlinear Schrödinger equation modeling laser propagation in a non-homogeneous medium are considered in this paper. Error and stability estimates for the semidiscrete solution are proven which are of infinite order depending upon regularity and decay of the initial data. A non-dissipative time stepping procedure is also described.

1. Introduction. Spectral methods for a nonlinear Schrödinger equation are considered in this paper. We study the following initial boundary value problem

$$
\begin{aligned}
&i \frac{\partial u}{\partial t} + \Delta u = \zeta u && \text{on } R^2 \times [0,T] \\
(1.1)\quad &u(x,t) \to 0 && \text{as } |x| \to \infty \\
&u(x,0) = u_0(x) && \text{for } x \text{ in } R^2
\end{aligned}
$$

Here Δ denotes the Laplace operator and i is the square root of minus one. The above equation is representative of the equations modeling thermal blooming in atmospheric laser propagation. The variable t points in the direction of propagation and x is a two dimensional vector in the plane transverse to the propagation direction. The irradience $I(x,t)$ is defined to be $|u(x,t)|^2$ and provides a measure of the energy in the beam. The function ζ is related to the density distribution in the atmosphere and may depend

* Applied Mathematics Dept., Brookhaven National Laboratory, Upton, New York 11973
The submitted manuscript has been authored under contract DE-AC02-76CH00016 with the U.S. Department of Energy.

nonlinearily on the history of the irradience.

Spectral methods have been extensively used in the numerical modeling of laser propagation [5,6,9,15,16,17]. These methods use trigonometric polynomials for spacial discretization and some type of time step discretization. Alternatively, this type of discretization can be motivated by using the discrete Fourier Transform (implemented by the Fast Fourier Transform Algorithm [8]) as an approximation to the Fourier Transform on R^2.

This paper provides some analysis for a semidiscrete spectral approximation of (1.1). To obtain any rigorous error estimates for these problems one needs regularity and decay results for the solution $u(x,t)$. Questions of existence, uniqueness, regularity and decay of solutions of (1.1) with nonlinear ζ are in general rather difficult and depend intrinsically on the properties of the nonlinearity. Existence, uniqueness, and regularity results for these equations have been investigated [1,2] for problems where the nonlinearity was given by

$$(1.2) \qquad \zeta(u(x,t)) = K\, |u(x,t)|^2$$

or

$$(1.3) \qquad \zeta(u(x,t)) = \int_{-\infty}^{x_1} |u(x,t)|^2 \, dx_1$$

To keep this paper from becoming too abstract, we shall only consider (1.1) with nonlinearity given by (1.2). The basic result of this paper (Theorem 5.1) shows that the approximate solution $U(x,t)$ exists, is unique and satisfies an error estimate of the form

$$(1.4) \qquad \|u(x,t) - U(x,t)\|_{L^2(R^2)} \leq C\{h^m + M^{-m}\}$$

under appropriate hypothesis. Here C is a positive constant which is independent of the mesh parameters h and M and m depends upon the regularity and decay of the initial data u_0.

The remainder of this paper is outlined as follows. Section 2 sets up the notation and definitions. In Section 3 decay results are proven for problem (1.1-1.2). Approximation properties of trigonometric polynomials are given in Section 4. The semidiscrete approximation to (1.1-1.2) is defined and analyzed in Section 5 and a nondissipative time stepping procedure is given in Section 6.

2. Notation. Let Ω be a domain in R^2 and N be the set of nonnegative integers. For $\alpha \varepsilon N^2$, let $D^\alpha u \equiv (\frac{\partial}{\partial x_1})^{\alpha_1} (\frac{\partial}{\partial x_2})^{\alpha_2} u$ denote the distributional derivative of u defined for u in $\mathcal{D}'(\Omega)$ (the space of distributions on Ω [14]). The Sobolev space $H^m(\Omega)$ is the set of functions $u \varepsilon \mathcal{D}'(\Omega)$ such that $D^\alpha u \varepsilon L^2(\Omega)$ for $|\alpha| \equiv \alpha_1 + \alpha_2 \leq m$. The space $H^m(\Omega)$ is a Hilbert space with norm

$$(2.1) \quad \|u\|^2_{H^m(\Omega)} = \sum_{|\alpha| \leq m} \|D^\alpha u\|^2_{L^2(\Omega)} .$$

Sobolev spaces on R^2 shall be denoted H^m with corresponding norm $\|\cdot\|_m$. Let $C^\infty(\Omega)$ (resp $C_0^\infty(\Omega)$) denote the space of infinitely differentiable functions (resp. infinitely differentiable functions with compact support) on Ω. Then for bounded domains, $H^m(\Omega)$ is the completion of $C^\infty(\Omega)$ in the norm $\|\ \|_{H^m(\Omega)}$ while H^m is the completion of $C_0^\infty(R^2)$.

We shall approximate the solution of problem (1.1) on large square domains

$$\Omega_M = \{(x,y) | -M \leq x,y \leq M\}$$

with periodic boundary conditions. A function f defined on R^2 is periodic with period 2M if

$$f(x + 2kM,\ y + 2jM) = f(x,y)$$

for all x and y and all integer pairs (k,j). Let $C_p^\infty(\Omega_M)$ be the set of infinitely differentiable functions which are periodic with period 2M and define $H_p^m(\Omega_M)$ to be the completion of $C_p^\infty(\Omega_M)$ in the $H^m(\Omega_M)$ norm.

To approximate the solution of a problem on an infinite

domain by discrete functions on finite domains, one must have some control over the decay of the solution at infinity. Thus, for nonnegative n and m, we introduce the norm

$$|u|^2_{m,n} \equiv \sum_{|\alpha| \leq m} \| (1+|x|^2)^{n/2} D^\alpha u(x)\|_0^2 .$$

Denote by $\Lambda_{m,n}$ the completion of $C_0^\infty(R^2)$ in the norm $|\cdot|_{m,n}$. The spaces $\Lambda_{m,n}$ measure regularity and decay. Indeed, if $u \varepsilon \Lambda_{m,n}$ and Ω' is a domain contained in the complement of the ball of radius R centered at the origin, then

$$(2.2) \qquad \|u\|_{H^m(\Omega')} \leq (1+R^2)^{-n/2} |u|_{m,n} .$$

The Fourier Transform shall be denoted F and is given by

$$Ff(\xi) = \int_{R^2} e^{ix\cdot\xi} f(x)\, dx$$

for $\xi \varepsilon R^2$ and $f \varepsilon L^1(R^2)$. F extends to an isometry of $L^2(R^2)$ onto $L^2(R^2)$ with inverse F^{-1}.

Remark 2.1. The Fourier Transform maps $\Lambda_{m,n}$ homeomorphically onto $\Lambda_{n,m}$. The above result is well known for n=0 and follows from a straightforward application of the Leibniz rule for n > 0.

For u in $C_0^\infty(R^2)$ define the norm

$$[u]^2_m = \sum_{j=0}^{m} |u|^2_{m-j,j}$$

and let Π^m denote the completion of $C_0^\infty(R^2)$ in $[\cdot]_m$. The space Π^m is obviously the intersection of $\Lambda_{m-j,j}$ for

$j=0,\dots,m$ and by Remark 2.1, the Fourier Transform maps Π^m onto Π^m.

Throughout this paper C will denote a generic positive constant which may assume different values in different places. The constant C will however always be independent of the mesh parameters h, N and M.

3. Decay of Problem (1.1). Decay estimates for Problem (1.1) with nonlinearity given by (1.2) are proven in this section. Regularity results for (1.1) with (1.2) or (1.3) were given by Baillon et al. in [1,2]. The analysis of decay given in this section proceeds along the same lines as the regularity analysis of [1,2]. The results of this section were essentially obtained in [4].

Let the nonlinearity (1.2) be denoted

$$Ju(x,t) \equiv K|u(x,t)|^2\, u(x,t).$$

For fixed t, $Ju(\cdot,t)$ is a function on R^2 which we shall often denote $Ju(t)$. The solution of (1.1) is given by Duhamel's principle

$$u(t) = E(t)\, u_0 + \int_0^t E(t-s)\, Ju(s)\, ds \tag{3.1}$$

where $E(t)$ is the solution operator for the linear Schrödinger equation, i.e. $E(t)w=v(t)$ where $v(t)$ solves the problem

$$i\,\frac{\partial v}{\partial t} + \Delta v = 0 \qquad \text{on } R^2 \times [0,T] \tag{3.2}$$

$$v(0,x) = w(x) \qquad \text{for } x\varepsilon R^2 \tag{3.3}$$

(3.4) $v(t) \in L^2(R^2)$

The solution of (3.2 - 3.4) for $w \in L^2(R^2)$ is given by

$$\widehat{V(t)}\ (\xi) = e^{-i|\xi|^2 t}\ \hat{w}(\xi)$$

where ^ denotes the Fourier Transform, i.e.

$$\hat{W} = Fw.$$

The decay spaces Π^m were defined so that the following result holds.

Lemma 3.1. The operator $E(t)$ is a bounded map of Π^m into Π^m. Furthermore, for $t \in [0,T]$ there is a constant C_T such that

(3.5) $[E(t)w]_m \leq C_T[w]_m.$

Proof. Let v be the solution of (3.2 - 3.4). By Remark 2.1, it is enough to show that $\hat{v} \in \Pi^m$. We consider $\hat{w}$ in $C_0^\infty(R^2)$ and get the result on all of Π^m by density. For $|\alpha| \leq j$, the Leibniz rule and the definition of the spaces Π^m give

$$\|(1+|\xi|^2)^{\frac{m-j}{2}} D^\alpha(e^{-i|\xi|^2 t}\hat{w})\|_0 \leq$$

$$C_t \sum_{|\tau| \leq |\alpha|} \|(1+|\xi|^2)^{\frac{m-j+|\alpha|-|\tau|}{2}} D^\tau \hat{w}\|_0 \leq C_t[\hat{w}]_m.$$

Thus summing over $|\alpha| \leq j$ and $j=0,\ldots,m$ gives

$$[E(t)w]_m \leq C_t[w]_m$$

and the lemma follows by density.

The following theorem gives existence, uniqueness, and regularity results for the solution of (1.1 - 1.2).

<u>Theorem 3.1</u>. [2] Let u_0 be $H^m(R^2)$, $T > 0$ and $m \geq 3$. If $K > 0$ or $|K| \, \|u_0\|_0^2 < 2$ then there exists a unique solution of (1.1 - 1.2) satisfying

$$u \varepsilon C^0(0,T;H^m(R^2)) \cap C^1(0,T;H^{m-2}(R^2)).$$

(The spaces $C^j(a,b;X)$ are spaces of vector valued functions defined on the interval (a,b) with range in X [10]).

We next prove decay results for (1.1 - 1.2).

<u>Theorem 3.2</u>. Let u be the solution of (1.1 - 1.2) with initial data u_0 in Π^m and $m \geq 3$. If $|K| \, \|u_0\|_0^2 \leq 2$ for $K < 0$ or $K > 0$ then u(t) is in Π^m for all positive t. Furthermore, there is a constant $C(T,[u_0]_m)$ satisfying

$$[u(t)]_m \leq C(T,[u_0]_m) \qquad \text{for } t \varepsilon [0,T].$$

<u>Proof</u>. By Duhamel's principle the solution u can be written

$$(3.6) \qquad u(t) = E(t)\, u_0 + \int_0^t E(t-s)\, Ju(s)\, ds.$$

To get local existence in Π^m, it is enough to show that the map $u \to Ju$ is Lipschitz continuous from a neighborhood of u_0 in Π^m into Π^m. To show that J is Lipschitz we must estimate terms of the form

$$(3.7) \qquad \|(1+|x|^2)^{\frac{m-j}{2}} D^{\alpha}(Ju - Jv)\|_0$$

for $|\alpha| \leqslant j$. We have that

$$(3.8) \qquad \begin{aligned} Ju(x) - Jv(x) &= [(u(x)-v(x))\ u^*(x)\ u(x) \\ &+ v(x)\ (u(x)-v(x))^*\ u(x) + v(x)\ v(x)^*\ (u(x)-v(x))]K \end{aligned}$$

where * denotes conjugation. To bound (3.7) we use (3.8) and the triangle inequality. By the Leibniz rule, one is left to estimate terms of the typical form

$$\|(1+|x|^2)^{\frac{m-j}{2}} D^{\alpha_1}(u-v)\ D^{\alpha_2}u^*\ D^{\alpha_3}u\|_0$$

where $|\alpha| = |\alpha_1| + |\alpha_2| + |\alpha_3|$. Let α_i denote the index with largest absolute value and let α_j, α_k denote the other two. Then

$$\begin{aligned} &\|(1+|x|^2)^{\frac{m-j}{2}} D^{\alpha_i} w_i\ D^{\alpha_j} w_j\ D^{\alpha_k} w_k\|_0 \leqslant \\ &\qquad \|D^{\alpha_j} w_j\|_{L^\infty(R^2)} \|D^{\alpha_k} w_k\|_{L^\infty(R^2)} \|(1+|x|^2)^{\frac{m-j}{2}} D^{\alpha_i} w_i\|_0 \\ &\qquad \leqslant C\ [w_i]_m\ [w_j]_m\ [w_k]_m . \end{aligned}$$

The last inequality follows from Sobolev inequalities and the fact that $|\alpha_j|, |\alpha_k| \leqslant m-2$. Thus

$$[Ju - Jv]_m \leqslant C\ \{[u]_m^2 + [v]_m^2\}\ [u-v]_m .$$

By the Contraction Mapping Theorem there exists $T_\varepsilon > 0$ depending upon a positive constant C such that the solution of (1.1 -1.2) exists in Π^m and is unique for $t \varepsilon [0, T_\varepsilon]$ as long as

$$[u(0)]_m \leq C.$$

We next extend the local result by applying Gronwall's Lemma. If the solution u(t) is in Π^m for all $t\varepsilon[0,T]$ then using (3.6) and the triangle inequality gives

$$(3.9) \qquad [u(t)]_m \leq [E(t)u_0]_m + \int_0^t [E(t-s)\, Ju(s)]_m ds$$

Using arguments similar to those above gives

$$(3.10) \qquad [Ju(s)]_m \leq C\|u\|_m^2 [u]_m.$$

Combining (3.9) and (3.10) and using Lemma 3.1 gives

$$[u(t)]_m \leq C\{[u_0]_m + \int_0^t [u(s)]_m\, ds\}.$$

By Gronwall's Lemma

$$(3.11) \qquad [u(t)]_m \leq C(T,[u_0]_m).$$

Using (3.11) and the local result on $(0,T_\varepsilon)$, $(T_\varepsilon, T_{2\varepsilon})$,... completes the proof of the theorem.

4. <u>Trigonometric Subspaces</u>. Approximating subspaces of trigonometric polynomials are defined in this section. The approximation properties of these subspaces are described approximating periodic functions as well as functions with decay. The approach used follows the one dimensional analysis presented in [13].

Let N be a positive integer and set

$$(4.1) \qquad h = \frac{2M}{(2N-1)}.$$

The spectral approximation subspace S_h is the linear space of trigonometric polynomials spanned by the functions

$$(4.2) \qquad \phi_{k,\ell}(x,y) = e^{i\pi kx/M} e^{i\pi \ell y/M}$$

where $-N+1 \leq k,\ell \leq N-1$. The functions in S_h are extended to all of R^2 by zero. The subspaces S_h obviously satisfy the inverse property

$$\|\psi\|_{H^j(\Omega_M)} \leq Ch^{k-j} \|\psi\|_{H^k(\Omega_M)} \quad \text{for } \psi \in S_h \text{ and } k \leq j.$$

The approximation properties of S_h are a consequence of the fact that the trigonometric functions are eigenfunctions for the elliptic boundary value problem

$$\begin{aligned} w - \Delta w &= f && \text{in } \Omega_M \\ w \text{ periodic} & && \text{on } \partial\Omega_M. \end{aligned}$$

Let $\lambda_j^k = \dfrac{1}{M^2} \sum_{n=0}^{k} (j/M)^{2n}$ and $\dot{H}^j(\Omega_M)$ be the set of functions f in $L^2(\Omega_M)$ such that

$$\sum_{(k,\ell)\in Z^2} \lambda^j_{|(k,\ell)|} \, |(f,\phi_{k,\ell})_{\Omega_M}|^2 < \infty$$

where $(\cdot,\cdot)_{\Omega_M}$ denotes the L^2 inner product on Ω_M. The space $\dot{H}^j(\Omega_M)$ has the natural norm

$$(4.3) \qquad \|f\|_{\dot{H}^j(\Omega_M)} \equiv \Big(\sum_{(k,\ell)\in Z^2} \lambda^j_{|(k,\ell)|} |(f,\phi_{k,\ell})_{\Omega_M}|^2 \Big)^{1/2}$$

In [12], it is shown that $\dot{H}^j(\Omega_M) = H_p^j(\Omega_M)$ and that the norm (4.3) is equivalent to the usual Sobolev norm on $H_p^j(\Omega_M)$. Furthermore, a dialation argument shows that the norm equivalence holds uniformly for $M \geq C > 0$.

Lemma 4.1. Let w be in $H_p^r(\Omega_M)$ and P_h denote the L^2 projection into S_h; then

$$\|w-P_h w\|_{H^s(\Omega_M)} \leq Ch^{r-s}\|w\|_{H^r(\Omega_M)}$$

for $0 \leq s \leq r$.

Proof. We prove this result using the equivalent norm (4.3). Then

$$\|w-P_h w\|^2_{\dot{H}^s(\Omega_M)} = \sum_{\substack{k,\ell\in Z^2 \\ |k|,|\ell|\geq N}} \lambda^s_{|(k,\ell)|} |(w,\phi_{k,\ell})_{\Omega_M}|^2$$

$$\leq h^{2r-2s} \sum_{k,\ell\in Z^2} \lambda^r_{|(k,\ell)|} |(w,\phi_{k,\ell})_{\Omega_M}|^2 \leq C_r h^{2r-2s}\|w\|^2_{H^r(\Omega_M)}$$

This completes the proof of the lemma.

In many applications, L^2 projections are replaced by collocation projections. A set of collocation points Γ_h for S_h is defined by

$$\Gamma_h = \{(kh,\ell h) \text{ where } -N+1 \leq k,\ell \leq N-1\}$$

Given the point values of w on Γ_h, the collocation projection is defined to be the unique function in S_h which interpolates w at the points of Γ_h. One can show (see [7], [10] or [12]) that the collocation projection attains the approximation properties of S_h on $H_p^r(\Omega_M)$ as described in the following lemma.

Lemma 4.2. Let $r > 1$ then for w in $H_p^r(\Omega_M)$ and $0 \leq s \leq r$

$$\|P_c w-w\|_{H^s(\Omega_M)} \leq C\, h^{r-s}\|w\|_{H^r(\Omega_M)}.$$

We next consider using trigonometric polynomials on finite domains to approximate functions on infinite domains. Let ψ_1 be a C^∞ function on R^2 satisfying

$$\psi_1 = \begin{cases} 0 & |x| \geq 1 \\ 1 & |x| < 1/2 \end{cases}$$

and let $\psi_M(x) = \psi_1(x/M)$.

<u>Lemma 4.3</u>. Let $r > 1$, u be in H^r and $\hat{u}$ be in H^s, then

$$\| u - P_c \psi_M u \|_0 \leq C \{h^r \|u\|_r + M^{-s} \| \hat{u} \|_s\}$$

<u>Proof</u>. For a function f in $C^\infty(R^2)$, the periodic extension of $\psi_M f$ is in $C_p^\infty(\Omega_M)$. It is not hard to show that multiplication by ψ_M maps H^r into $H_p^r(\Omega_M)$ uniformly for $M \geq C$. In addition

$$(4.4) \qquad \|\psi_M u - u\|_0^2 \leq C \int_{|x| \geq M/2} |u|^2 \, dx \leq C M^{-2s} \|\hat{u}\|_s^2 .$$

By the triangle inequality

$$\|u - P_c \psi_M u\|_0 \leq \|u - \psi_M u\|_0 + \|(I - P_c)\psi_M u\|_0$$

The lemma follows immediately from (4.4) and Lemma 4.2.

5. <u>Semidiscrete Estimates</u>. Semidiscrete (time continuous) approximation to (1.1 - 1.2) shall be considered in this section. The idea is to first define a discrete approximation $E_h(t)$ to $E(t)$. Then motivated by (3.1), we define the semidiscrete approximation to (1.1 - 1.2) by

$$(5.1) \qquad U(t) = E_h(t)(P_c \psi_M u_0) + \int_0^t E_h(t-s)(P_c \psi_M J U(s)) ds .$$

The operator $E_h(t)$ is defined to be the solution operator for the problem

$$i \frac{\partial V}{\partial t} + \Delta V = 0 \qquad \text{in } [0,T] \times \Omega_M$$

$$V(t) \text{ periodic on } \partial\Omega_M$$

$$V(0) = W \qquad \text{for } W \in L^2(\Omega_M)$$

that is $E_h(t)W = V(t)$. As usual $V(t)$ is defined to be zero outside of Ω_M. Since the trigonometric polynomials are eigenfunctions for the Laplacian with periodic boundary conditions, $E_h(t)$ leaves S_h invariant. Indeed if W is in S_h and is given by

$$(5.2) \qquad W = \sum_{|j|,|k| < N} C_{j,k}\, \phi_{j,k}$$

then

$$E_h(t)\, W = \sum_{|j|,|k| < N} C_{j,k}\, e^{-i\pi^2(j^2+k^2)/M^2}\, \phi_{j,k}.$$

Note that the coefficients in (5.2) can be calculated given the functional values on Γ_h (and vice versa) by the Fast Fourier Transform Algorithm. Thus the operators $E_h(t)P_c$ in (5.1) can be efficiently evaluated. The main results of this section are given in the following theorem.

Theorem 5.1. Let u_0 be in Π^m for $m \geq 3$ and let $T > 0$ be given. Then for γ with $0 < \gamma < 1$, there exists $\varepsilon > 0$ such that if $\{h + M^{-1}\}^{m-2-\gamma} < \varepsilon$ then the solution of (5.1) exists on $[0,T]$ and satisfies

$$\| U(t) - u(t) \|_0 \leq C\,\{h^m + M^{-m}\}$$

The constants ε and C depend upon T, γ and $[u_0]_m$ but are independent of M and h.

Corollary 5.1. $U(t)$ is a differentiable vector valued function on $[0,T]$ and solves

$$(5.3)\qquad i\frac{d}{dt}U + \Delta U = P_c\psi_M JU \qquad \text{on } \Omega_M$$

$$(5.4)\qquad U(t) \text{ periodic on } \partial\Omega_M$$

$$(5.5)\qquad U(0) = P_c\psi_M u_0 .$$

To prove Theorem 5.1 we shall first need some auxiliary lemmas. Let F_M denote the Finite Fourier Transform on Ω_M, i.e. for f in $L^2(R^2)$

$$(5.6)\qquad F_M f(\zeta) = \int_{\Omega_M} \exp(ix\cdot\zeta)\, f(x)dx \text{ for } \zeta \in R^2.$$

Let $Z_M = \{(\pi k/M, \pi j/M) \text{ where } j,k \in Z\}$ then F_M is an isometry of $L^2(\Omega_M)$ onto $L^2(Z_M)$ with inverse denoted F_M^{-1}. Note, however, that we have not restricted the domain of F_M to $L^2(\Omega_M)$ and we have not restricted the domain of $F_M f$ to Z_M in the definition (5.6). We quote the following lemma.

Lemma 5.1. [3, Thm. 5]. Let f be in H^j for $j > 1$; then

$$\|(F_M^{-1} - F^{-1})\, f\|_0 \leq CM^{-j}\|f\|_j$$

Here F_M^{-1} is defined first from $L^2(Z_M)$ into $L^2(\Omega_M)$ and then extended to all of R^2 by zero.

Lemma 5.2. Let w be in Π^m; then

$$(5.7)\qquad \| E(t)w - E_h(t)\, P_c\psi_M w\|_0 \le C\,\{h^m + M^{-m}\}\,[w]_m$$

Proof of Lemma 5.2. First we observe that the discrete operator $E_h(t)$ is given by

$$(5.8)\qquad E_h(t)\,u = F_M^{-1}\,[e^{-i|\zeta|^2 t}(F_M u)(\zeta)]$$

Then by the triangle inequality

$$\| E(t)w - E_h(t)\, P_c\psi_M w\|_0 \le \|E(t)\,(I-\psi_M)w\|_0$$

$$+ \|(E(t) - E_h(t))\,\psi_M w\|_0 + \|E_h(t)\,(I-P_c)\psi_M w\|_0$$

Using (4.6) and the fact that $E(t)$ is a bounded operator from $L^2(R^2)$ into $L^2(R^2)$ gives

$$\|E(t)\,(I-\psi_M)\,w\|_0 \le CM^{-m}\,[w]_m .$$

$E_h(t)$ is also a bounded operator from $L^2(\Omega_M)$ onto $L^2(\Omega_M)$; thus Lemma 4.2 implies

$$\|E_h(t)\,(I-P_c)\,\psi_M w\|_0 \le C\,h^m\,[w]_m .$$

Using (5.8) and Lemma 5.1 gives

$$\|(E(t) - E_h(t))\,\psi_M w\|_0 = \|(F^{-1}-F_M^{-1})\,e^{-i|\zeta|^2 t}F(\psi_M w)\|_0$$

$$\le CM^{-m}\|e^{-i|\zeta|^2 t}F(\psi_M w)\|_m$$

Applying the Leibniz rule and Remark 2.1 gives

$$\|e^{-i|\zeta|^2 t}F(\psi_M W)\|_m \le C\,[\psi_M w]_m \le C[w]_m .$$

The proof of the lemma follows by combining the above estimates.

<u>Proof of Theorem 5.1</u>. Let us assume for the moment that the solution of (5.1) exists for $t \varepsilon [0,T]$ and satisfies the estimate

$$(5.9) \qquad \|U(t)\|_{H^{2+\gamma}(\Omega_M)} \leq C_0 \text{ for } \gamma > 0 \text{ and } t \varepsilon [0,T]$$

Let $e(t) \equiv u(t) - U(t)$ denote the error. Subtracting (3.1) and (5.1) and using the triangle inequality gives

$$(5.10) \qquad \begin{aligned} \|e(t)\|_0 &\leq \|E(t)u_0 - E_h(t)\, P_c \psi_M u_0\|_0 \\ &\quad + \int_0^t \|E(t-s)Ju(s) - E_h(t-s)P_c \psi_M Ju(s)\|_0 \\ &\quad\quad + \| E_h(t-s)\, P_c \psi_M\, (Ju(s)-JU(s)\|_0 ds \\ &\equiv T_0 + \int_0^t T_1 + T_2 \, ds. \end{aligned}$$

We first estimate by the triangle inequality

$$(5.11) \qquad \begin{aligned} \|(P_c-I)\psi_M(JU-Ju)\|_0 &\leq \|(P_c-I)\psi_M(JU-JP_h u)\|_0 \\ &\quad + \| (P_c-I)\, \psi_M\, (JP_h u - Ju)\|_0 \end{aligned}$$

By (3.8), Sobolev inequalities and the Leibniz rule

$$(5.12) \| Jf-Jg \|_{H^p(\Omega_M)} \leq C\{\|f\|^2_{H^{2+\gamma}(\Omega_M)} + \| g \|^2_{H^{2+\gamma}(\Omega_M)}\} \| f-g \|_{H^p(\Omega_M)}$$

for $p=0,1,2,3$. The boundedness properties of P_h, Lemma 4.2, the inverse properties of S_h and (5.11) imply that

$$\|(P_c - I)\ \psi_M(JU - Ju)\|_0 \leq Ch^2\{\|U - P_h u\|_{H^2(\Omega_M)} + \|P_h u - u\|_{H^2(\Omega_M)}\}$$
(5.13)
$$\leq C\{\|U - P_h u\|_{L^2(\Omega_M)} + h^m\,[u]_m\} \leq C\{\|e\|_0 + h^m[u]_m\}$$

We next use (5.13) to bound T_2. By the triangle inequalty T_2 can be bounded by

$$T_2 \leq \|(P_c - I)\ \psi_M(JU - Ju)\|_0 + \|\psi_M(JU - Ju)\|_0$$

Using (5.13) gives

$$T_2 \leq C\ \{\|e(s)\|_0 + h^m[u(s)]_m\} \tag{5.14}$$

Combining (5.10), (5.14) and Lemma 5.2 implies

$$\|e(t)\|_0 \leq C\ \{h^m + M^{-m}\} + \int_0^t \|e(s)\|_0\ ds.$$

Gronwall's Lemma gives that there is a constant C_1 (depending on C_0 and T but not on h or M) such that

$$\|e(t)\|_0 \leq C_1\{h^m + M^{-m}\} \quad \text{for all } t\varepsilon[0,T]. \tag{5.15}$$

Note that the constant C_1 is well defined irregardless of whether the solution of (5.1) exists.

Now by Theorem 3.1 there is a constant C_2 such that

$$\|u(t)\|_{H^{2+\gamma}(\Omega_M)} \leq C_2. \tag{5.16}$$

Define $C_0 = 2C_2$. We use a bootstrap argument to show that the solution of (5.1) exists on [0,T], is unique, and satisfies (5.9) and (5.14). By (5.12) and the inverse properties of S_h it is not hard to show local existance and uniqueness for solutions of (5.1) by the contraction mapping theorem.

Now, by the triangle inequality

$$(5.17)\quad \begin{aligned} \|U(t)-u(t)\|_{H^{2+\gamma}(\Omega_M)} &\leq \|U(t)-P_h\psi_M u(t)\|_{H^{2+\gamma}(\Omega_M)} \\ &+\|(I-\psi_M)\ u(t)\|_{H^{2+\gamma}(\Omega_M)} +\|\ (I-P_h)\psi_M u(t)\|_{H^{2+\gamma}(\Omega_M)} \end{aligned}$$

One can show that

$$\|(I-\psi_M)\ u(t)\|_{H^j(\Omega_M)} \leq CM^{j-m}[u]_m$$

and thus by interpolation, the second term of (5.17) is bounded by $CM^{2+\gamma-m}$. If (5.15) holds then the inverse properties of S_h and Lemma 4.1 give

$$\begin{aligned} \|U(t)-u(t)\|_{H^{2+\gamma}(\Omega_M)} &\leq C\ \{h^{-2-\gamma}\|U(t)-P_h\psi_M u(t)\|_{L^2(\Omega_M)} \\ &\qquad + M^{-m+2+\gamma} + h^{m-2-\gamma}\} \\ &\leq C_3\{h^{-2-\gamma}(M^{-m}+h^m) + M^{-m+2+\gamma}\} \end{aligned}$$

Thus if h and M are chosen so that

$$C_3\{h^{-2-\gamma}(M^{-m}+h^m) + M^{-m+2+\gamma}\} < C_0/4$$

then (5.15) and (5.16) implies (5.9). The theorem follows from local existence and (5.15).

6. <u>Time Step Discretization</u>. To describe an algorithm approximating the solution of (1.1 - 1.2) one must discretize (5.1) or (5.3-5.5) in time. Note that the L^2

norm of the solution of (1.1-1.2) remains constant independent of t as can be seen integrating (1.1-1.2) against u and taking imaginary parts. We will consider a fractional step time discretization of (5.3-5.5) which generates approximate solutions with constant L^2 norms. This discretization does not introduce numerical dissipation as may be the case with Crank-Nicolson and similar time discretization methods based on rational approximations to the exponential.

The fractional step discretization of these equations was, I believe, first used by J. Wallace [14] and can be described by

(i) $U^0 = P_c \psi_M u_0$

(ii) For i=0,1,2...

$$U^{i+1/2} = P_c(e^{-i\zeta(U^i)k}U^i) \qquad \text{where } \zeta(U)U \equiv JU$$

$$U^{i+1} = E_h(k)\, U^{i+1/2}$$

Here k>0 is the time step size and U^j approximates u(kj). I do not know of any analysis for fractional step methods applied to nonlinear problems although linear fractional step methods are considered in [18] and the references therein. The fractional step method given above satisfies

$$\| U^j \|_0 = \| P_c \psi_M u_0 \|_0$$

Numerical results for the spectral discretization of (1.1) with time step defined by (i) and (ii) are given in Figure 1. The results of Fig. 1 are for calculations modeling the propagation of a focused pulsed high power laser beam through an absorbing medium. The beam heats the medium which generates the density gradients defining ζ. The beam is initially a Gaussian distribution. The plots give contours of constant irradience as a function of T (the propagation distance) and the pulse number. Computations of this sort can be used to predict beam distortion patterns and overall power dissipation properties.

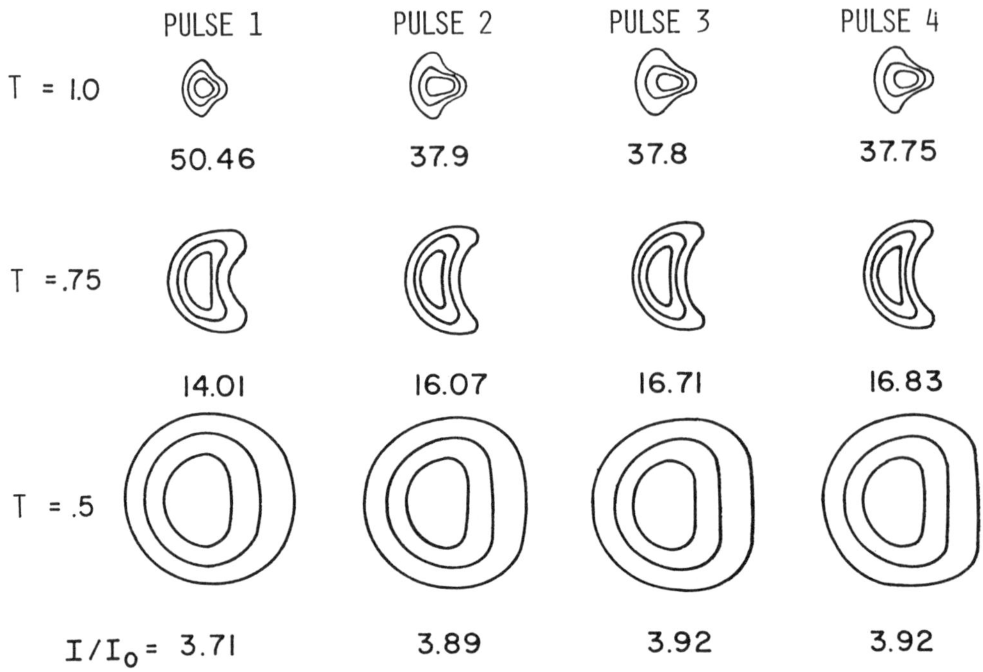

FIGURE 1. PULSED PROPAGATION THRU AN ABSORBING MOVING MEDIUM. (COURTESY OF FAR FIELD INC., SUDBURY, MASS.)

REFERENCES

[1] J. B. BAILLON., T. CAZENAVE, AND M. FIGUEIRA, Équation de Schrödinger avec non-linéarité intégrale, C. R. Acad. Sc., Paris Sér. A 284(1977), pp. 939-942.

[2] J. B. BAILLON, T. CAZENAVE AND M. FIGUEIRA, Équation de Schrödinger non linéaire, C. R. Acad. Sc., Paris Ser. A, 284(1977), pp. 869-872.

[3] J. H. BRAMBLE AND S. R. HILBERT, Estimation of linear functionals on Sobolev spaces with application to Fourier transforms and spline interpolation, Math. Comp., 7(1970), pp. 112-124.

[4] J.H. BRAMBLE AND J.E. PASCIAK, High order time stepping methods for a nonlinear Schrodinger equation, unpublished notes.

[5] H. J. BREAUX, An analysis of mathematical transformations and a comparison of numerical techniques for computation of high energy cw laser propagation in an inhomogeneous medium, Ballistic Research Laboratories, Rep BRLR 1723(1974).

[6] W. P. BROWN, JR., Computer simulation of adaptive optical systems, Hughes Research Lab., Rep 60921-74-C-0249(1975).

[7] C. CANUTO AND A. QUATERONI, Approximination results for orthogonal polynomials in Sobolev spaces, Instituto Di Analisi Numerica del Consiglio Nazionale delle Ricerche, 251(1980).

[8] J. W. COOLEY AND J. W. TUKEY, An algorithm for the calculation of complex Fourier series, Math. Comp., 19 (1965), pp. 297-301.

[9] J. A. FLECK, JR., J. R. MORRIS and M. J. FERT, Time dependent propagation of high energy laser beams through the atmosphere, UCRL Rep. 51826(1975).

[10] H. O. KREISS and J. OLIGER, Stability of the Fourier method, SIAM J. Numer. Anal., 16 (1979), pp. 421-433.

[11] J. L. LIONS, Non Homogeneous Boundary Value Problems and Applications, Springer-Verlag, N.Y.,(1972).

[12] J. E. PASCIAK, Spectral and pseudo spectral methods for advection equations, Math. Comp., 35(1980), pp. 1081-1092.

[13] J. E. PASCIAK, Spectral methods for a nonlinear initial value problem involving pseudo differential operators, SIAM J. Numer. Anal., 19(1982), pp. 142-154.

[14] V. S. VLADIMIROV, Equations of Mathematical Physics, Marcel Dekker, N.Y. (1971).

[15] J. WALLACE and J. LILLY, Thermal blooming of repetitively pulsed laser beams, J. Opt. Soc. Am., 64(1974), pp. 1651-1655.

[16] J. WALLACE and J. PASCIAK, Compensating for thermal blooming of repetitively pulsed lasers, J. Opt. Soc. Am., 65(1975).

[17] K. G. WHITNEY, G. L. MADER, and P. B. ULRICH, SSPARAMA: A nonlinear wave optics multipulse (and cw) steady state propagation code with adaptive coordinates, Naval Research Lab.

[18] N. N. YANENKO, The Method of Fractional Steps, Springer-Verlag, N.Y.(1971).

SPECTRAL METHODS FOR SPATIALLY EVOLVING HYDRODYNAMIC FLOWS

A. T. PATERA*

Abstract. An explicit hybrid Fourier spectral technique is developed for the numerical simulation of spatially-evolving (inflow/ outflow) hydrodynamic flows. The hybrid scheme relies on representation of a non-periodic function by low-order polynomials in conjunction with a Fourier expansion, the former used to satisfy the boundary constraints required for rapid convergence of the latter. The method is applied to the small-amplitude and nonlinear evolution of vibrating-ribbon disturbances to plane Poiseuille flow. Good agreement is obtained with both spatial linear theory and previously published work. The hybrid Fourier techniques presented here apply with little modification to general flow problems with rigid boundary conditions.

Introduction. Spectral numerical methods have proven invaluable in the study of transition and turbulence in fluid flows. A particular area in which spectral methods have had a marked effect is that of transition in wall-bounded shear flows. The accuracy and efficiency of spectral methods as applied to these flows has led in the past ten years to:

(i) accurate solution of the Orr-Sommerfeld equation [1];

(ii) calculation of two-dimensional nonlinear travelling waves [2];

(iii) calculation of the linear three-dimensional instability of two-dimensional nonlinear states by direct numerical simulation [3], [4], [5];

(iv) calculation of the linear three-dimensional instability of two-dimensional nonlinear states by spectral solution of non-separable perturbation equations [6], [7];

(v) investigation of fully nonlinear effects on transition and turbulence [8], [9], [10].

*Department of Mechanical Engineering, Massachusetts Institute of Technology, Cambridge, MA 02139

The major disparity between transition spectral simulations and reality has been the assumption in the former of temporally growing perturbations and associated spatial periodicity downstream. Although such assumptions have proven less restrictive than might be imagined, they nevertheless preclude direct comparison with experiment, especially in the later stages of transition. The assumption of downstream periodicity becomes even more bothersome when the base flow is not strictly parallel (e.g. boundary layers, jets), and when linear growth rates are large and hence temporal and spatial stability theory cannot be expected to give similar results.

Lack of periodicity in the downstream direction has led previous investigators to a finite-difference representation in this direction [11], [12]. The advective instabilities and large artificial viscosities usually associated with difference techniques are elminated by use of implicit (iterative) high-order schemes for the nonlinear terms in the Navier-Stokes equations. It is our purpose in this paper to present an explicit hybrid Fourier technique as a viable alternative (in terms of accuracy, ease of implementation, and efficiency) to these implicit difference formulations.

To test our inflow/outflow methods we will examine the two-dimensional linear and nonlinear spatial stability characteristics of plane Poiseuille flow. The real application of these techniques will be in investigating three-dimensional spatial instability (in channels and boundary layers) resulting from both vibrating-ribbon and point disturbances, the latter corresponding to transition spots. (It should be noted that there is no pretense in this paper of providing any new physical insights into the instability process.)

The techniques presented in this paper are in fact general algorithms for extending Fourier techniques to "rigid" boundary conditions, however they are clearly only useful when high resolution/accuracy is not required near the boundary. The methods have already been successfully used for flow over a backward-facing step and for the developing flow in the entrance region of a channel. These calculations will be reported elsewhere.

Fourier Series with Polynomial Subtraction. The accuracy of spectral techniques is largely determined by the choice of spectral expansion. As is well known, expansions based on eigenfunctions of singular Sturm-Liouville problems converge faster than algebraically for "smooth" functions, independent of the boundary conditions. On the basis of these considerations, Chebyshev polynomial expansions present themselves as ideal for the downstream direction of a spatially evolving flow. However, on three points, Chebyshev representation is less than ideal for inflow/outflow situations. First, they provide the most resolution near inflow and outflow, where very little of physical import is occurring. Second, this increased (undesired) resolution near inflow and outflow drastically reduces the maximum allowable time step on the basis of the Courant condition. Lastly, Chebyshev operators are, in general, difficult and time-consuming to invert. This leads one to consider Fourier representations, as here the collocation points are evenly spaced and the operators diagonal.

The drawback to Fourier series representations is that the convergence of the expansion depends on the boundary conditions of the problem. The idea of polynomial subtraction to satisfy the boundary constraints required for rapid convergence of the Fourier series is not particularly new [13], and has already been considered in the context of modern spectral methods [14]. A trivial example of how such a technique works is given below.

Consider the forced heat equation

$$\frac{\partial\phi}{\partial t} = \frac{\partial^2\phi}{\partial x^2} + W(t) \qquad \phi(0,t) = \phi(1,t) = 0 \tag{1}$$

with some specified initial condition. An expansion of the form

$$\phi(x,t) = \sum_{n=1}^{N} a_n(t) \sin n\pi x \tag{2}$$

will lead to a third order (in space) scheme (i.e. $a_n \sim 0\,(n^{-3})$ as $n \to \infty$). However, the trivial modification

$$\phi(x,t) = \frac{W(t)}{8}\left\{1 - 4(x - \tfrac{1}{2})^2\right\} + \sum_{n=1}^{N} b_n(t) \sin n\pi x \tag{3}$$

directly leads to a fifth order scheme ($b_n \sim 0\,(n^{-5})$, $n \to \infty$), and in the case where $W(t) = W_0$, an "infinite" order scheme. (A less trivial example of polynomial subtraction for a hyperbolic equation is given in [14].)

In many cases it is not clear or easily determined what polynomials must be subtracted. For instance, if a function ϕ is specified on the interval $x \in [0,1]$, with $\phi(0) = \phi(1) = 0$, a priori the subtractions required for a fifth-order Fourier representation of ϕ can not be known. However, if we denote a jth-order approximation (say finite-difference) to $d^k\phi(.)/dx^k$ as $D_j^k\phi(.)$, then

$$\phi + \left\{\frac{1 - 4(x - \frac{1}{2})^2}{8}\right\}\left[D_j^2\,\phi(0) + \tfrac{1}{3}(1+x)(D_j^2\,\phi(1) - D_j^2\phi(0))\right] \tag{4}$$

$$= \sum_{n=1}^{N} c_n \sin n\pi x + 0(1/N^5) \qquad (N \to \infty),\quad j \geq 2 \quad .$$

Although the expansion (3) (which corresponds to (4) with $j = \infty$, $D_\infty^2\phi(0) = D_\infty^2\phi(1) = -W(t)$) is clearly cleaner than (4), it should be noted that the order of the spatial error in the solution of (1) - (2) using (4) is formally independent of j for $j \geq 2$.

Thus it is seen that high order Fourier representations can be constructed in many cases, either on the basis of exact information in the equations/boundary conditions or from low-order approximations to boundary derivatives.

The Inflow/Outflow Problem. The geometry for the inflow/outflow problem as applied to two-dimensional plane Poiseuille flow is shown in Fig. 1. The governing equations for the velocity $\vec{v}(x, z, t) = u\hat{x} + w\hat{z}$ are

$$\frac{\partial \vec{v}}{\partial t} = \vec{v} \times \vec{\omega} - \nabla \pi + \frac{1}{R} \nabla^2 \vec{v} \tag{5a}$$

$$\nabla \cdot \vec{v} = 0 \tag{5b}$$

where $\vec{\omega} = \nabla \times \vec{v}$ $(= \zeta\hat{y})$, π is the pressure head, $p + \frac{1}{2}|\vec{v}|^2$, and R is the Reynolds number based on channel half-width, h, and centerline velocity at the inflow boundary (x = 0), U. The equations are non-dimensionalized with respect to length, velocity, and time scales of h, U, and h/U, respectively. In general, we decompose v into the basic flow and a perturbation, $\vec{v} = \vec{v}'(x, z, t) + (1 - z^2)\hat{x}$.

The boundary conditions on the rigid walls are no-slip,

$$\vec{v}(x, z = \pm 1, t) = 0 \tag{6a}$$

In our work here we are simulating vibrating-ribbon experiments, and the boundary conditions at inflow are therefore taken as

$$\vec{v}(x{=}0, z, t) = A\,\mathrm{Re}\left\{\vec{v}_\ell(z)e^{-i\omega_R t}\right\} + (1-z^2)\hat{x} \tag{6b}$$

where $\vec{v}_\ell(z)$ is the spatial eigenfunction of the Orr-Sommerfeld equation corresponding to the temporal frequency ω_R and spatial eigenvalue $\alpha_R + i\alpha_I$ with smallest imaginary part. A is an amplitude parameter: in general $u'_{t,o} = \max_z \langle u'^2(x = 0, z, t)\rangle^{1/2}$ (where $\langle\ \rangle$ denotes time-averaging) is used as a measure of the magnitude of the vibrating-ribbon disturbance.

Much attention has been given the outflow (x = L) boundary conditions. In general it appears that these conditions (which are mathematically required but do not reflect any physical condition) do not "contaminate" the actual solution appreciably upstream. For our purposes here we choose the simple conditions

$$\text{(6c)} \qquad \frac{\partial u}{\partial x}(x = L,\ z,\ t) = w(x = L,\ z,\ t) = 0 \quad ,$$

although improvements on (6c) are clearly possible.

To complete the problem statement the initial conditions must be specified. To preclude any bias towards the linear-theory solution, the initial state is taken to be the basic flow,

$$\text{(7)} \qquad \vec{v}(x,\ z,\ t = 0) = (1 - z^2)\hat{x}$$

i.e. $\vec{v}'(x,\ z,\ t = 0) = 0$.

Note that for $A \to 0$ the solution to eqns. (5) - (7) as $t \to \infty$ (without the outflow condition (6c)) is given by linear theory as

$$\text{(8)} \qquad \vec{v}'(x,\ z,\ t) = A\, e^{-\alpha_I x}\, \mathrm{Re}\left\{ \vec{v}_\ell(z)\, e^{i(\alpha_R - \omega_R t)} \right\} \quad ,$$

where $(-\alpha_I)$ is the growth rate of the mode.

<u>Numerical Methods--The Hybrid Spectral Technique</u>. The numerical method to be presented here is a fractional-step method [9], [15]. Each time step consists of three partial steps; a nonlinear step, a pressure step, and a viscous step

$$\text{(9)} \qquad \frac{\hat{\vec{v}}^{n} - \vec{v}^{n-1}}{\Delta t} = \frac{3}{2}(\vec{v} \times \vec{\omega})^{n-1} - \frac{1}{2}(\vec{v} \times \vec{\omega})^{n-2} \quad ,$$

$$\text{(10a)} \qquad \frac{\hat{\hat{\vec{v}}}^{n} - \hat{\vec{v}}^{n}}{\Delta t} = -\nabla \pi^{n}$$

$$\text{(10b)} \qquad \nabla \cdot \hat{\hat{\vec{v}}}^{n} = 0 \quad ,$$

and

$$\text{(11)} \qquad \frac{\vec{v}^{n} - \hat{\hat{\vec{v}}}^{n}}{\Delta t} = \frac{1}{R}\nabla^2 \vec{v}^{n} + \frac{2}{R}\hat{x} \quad ,$$

respectively. (Here superscript denotes time step.) In the first step, (9), no boundary conditions are imposed. In the pressure step, (10), the velocity <u>normal</u> to the domain boundaries is specified per eqns. (6a) - (6c). In the viscous step, (11), both normal <u>and</u> tangential velocity components are made to satisfy eqns. (6a) - (6c). Al-

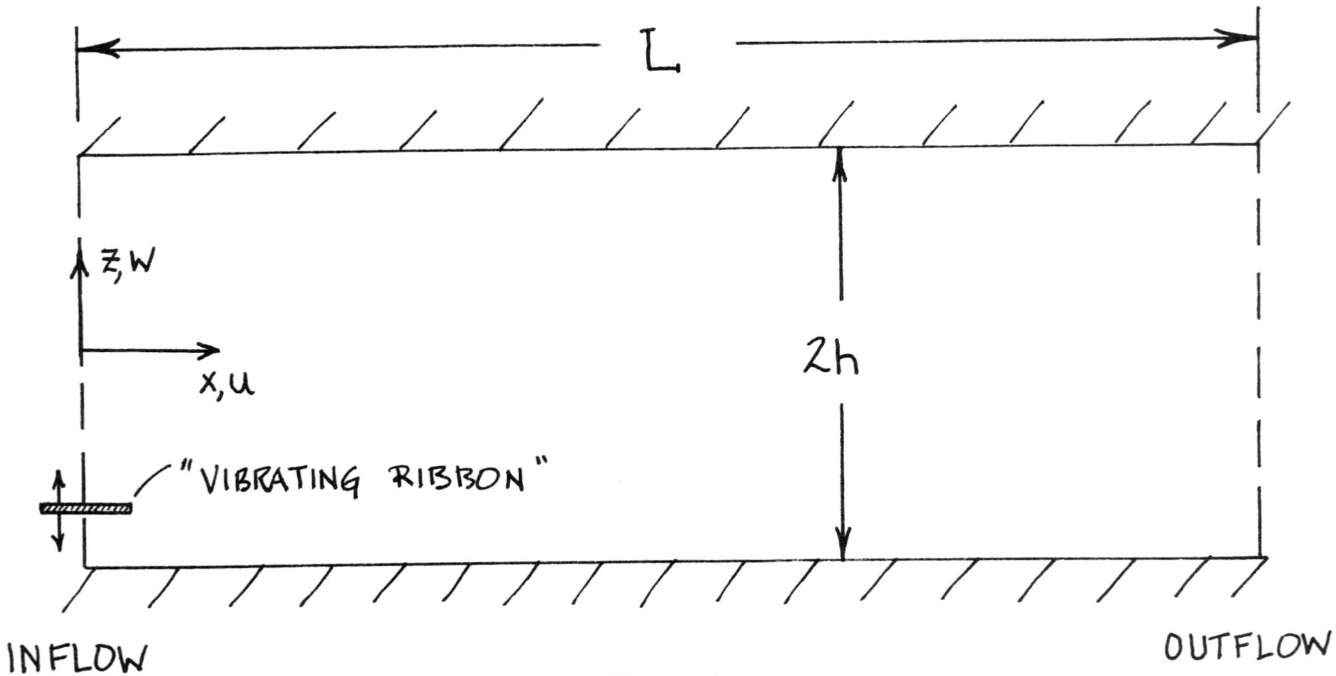

Figure 1

Inflow/outflow geometry for plane Poiseuille flow. Disturbances are introduced into the flow at the inflow boundary, $x = 0$ and propagate out at the outflow point $x = L$. The outflow boundary is artificial in the sense that, in reality, no physical condition is imposed there.

though the explicit Adams-Bashforth step, (9), is second-order accurate in time, even if Crank-Nicolson rather than Euler were used for (11), the total step would still not be second order due to splitting errors [16]. (The $2/R\ \hat{x}$ term in (11) is not required, however it eliminates a trivial splitting error.)

The basic premise behind the hybrid Fourier technique is that, although the viscous step, (11), can be adequately represented in x by finite-differences due to the high Reynolds number and absence of significant shear in the x-direction (relative to the z-direction), the advection and pressure steps are susceptible to phase errors and numerical instabilities and hence are better treated by a hybrid Fourier technique. Although such a Fourier technique cannot be "infinite" order, interior errors can be made very small by polynomial subtractions, and even the highest modes accurately preserve phase information. From a practical point of view, it is important to note that existing periodic codes require very little modification to handle inflow/outflow problems with the hybrid technique to be presented here.

We now examine in detail the spatial representation of $\vec{v}$ in the x-direction at each partial step. For now we treat z as a continuous variable: its (standard) discretization into a Chebyshev basis will be summarized at the end of this section. Furthermore, to simplify the nomenclature, all variables are implicitly assumed to depend on z, and the dependence is therefore not explicitly indicated at each step.

In the first step, (9), we write

$$u^{n-1}(x) - u^{n-1}(0) = \sum_{j=0}^{N} u_j^{n-1} \sin \alpha_j x \tag{12a}$$

$$w^{n-1}(x) - D_2^1\, w^{n-1}(0) \left\{ x - L \right\} = \sum_{j=0}^{N} w_j^{n-1} \cos \alpha_j x \tag{12b}$$

where

$$\alpha_j = \frac{(j + \frac{1}{2})\pi}{L} \quad . \tag{13}$$

For future reference we indicate the physical space representation of variables as

$$\tilde{u}_k^{n-1} = u^{n-1}(x_k) \ , \ . \ . \ . \tag{14a}$$

where

(14b) $$x_k = \frac{k}{N} L , \quad k = 0, \ldots, N$$

From (12) it follows that, given that the boundary conditions (6a) - (6c) have been satisfied on the previous viscous step, $u_j^{n-1}, w_j^{n-1} \sim 0\,(j^{-3})$ as $j \to \infty$. From (12) the vorticity is easily constructed,

(15) $$\zeta^{n-1}(x) - \frac{d}{dz} u^{n-1}(0) + D_2^1 w^{n-1}(0) = \sum_{j=0}^{N} \left(\frac{d}{dz} u_j^{n-1} + \alpha_j w_j^{n-1} \right) \sin\alpha_j x .$$

From (15) ζ^{n-1} is transformed to physical space (a FFT has been constructed for half sine/cosine expansions: see [17]), the nonlinear term is formed by collocation, and $\hat{\vec{v}}^n$ is computed. (The approximation we use here for $D_2^1 f(0)$ is $(-3/2\ \tilde{f}_0 + 2\tilde{f}_1 - 1/2\ \tilde{f}_2)/\Delta x$, where $\Delta x = L/N$).

The pressure step is by far the most critical step in terms of both accuracy and stability. The velocity is represented as

(16a) $$\hat{u}^n(x) - \hat{u}^n(0) = \sum_{j=0}^{N} \hat{u}_j^n \sin \alpha_j x$$

(16b) $$\hat{w}^n(x) - \hat{w}^n(L) - D_2^1\, \hat{w}^n(0) \left\{ x - L \right\} = \sum_{j=0}^{N} \hat{w}_j^n \cos \alpha_j x$$

(16c) $$\pi^n(x) - a - b \left\{ x - L \right\} = \sum_{j=0}^{N} \pi_j^n \cos\alpha_j x$$

and

(17a) $$\hat{\hat{u}}^n(x) - \hat{\hat{u}}^n(0) = \sum_{j=0}^{N} \hat{\hat{u}}_j^n \sin \alpha_j x$$

(17b) $$\hat{\hat{w}}^n(x) = \sum_{j=0}^{N} \hat{\hat{w}}_j^n \cos \alpha_j x .$$

There are two points to note. First, to make (16a) formally third-order there should be a term of the form $D_2^1 \hat{u}^n(L) \left\{ x \right\}$. However, $\frac{\partial u^{n-1}}{\partial x}(L) = 0$ and hence this term is typically small: experimentation indicates that addition of this derivative correction to (17a) does not affect the solution. Second, the form of (17) is determined by imposing the outflow boundary conditions (6c) on both $\hat{u}^n$ and $\hat{w}^n$, but the inflow boundary condition (6b) only on $\hat{u}^n$ (i.e. $\hat{u}^n(0)$ is specified). We do not allow a linear polynomial in (17b) in order that incompressibility can be satisfied "exactly".

Inserting (16) - (17) into (10) gives (absorbing the Δt in π)

$$\hat{u}^n(0) - \hat{u}^n(0) + \sum_{j=0}^{N} \hat{u}_j^n \sin\alpha_j x - \sum_{j=0}^{N} \hat{u}_j^n \sin\alpha_j x \tag{18a}$$

$$= \sum_{j=0}^{N} \alpha_j \pi_j^n \sin\alpha_j x \quad - b$$

$$-\hat{w}^n(L) - D_2^1 \hat{w}^n(0) \left\{x - L\right\} + \sum_{j=0}^{N} \hat{w}_j^n \cos\alpha_j x - \sum_{j=0}^{N} \hat{w}_j^n \cos\alpha_j x \tag{18b}$$

$$= - \frac{da}{dz} - \frac{db}{dz}\left\{x - L\right\} - \sum_{j=0}^{N} \frac{d}{dz} \pi_j^n \cos\alpha_j x$$

$$\sum_{j=0}^{N} \alpha_j \hat{u}_j^n \cos\alpha_j x + \sum_{j=0}^{N} \frac{d}{dz} \hat{w}_j^n \cos\alpha_j x = 0 \quad . \tag{18c}$$

We now choose a, b such that the Fourier expansions in (18) converge rapidly. Using the facts that

$$1 = \frac{2}{L} \sum_{j=0}^{\infty} \frac{(-1)^j}{\alpha_j} \cos\alpha_j x \tag{19a}$$

and

$$\left\{x - L\right\} = \frac{-2}{L} \sum_{j=0}^{\infty} \frac{1}{\alpha_j^2} \cos\alpha_j x \tag{19b}$$

it then follows that

$$b = \hat{u}^n(0) - \hat{u}^n(0) \tag{20a}$$

$$\frac{da}{dz} = \hat{w}^n(L) \quad . \tag{20b}$$

The Galerkin (in x) approximation to (18) then becomes

(21a) $$\hat{\hat{u}}_j^n - \hat{u}_j^n = \alpha_j \pi_j^n$$

(21b) $$\hat{\hat{w}}_j^n - \hat{w}_j^n = -\frac{d}{dz}\pi_j^n - \frac{2}{\alpha_j^2 L}\left(D_2^1 \hat{w}^n(0) - \frac{db}{dz}\right)$$

(21c) $$\alpha_j \hat{\hat{u}}_j^n + \frac{d}{dz}\hat{\hat{w}}_j^u = 0 \quad .$$

Eliminating $\hat{\hat{u}}_j^n$ and π_j^n from the above equations gives the single equation for $\hat{\hat{w}}_j^n$,

(22a) $$\left(\frac{d^2}{dz^2} - \alpha_j^2\right)\hat{\hat{w}}_j^n = -\alpha_j^2 \hat{w}_j^n - \alpha_j \frac{d}{dz}\hat{u}_j^n + \frac{2}{L}\left(D_2^1 \hat{w}^n(0) - \frac{db}{dz}\right)$$

with boundary conditions

(22b) $$\hat{\hat{w}}_j^n(z = \pm 1) = 0$$

The calculation of $\hat{\hat{u}}_j^n$ follows from (21c) once (22) has been solved.

To compute the final step, we transform into physical space in the x-direction and discretize (11) using second-order differences,

(23a) $$\frac{\tilde{u}_j^n - \hat{\hat{u}}_j^n}{\Delta t} = \frac{1}{R}\left(\frac{d^2\tilde{u}_j^n}{dz^2} + \frac{\tilde{u}_{j+1}^n - 2\tilde{u}_j^n + \tilde{u}_{j-1}^n}{\Delta x^2}\right) + \frac{2}{R}$$

$$\tilde{u}_j^n(z = \pm 1) = 0$$

(23b) $$\frac{\tilde{w}_j^n - \hat{\hat{w}}_j}{\Delta t} = \frac{1}{R}\left(\frac{d^2\tilde{w}_j^n}{dz^2} + \frac{\tilde{w}_{j+1}^n - 2\tilde{w}_j^n + \tilde{w}_{j-1}^n}{\Delta x^2}\right)$$

$$\tilde{w}_j^n(z = \pm 1) = 0 \quad .$$

The boundary conditions in x on u^n, w^n are given by (6b) and (6c). To handle zero-derivative boundary conditions in x (i.e. $\partial/\partial x = 0$) at a point j_* we simply set $\tilde{u}_{j_*+1}^n = \tilde{u}_{j_*-1}^n$. To handle Dirichlet conditions we set $\tilde{w}_{j_*}^n$ (or $\tilde{u}_{j_*}^n$) to its boundary value and condense the matrix.

We will not dwell at any length here on the z-discretization, identical to that used in previous periodic-in-x calculations [4]. In the z-direction the variables are represented in spectral space by a Chebyshev polynomial expansion, and in physical space at the Chebyshev collocation points [14]. Derivatives are calculated by transforming to Chebyshev space, computing the derivative, and returning to physical space representation. Collocation (pseudo-spectral) approximation is used for both the nonlinear terms and the Poisson equations, (22) - (23). The Poisson operator in (22) is inverted using the eigenfunction technique described in [15]. The same technique is used for the Poisson operators in the viscous step, (23), however here the diagonalization in the z-direction results in a (easily inverted) block tridiagonal system where each block is diagonal.

The operation count for an entire time step is then

$$\text{(24)} \qquad \text{operations} \sim O(P^2N) + O(PN\log_2 N) + O(NP\log_2 P)$$

where P is the number of Chebyshev polynomials and N the number of x collocation points. The first term in (24) results from the Poisson solves, the last two from the Fourier transforms between physical and spectral space. With respect to accuracy in the x-direction, it is difficult to estimate the errors in a useful manner. It would appear that the scheme is at least third-order in the interior and lower-order at the boundaries.

We will comment only briefly on the numerical methods for generating the inflow spatial eigenfunctions. A Chebyshev collocation technique is used with a matrix eigenvalue solver: the temporal problem is solved (using a method similar to that of [1]) and used as an initial guess [18] for the spatial iteration.

Results. (Linear Theory) We run here a sub-critical case similar to that investigated in [12]. $\vec{v}_\ell(z)$ is taken to be the eigenfunction at $R = 5000$, $\omega_R = .33017$ corresponding to $\alpha_R = 1.1557$, $\alpha_I = .0106$. The parameters for the simulations to be reported here are $P = 32$, $N = 32$, $L = 16$, $\Delta t = T/320$ (where T is the period of oscillation, $2\pi/\omega_R$). The excessively small time step is due to the Courant condition on the base parallel flow and could be easily alleviated by a semi-implicit treatment of the nonlinear terms.

In Fig. 2 we plot $u'_{max}/u'_{t,o}$ (where $u'_{max} = \max_z u'$) versus x for the exact (linear) and numerical solutions at a time $t = 6T$ when the "steady-state" has been achieved. ($u'_{t,o}$ is taken to be 10^{-8} for these linear runs: its value is clearly irrelevant if taken sufficiently small). The wavelength (α_R), the phase, the attenuation (α_I), and the "shape" of the numerical solution are all seen to be extremely accu-

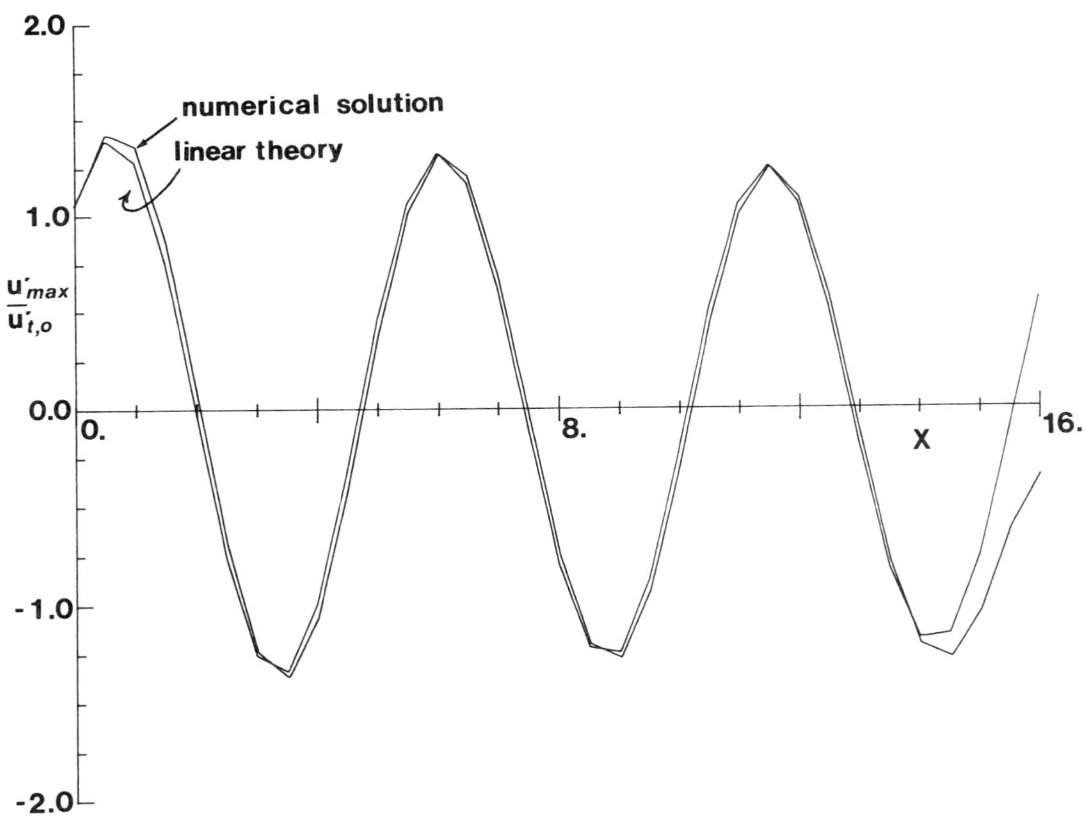

Figure 2

A plot of $u'_{max}/u'_{t,o}$ (where $u'_{max} = \max_z u'$, $u'_{t,o} = \max_z \langle u'^2(0, z, t)\rangle^{1/2}$) versus x at t = 6T. The numerical solution is seen to compare very well with the exact (linear) solution, especially away from the inflow and outflow boundaries.

rate. Even when the x-resolution is halved (N = 16 corresponds to only 5 points/wavelength) the solution remains fairly accurate. The solution in the interior is markedly more accurate than at the boundaries: although this is to be expected at outflow, it is slightly more bothersome at inflow.

To better see the errors in growth rate we plot in Fig. 3 $\ln(u'_{t,max}/u'_{t,o})$ versus x (in the steady-state) for the exact and numerical solutions. In the interior of the domain the agreement with linear theory is excellent. At inflow $u'_{t,max}$ actually rises slightly before descending: increasing the resolution indicates that this aberration does vanish for sufficiently large N, however it appears that the error in $u'_{t,max}$ and the extent in x over which it occurs decrease only as 1/N (a typical Fourier boundary error).

As a last verification of the scheme's ability to accurately reproduce linear theory, we plot in Fig. 4 u'_t versus z at x = 0 and at x = L/2. It is seen that the eigenfunction structure is preserved almost exactly as the perturbation proceeds downstream.

In summary, the hybrid Fourier method accurately predicts the phase, wavelength, and growth rate of linear modes as well as their spatial structure. Low-order accuracy at inflow can shift overall magnitudes slightly, however it does not appear to significantly alter the solution in the interior in any other way.

(Nonlinear Theory) As previously stated, it is not our purpose here to investigate the physics of nonlinear spatially evolving disturbances, but rather to demonstrate the viability of the hybrid Fourier scheme. To this end, we run an experiment similar to that reported in Fig. 9b of [12] for the R = 5000, ω_R = .33017 case described above (note that although the perturbations are nonlinear here, the form of the inflow disturbance is still taken to be that of a linear eigenfunction).

Fig. 5 shows a plot of $\zeta'_{t,c}$ (centerline (z = 0) r.m.s. perturbation vorticity) versus x for several values of $u'_{t,o}$. We do not plot the initial fluctuation in $\zeta'_{t,c}$ due to the low-order accuracy of the scheme at inflow. The calculation was run with L = 24, however in Fig. 5 we plot only to x = 20. Our Fig. 5 compares well with the previously published report: a threshold for nonlinear growth of $u'_{t,o} \cong .021$ is clearly indicated. On the basis of this result we conclude that the accuracy obtained for the linear case extends to the nonlinear regime.

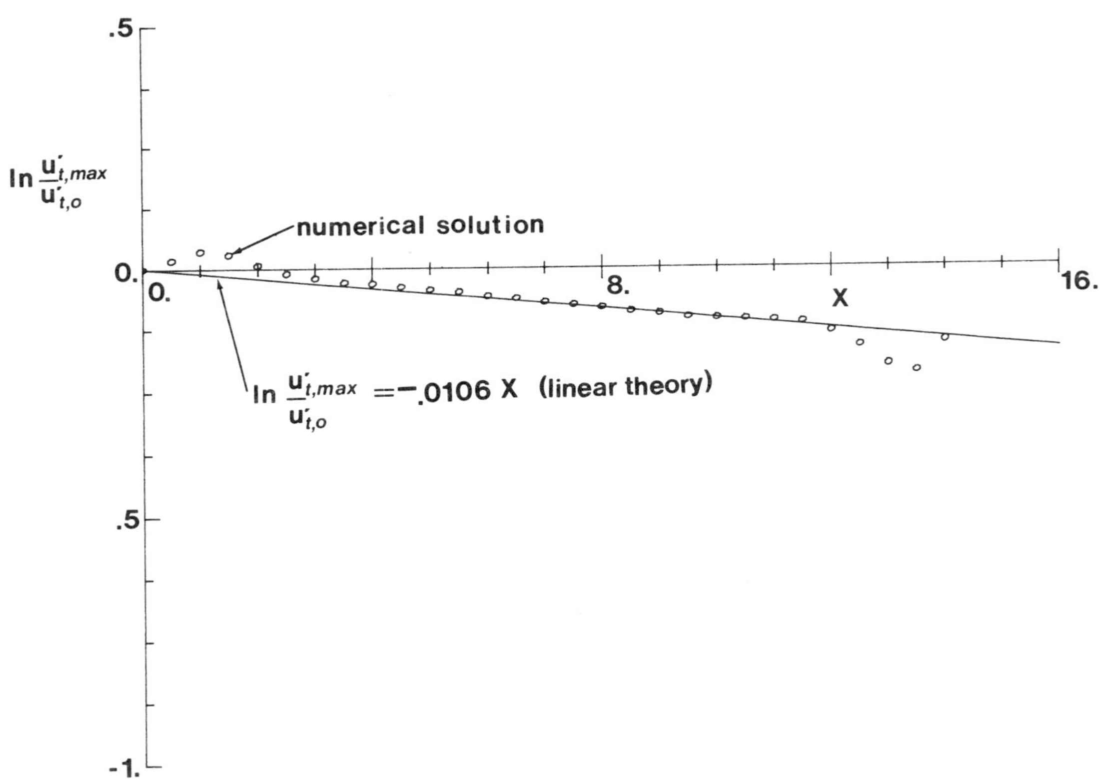

Figure 3

A plot of $\ell n[u'_{t,max}/u'_{t,o}]$ (where $u'_{t,max} = \max_z \langle u'^2(x, z, t)\rangle^{1/2}$) versus x. The time-averaging is performed over the time interval $5T \leq t < 6T$. The agreement between the numerical solution and linear theory is seen to be very good: in the interior the growth rate predicted by the numerical solution is between -.0102 and -.0112.

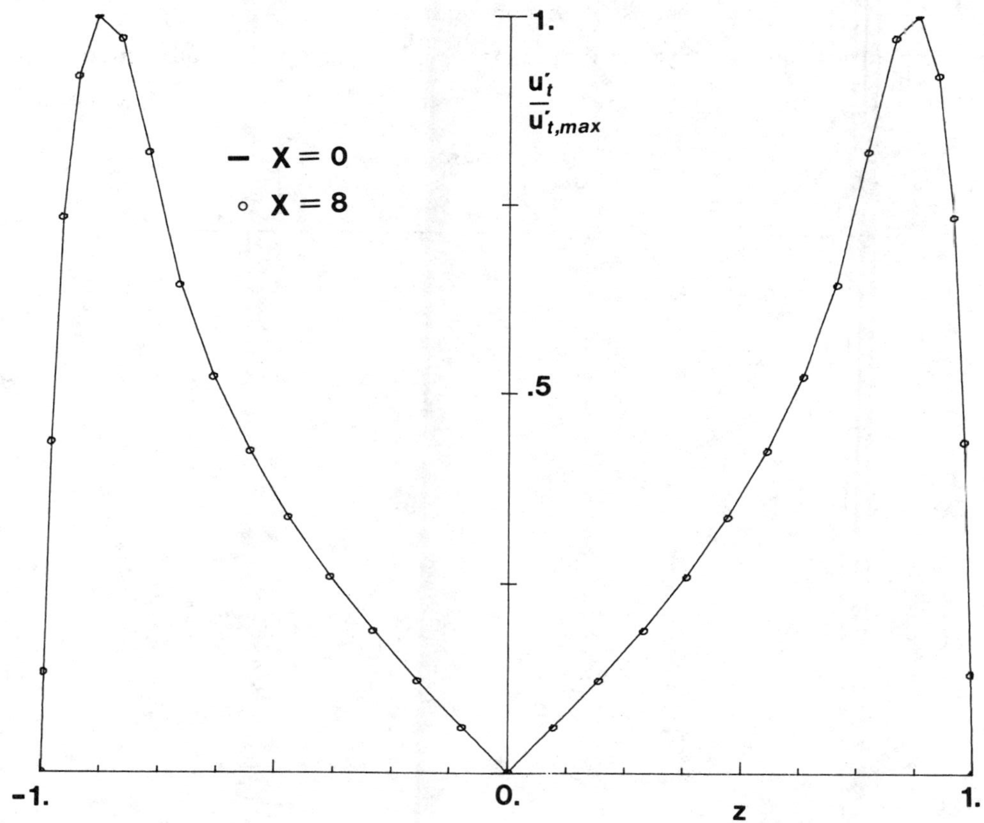

Figure 4

A plot of $u'_t/u'_{t,max}$ ($u'_t = \langle u'^2(x, z, t)\rangle^{1/2}$) versus z at two downstream locations, x = 0 and x = 8. The time-averaging is performed over the time interval $5T \le t < 6T$. The shape of the linear eigenfunction is preserved almost exactly by the hybrid Fourier technique.

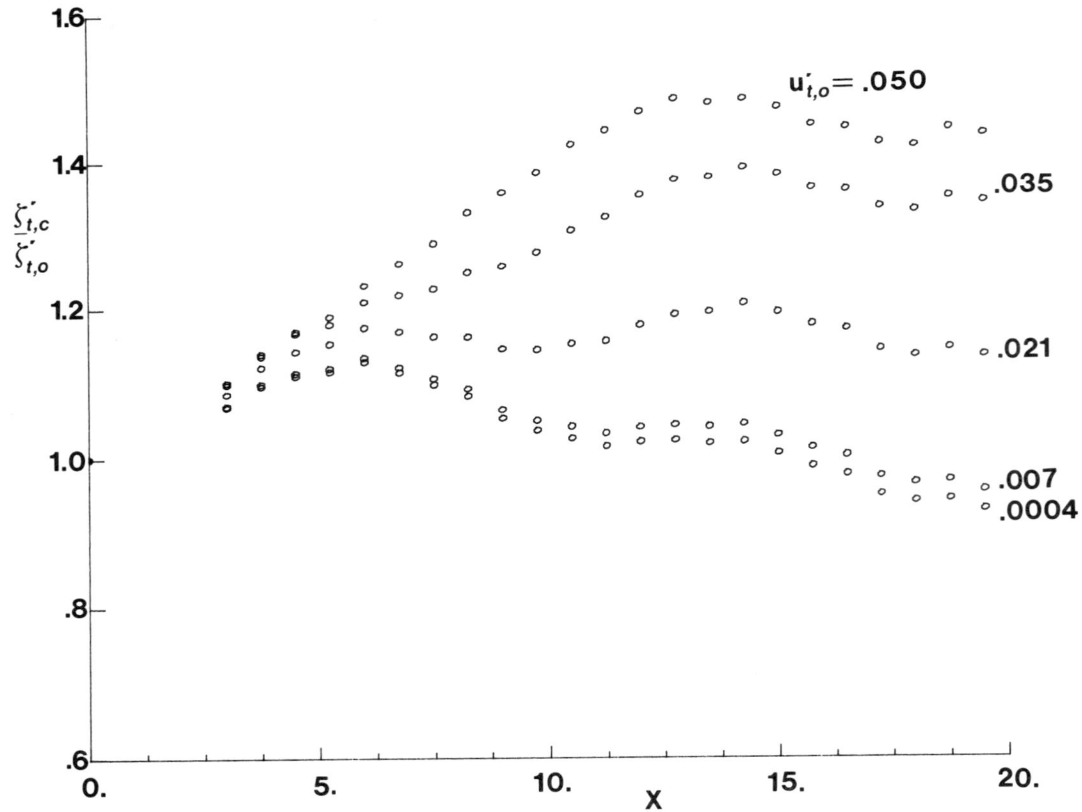

Figure 5

A plot of $\zeta'_{t,c}/\zeta'_{t,o}$ (where $\zeta'_{t,c} = \langle\zeta'^2(x, 0, t)\rangle^{1/2}$, $\zeta'_{t,o} = \langle\zeta'^2(0, 0, t)\rangle^{1/2}$) versus x for various values of the inflow disturbance strength, $u'_{t,o}$. At this subcritical Reynolds number, R = 5000, it is seen that for $u'_{t,o} \gtrsim .021$ nonlinear disturbances will grow.

ACKNOWLEDGEMENTS

I would like to thank Prof. S. A. Orszag for numerous helpful discussions and suggestions concerning this work. Thanks also to Ms. Jane Chung for programming assistance, and Ms. Eva Kish for help in preparing the manuscript.

REFERENCES

[1] Orszag, S. A. 1971 Accurate solution of the Orr-Sommerfeld stability equation. J. Fluid Mech., 50, p. 689.

[2] Herbert, T. 1976 Periodic secondary motions in a plane channel. In Proc. 5th Int. Conf. on Numerical Methods in Fluid Dynamics (ed. A. I. van de Vooren and P. J. Zandbergen), p. 235, Springer.

[3] Patera, A. T. & Orszag, S. A. 1980 Transition and turbulence in plane channel flows. In Proc. 7th International Conference on Numerical Methods in Fluid Dynamics (ed. R. W. MacCormack and W. C. Reynolds), Springer, pp. 329-335.

[4] Patera, A. T. & Orszag, S. A. 1982 Secondary instability in wall-bounded shear flows. J. Fluid Mech., to appear.

[5] Kleiser, L. 1982 Numerische Simulationen zum laminar-turbulenten Umschlagsprozeß der ebenen Poiseuille-Strömung. Ph.D. Thesis, Universität Karlsruhe.

[6] Orszag, S. A. & Patera, A. T. 1980 Subcritical transition to turbulence in plane channel flows. Phys. Rev. Lett. 45, pp. 989-983.

[7] Orszag, S. A. & Patera, A. T. 1981 Subcritical transition to turbulence in planar shear flows. In Transition and Turbulence (ed. R. E. Meyer), Academic Press, pp. 127-146.

[8] Moin, P., Reynolds, W. C., & Ferziger, J. H. 1978 Large eddy simulation of incompressible turbulent channel flow. Report No. TF-12, Dept. of Mech. Engrg., Stanford University.

[9] Orszag, S. A. & Kells, L. C. 1980 Transition to turbulence in plane Poiseuille flow and plane Couette flow. J. Fluid Mech., 96, p. 159.

[10] Orszag, S. A. & Patera, A. T. 1981 Calculation of Von Karman's constant for turbulent channel flow. Phys. Rev. Lett. 47, p. 832.

[11] Fasel, H. 1976 Investigation of the stability of boundary layers by a finite-difference model of the Navier-Stokes equations. J. Fluid Mech., 78, p. 355.

[12] Fasel, H. & Bestek, H. 1979 Investigation of nonlinear, spatial disturbance amplification in plane Poiseuille flow. In: Laminar-Turbulent Transition (eds. R. Eppler, H. Fasel), Springer-Verlag.

[13] Lanczos, C. 1956 Applied Analysis. Prentice-Hall, Englewood Cliffs, N.J.

[14] Gottlieb, D. & Orszag, S. A. Numerical Analysis of Spectral Methods: Theory and Applications. NSF-CBMS Monograph no. 26, Soc. Ind. App. Math., Philadelphia.

[15] Patera, A. T. & Orszag, S. A. 1981 Finite amplitude stability of axisymmetric pipe flow. J. Fluid Mech., 112, p. 467.

[16] Deville, M. O. & Orszag, S. A. To be published.

[17] Brachet, M. E., Meiron, D. I., Orszag, S. A., Nickel, B. G., Morf, R. H., & Frisch, U. 1982 Small-scale structure of the Taylor-Green vortex. J. Fluid Mech., to appear.

[18] Gaster, M. 1962 A note on the relations between temporally-increasing and spatially-increasing disturbances in hydrodynamic stability. J. Fluid Mech., 14, p. 222.

RECENT ADVANCES IN PSEUDO-SPECTRAL METHODS

T. D. TAYLOR*

Abstract. This paper summarizes some advances made in pseudo-spectral methods for computation of fluid flow. In particular, progress made with finite difference predictors and spectral correctors is outlined.

1. Introduction. This paper is a brief summary of recent developments in pseudo-spectral methods applied to flow problems. The work described is an accumulation of progress made while the author was at Aerospace Corporation and currently at Johns Hopkins University/Applied Physics Laboratory. The efforts of Dr. Richard Hirsh of the Applied Physics Laboratory, as well as Dr. John Murdock and Dr. Robert Myers at Aerospace, made much of the progress possible.

The advances discussed focus largely on improvements in the efficiency of application of spectral methods. They include the use of a finite difference predictor and spectral corrector approach, as well as non-FFT techniques for matrix inversions. Techniques for calculating the pressure for incompressible flows using pseudo-spectral methods are also addressed. In the paper example results are presented which demonstrate the main points.

2. The Pseudo-Spectral Methods. In the application of spectral methods to problem solutions one usually tries to develop an expansion which represents the solution, f, in one dimension in the form

$$f = \sum_{n=0}^{N} a_n(t)\, F_n(x) \tag{1}$$

*Applied Physics Laboratory, Laurel, MD 20707

and then attempts to determine the values of the a_n's. For a simple 1-D equation of the form

(2) $$\frac{\partial f}{\partial t} + \frac{\partial f}{\partial x} = \alpha \frac{\partial^2 f}{\partial x^2}$$

this can be attempted by direct substitution with the result

(3) $$\sum_{n=0}^{N} a_n' F_n(x) + \sum_{n=0}^{N} a_n (F_n' - \alpha F_n'') = 0$$

If now one can relate $(F_n' - \alpha F_n'')$ to F_n then equations for each a_n can be developed. If this is not straightforward then it is more practical to attack the problem in real space by integrating equation (2) to obtain

(4) $$f(t_2,x) - f(t_1,x) = \int_{t_1}^{t_2} (\alpha \frac{\partial^2 f}{\partial x^2} - \frac{\partial f}{\partial x}) dt$$

This equation can be used to determine the values of $f(x)$ at time t_2 provided values are known at t_1. The procedure is to evaluate the right hand side of (4) as accurately as possible using equation (1) as an interpolation formula for spacial derivatives. Clearly, since the $a_n(t)$ are unknown except at time t_1 and earlier times this is possible only by an explicit predictor corrector approach. If, however, one uses a finite difference predictor and a spectral corrector on the right hand side then implicit finite difference schemes can be used to stabilize the predictor and enhance the time step. The corrector can still be explicit and spectral to retain the accuracy of the scheme and it can involve both the t_1 and t_2 time levels. For the predictor corrector approach one then has

(5a) $$\bar{f}(t_2,x) - f(t_1,x) = \int_{t_1}^{t_2} (\alpha \frac{\partial^2 f}{\partial x^2} - \frac{\partial f}{\partial x})_{\text{F.D.}} dt$$

(5b) $$f(t_2,x) - f(t_1,x) = \int_{t_1}^{t_2} (\alpha \frac{\partial^2 f}{\partial x^2} - \frac{\partial f}{\partial x})_{\text{spectral}} dt$$

with the R.H.S. derivatives being evaluated from

$$\bar{f}(t_2,x) = \sum_{n=0}^{N} a_n(t_2) F_n(x)$$

$$f(t_1,x) = \sum_{n=0}^{N} a_n(t_1) F_n(x).$$

where the a_n's are found by fitting the series to the point of values of f. The time integral is then evaluated by some weighted average of the derivatives at times t_1 and t_2. A higher order scheme using more time terms could also be employed.

The finite difference predictor and the spectral corrector can also be used to solve the Poisson equation for the pressure or the two dimensional stream function equation of incompressible flows. The use of the finite difference predictor approach lowers the density of the matrix that must be inverted to solve the equation. The procedure is to employ the equation (Orszag, Ref. 7)

$$\nabla^2_{F.D.} f^{m+1} = \nabla^2_{F.D.} f^{m} + \alpha(g - \nabla^2_{sp} f^{m}) \tag{6}$$

where now α is a relaxation parameter, m is the number of the iteration and g is the right hand side of the Poisson equation

$$\nabla^2 f = g \tag{7}$$

Standard techniques such as the conjugate gradient method or direct solvers can be used to solve equation (6).

The concept of the finite difference predictor and spectral corrector can be generalized in the context that one basically employs low order interpolation formulas for the predictor in order to simplify the matrices. Then a higher order interpolation formula is employed in the predictor step. This general concept permits one to derive all types of approaches for pseudo-spectral solution techniques. It permits one to consider a problem with one element or many elements (Ref. 6) with multiple grid systems. The only requirement is that one be able to develop interpolation approximations which apply at points connecting the elements. These of course can be developed to the order of accuracy desired.

The advantages of the predictor-corrector approach over a direct spectral approach are

1) boundary conditions are easily applied
2) larger time steps can be used

3) finite difference knowledge can be used to guide pseudo-spectral solutions

4) matrix inversions are usually simplified.

For viscous problems the ability to take longer time steps has a large impact. For example using Chebyshev expansion in space the time step is typically limited to $0\left(\frac{1}{N^4}\right)$ where N is the number of terms employed. By employing an implicit finite difference predictor and modified Euler corrector computer experiments have shown that the time step can be increased to $0\left(\frac{1}{N}\right)$. This test was conducted on the Blasius boundary layer equation since the solution was known and could be compared in detail.

The method which has been described has been applied to various problems. Some of these applications and the testing utilized will now be described.

3. Applications and Test Examples. The first area to be discussed are the tests that have been performed to determine an optimum way to compute the a_n's from point values of the function f for the expansion

$$f = \sum_{n=0}^{N} a_n F_n(x)$$

For the tests F_n was chosen to be a Chebyshev polynomial or a sine.

The primary purpose of the evaluation was to determine if one was confined to the Fast Fourier Transform, FFT, to maintain an efficient matrix inversion. If this restriction could be removed then one would not be confined to powers of two for selection of the grid point count and also functions that do not lend themselves to the FFT could be employed as needed. In order to perform the test both one and two dimensional cases were studied. The coefficients of the series expansions were determined by employing the FFT, the conjugate gradient method and direct solution by Crout reduction to invert the matrix formed by evaluating the series at N Chebyshev nodal points. The tests were made for

1) a Chebyshev fit to $e^{-\alpha x^2}$ with $\alpha = 0.5$ and 50

2) a Chebyshev fit to $e^{-(x^2+y^2)}$, and

3) a sine series solution to the 1-D Burgers equation $u_t+uu_x = \frac{1}{50} u_{xx}$

for the initial condition

$$u = 1.0 \text{ for } x \leq 5.0$$
$$u = \frac{1}{2} - \frac{x}{10} \text{ for } -5.0 \leq x \leq 5.0$$
$$u = 0 \text{ for } x \geq 5.0$$

and boundary conditions

$$u = 1 \text{ as } x \to -\infty$$
$$u = 0 \text{ as } x \to \infty$$

Tables 1, 2 and 3 show the timing results using a VAX 11/780. Note that FFT2 denotes the fastest results from three FFT routines. The other FFT routines were a factor of two slower and it should be mentioned that they were taken from conventional operating system subroutine packages. The fast FFT was specially written for signal processing. DIM2 denotes the Crout reduction timing and CGM the conjugate gradient method.

From the results for the 1-D cases the FFT is clearly better as N exceeds 64 but for N of 64 or less the penalty for not using the FFT is small and as a consequence it is clear that for $N \leq 64$ both CGM and a direct matrix inversion can be used with any number of points or any functions.

The 2-D test was more significant since it showed that the Crout reduction method was very competitive with the FFT up to N=128. Even more significant is the value of this result for use in unsteady calculations. The result shows that direct solution at each time step can be competitive with the FFT but it also implies that it can even be more efficient in two and three dimensional time integration solutions of the flow equations. The reason is that since the matrix coefficients are the same of each time step the direct solution method with LU decomposition can be used to upper triangularize the matrix

Table 1. Matrix Inversion Timings for $e^{-\alpha x^2}$ fit with Chebyshev at Nodal Points

$\alpha = 0.5$

	N			
	16	32	64	128
CGM	3.67	3.82	4.67	7.24
FFT2	1.22	1.29	1.70	2.51
DIM2	1.10	1.53	3.63	19.12

$\alpha = 50.0$

	16	32	64	128
CGM	3.67	3.88	4.56	6.90
FFT2	1.11	1.31	1.61	2.52
DIM2	1.18	1.57	3.65	19.23

Table 2. Matrix Inversion Timings for Two Dimensional Interpolation

	4x4	8x8	16x16	32x32	64x64	128x128
DIM2	1.20	1.42	2.32	6.29	15.49	117.03
FFT2	1.68	1.91	3.08	8.31	15.36	104.38

Table 3. Timings for the Solution of Burgers Equation

	16	32	64
FFT2	15.9	30.4	55.8
DIM2	13.3	50.5	202.0
CGM	18.1	72.0	284.4

for each dimension once and then for each time step thereafter only a sweep is needed to compute the series coefficients. The FFT must be applied completely, however, at each time step. The details of this approach are available in Reference (1). The end result is that for a large number of time steps the direct method can be made to be faster than the FFT for multidimensional problems.

One other point worth mentioning which was learned in conducting the timing tests was that a sine series representation took about four times as many modes as a Chebyshev expansion to fit the function

$$f(x) = e^{-\alpha x^2} - [1 - (1+e^{-\alpha})x] \quad 0 \leq x \leq 1$$

In fact it took 128 terms to show significant convergence. This means that one must be careful in assuming that a sine series is yielding an accurate representation of a function in a flow problem.

4. Incompressible Flow Applications. The finite difference predictor spectral corrector approach is just beginning to be applied to solution of the Navier-Stokes equations. The advantage as pointed out is the increased ease of solving a Poisson equation and the increased time step possibility. In two dimensions the equations for flow past a wall can be solved by pseudo-spectral methods in the form

$$u(t_2,x,y) = u(t_1,x,y) + \int_{t_1}^{t_2} \left(\frac{1}{Re} \nabla^2 u - p_x - uu_x - vu_y \right) dt \tag{8}$$

$$v(t_2,x,y_2) = - \int_{y_1}^{y_2} (u_x)_{t_2} \, dy + v(t_2,x,y_1) \tag{9}$$

$$\nabla^2 p = - (u_x^2 + 2u_y v_x + v_y^2) = f \tag{10}$$

$$\text{or} \quad p = - \int_{y_1}^{y_2} \left(v_t + uv_x + vv_y - \frac{1}{Re} \nabla^2 v\right) dy. \tag{11}$$

In Reference 2 these equations were solved by a pseudo-spectral approach for flow over a plate and the Poisson equation for the pressure was used. However, the finite difference predictor was not employed so the solution was somewhat inefficient. By using the predictor-corrector approach outlined earlier, this solution could be made much more efficient both in improving the size of the time step and in solving the Poisson equation.

The predictor-corrector scheme has been tested on two problems with success. The first was solution of the stream function vorticity equations for a vortex in a box with solid walls and the second was for flow over a flat plate using equations (8) and (9) noting that $\partial p/\partial x$ is zero. The details of the vortex calculation are available in Ref. (3) and the flat plate results are presented in Ref. (4). They are straightforward except for the important finding that the predictor corrector time step could be increased to $0\left(\frac{1}{N}\right)$ instead of $0\left(\frac{1}{N^4}\right)$ for a simple explicit pseudo-spectral Chebyshev solution approach.

The results mentioned thus far have been for incompressible flow but a few tests have been conducted for compressible flow (Ref. 5). A simple one dimensional problem was chosen and tests were conducted for calculation of flow through a shock, rarefaction and contact surface. The solution was developed by expanding the density, ρ, velocity, u, and energy e, in Chebyshev series. These series were then used to solve the equations

$$\frac{\partial \rho}{\partial t} + \frac{\partial \rho u}{\partial x} = 0$$

$$\frac{\partial \rho u}{\partial t} + \frac{\partial (p+\rho u^2)}{\partial x} = 0$$

$$\frac{\partial \rho e}{\partial t} + \frac{\partial u(\rho e+p)}{\partial x} = 0$$

for constant but different conditions at each boundary. The calculations were accomplished initially by a direct pseudo-spectral approach and Adams Bashforth time integration. The results required a filter to remove oscillations induced by the steep gradients. Some typical results are shown in Figure 1. Here the Boris and Book Flux Corrector Scheme was used to filter in real space after each time step. This is probably not the best approach since it yields plateaus in the rarefaction wave results. By employing a finite difference predictor with any of the well known shock capturing schemes, and a spectral corrector instead of the direct pseudo-spectral approach, the results could be improved considerably.

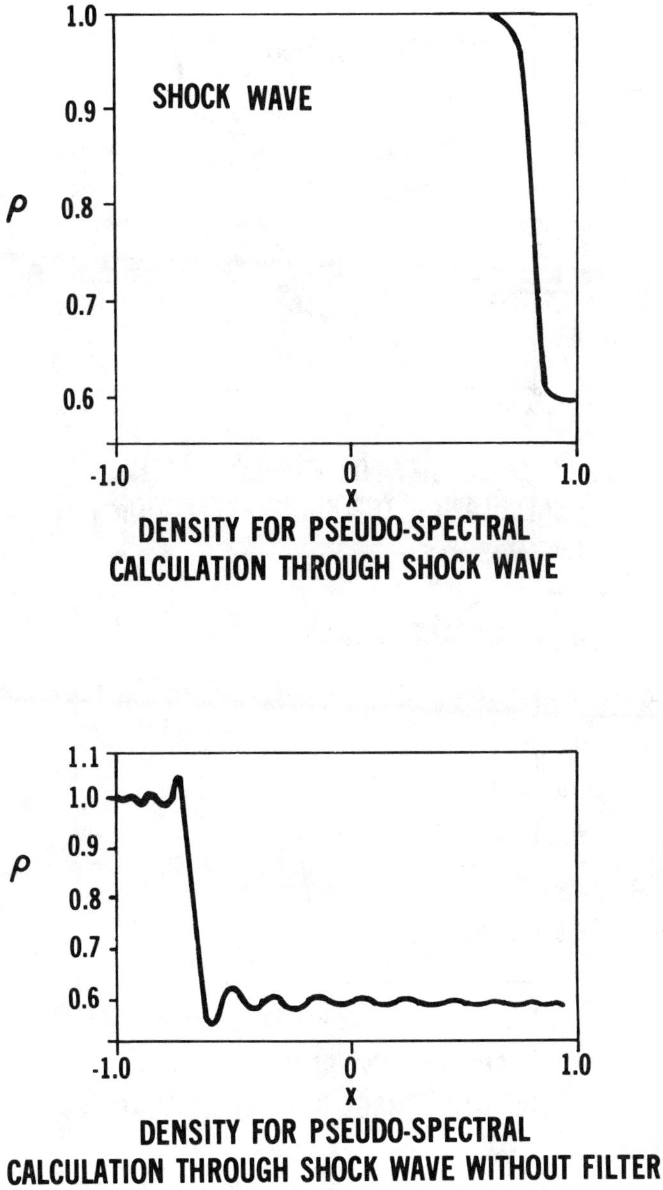

Figure 1a. Shock Wave Calculations

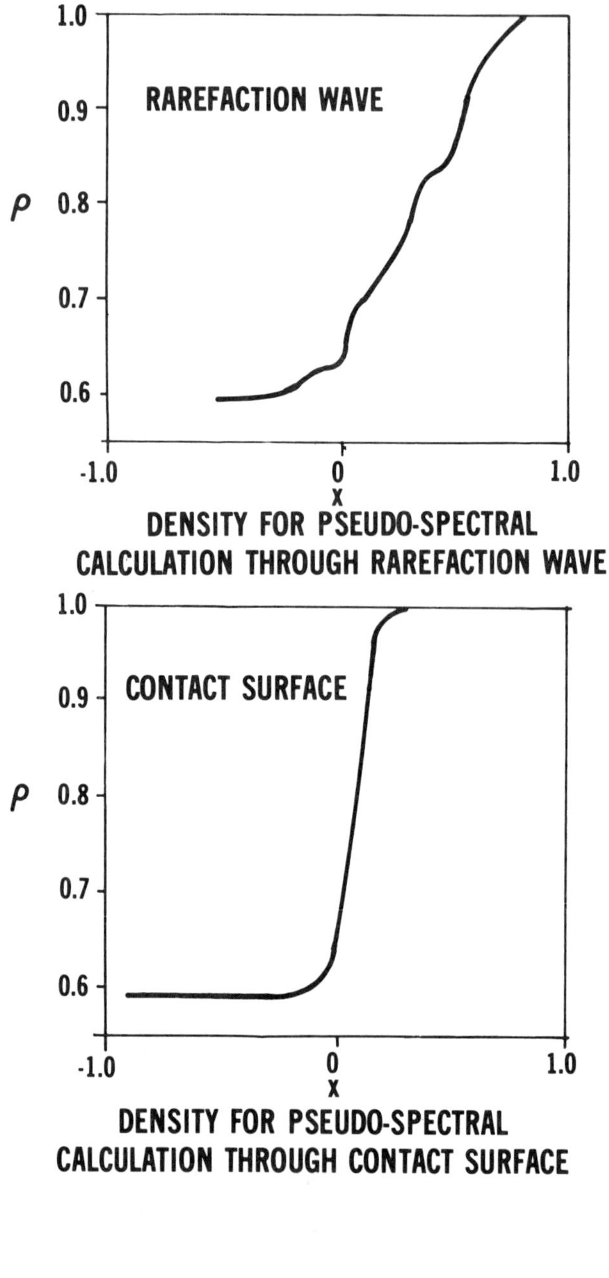

Figure 1b. Rarefaction and Contact Surface Calculations

5. Summary. In this overview paper the concept of finite difference predictor and a spectral corrector has been discussed for solving flow problems. The most significant advance of this approach is that it allows one to apply Chebyshev series near boundaries with time integration steps much greater than $0\left(\frac{1}{N^4}\right)$. It also permits the adaptation of the shock capturing work in finite differences to pseudo-spectral calculations, as well as finite difference Poisson solvers to pseudo-spectral problem solutions. All these advances are of considerable help in solving flow problems.

REFERENCES

(1) Taylor, T.D., Hirsh, R.S. and Nadworny, M.M., "Comparison of FFT, Direct Inversion and Conjugate Gradient Methods for Solution of Flow Equations by Pseudo-Spectral Methods", submitted to Computers and Fluids, also proceedings of 4th GAMM Conf. on Numerical Methods, Paris, France, Oct (1981).

(2) Taylor, T.D. and Murdock, J.W., Comp. and Fluids 9, pp.-255-263, (1981).

(3) Hirsh, R.S., Taylor, T.D., Nadworny, M.M., and Kerr, J.L., "Techniques for Efficient Implementation of Pseudo-Spectral Methods and Comparison with Finite Difference Solutions of the Navier-Stokes Equations", proc. of 8th Int. Conf. on Num. Meth., Aachen, 1982 to be published by Springer Verlag (1982).

(4) Hirsh, R.S. and Taylor, T.D., "Recent Developments at APL in the Use of Pseudo-Spectral Techniques", Proc. of EUROMECH Conf. on Spectral Methods, Nice, France (1982).

(5) Taylor, T.D., Myers, R.B. and Alber, J.H., Comp. and Fluids, 9, pp.-469-473, (1981).

(6) Metivet, B. and Morchoisne, Y. "Resolution of Unsteady Navier-Stokes Equations by Multi-domain Spectral Method", Proc. of EUROMECH Conf. on Spectral Methods, Nice, France (1982).

(7) Orszag, S., JCP, 37 pp.-70-92 (1980).